# ORGANIC SPECTROSCOPY

*HARPER'S CHEMISTRY SERIES*

*Under the Editorship of Stuart Alan Rice*

# ORGANIC SPECTROSCOPY

PRINCIPLES
AND
APPLICATIONS

*PIERRE LASZLO* *Liège University, Belgium*

*PETER J. STANG* *The University of Utah*

*HARPER & ROW, PUBLISHERS*

*New York, Evanston, San Francisco, London*

TO M. L. and C. S. S.

ORGANIC SPECTROSCOPY: Principles and Applications


Standard Book Number: 06–043852–5

Library of Congress Catalog Card Number: 75–148447

# CONTENTS

# PREFACE

Modern organic chemistry depends heavily upon physical methods, particularly spectroscopy of one sort or another, for structure elucidation and evaluations of reactivity. On the pedagogic level this is reflected in the increasing movement away from the traditional "qualitative organic" courses with their "wet" chemistry to courses in spectroscopy. Unfortunately, current books are either monographs about a single spectroscopic method directed to the specialist, or collections of data in tabular form with little or no reference to the basic principles involved. The need is evident for a compact, unified text, on the beginning graduate and senior undergraduate level, emphasizing a logical approach to spectroscopy and its applications to modern organic chemistry, with a pedagogic emphasis including illustrative examples and problems. We hope this text will satisfy this need. We have chosen to write it from an angle which does not appear to have been explored at an elementary level: namely, that it is intellectually more satisfying to obtain useful results about an organic molecule by deriving them from basic principles, so that memorization by the student of numerous spectral-group frequencies then becomes a useful adjunct rather than a mandatory burden.

In such an undertaking, one is forced to make some judgment about what is going to be most useful to the greatest number of readers. This judgment rests on our joint estimate of the present thrust of organic chemistry, an estimate whose subjectivity does not have to be stressed. The selection of the various topics discussed, and the amount of space devoted to each, rests on this subjective estimate and accounts, for instance, for the emphasis on the various resonance techniques.

Since it is impossible to be comprehensive about each of the current spectroscopic techniques, we have tried to emphasize the unity that underlies spectroscopy as a whole and to provide an understanding of the main principles operative in the various individual methods. We have noticed in our teaching how important it is for the student to "get a feel" for the reason behind a particular fact, even if he is not exposed to a complete theory. We have attempted to be simple but rigorous, and to provide an understanding of physical phenomena above and beyond a mere tabulation of data. Moreover, because in actual practice and application the various spectroscopies are always used jointly (since they complement each other nicely), we have reflected this observation in our approach. We have provided the reader with an integrated text so that he will have as a constant guide the unity provided by the object, molecular structure, and by the agent, the radiation used to induce transitions between quantized energy levels.

Spectroscopic tools are often likened to a camera, taking pictures of the molecular assemblies in which the chemist is interested. The aperture of the lenses is one of the outstanding characteristics that the user of a camera will select carefully. We have oriented this book along the current trend, that is, to reconstruct molecules from their constituent parts once the structure of the parts has been determined accurately by mass spectrometry or by one of the resonance spectroscopies. These methods focus on a small part of the molecule at a time and yield detailed and precise information about it, whereas the optical spectroscopies "see" the molecule as a complex set of vibrations or transitions for the entire molecular assembly and play a lesser role. We have not attempted to give a comprehensive coverage of the field; this would be an impossible task within such a format. But we have not avoided any of the present existing facets of organic spectroscopy: nuclear magnetic, electron spin, nuclear quadrupole, and ion cyclotron resonances; mass spectrometry, photoelectron spectroscopy, infrared and ultraviolet spectroscopy, optical rotatory dispersion, and circular dichroism.

We have chosen to subdivide the subject into three major sections: (1) basic principles and properties of spectroscopy, (2) molecular parameters derived from spectroscopy, and (3) applications. Each section has a number of worked-out sample problems to illustrate the main points further; ample problems are also provided for the student's own practice and benefit. We have also provided suggested readings at the end of each section, chosen deliberately to be short and simple to encourage the student to consult them. At the end of the text we have provided a comprehensive bibliography of current monographs and texts.

We hope this text will prove to be a valuable guide to the teacher and experienced scientist and a significant tool for the student in the mastery of spectroscopy.

We are grateful to our colleagues at Princeton and Utah for many valuable discussions and suggestions and to all our colleagues and various reviewers who have offered helpful comments and criticisms. We are also grateful to our students for their assistance and to Marjorie Bhavnani and Joyce Bennett for their technical services in the preparation of the manuscript.

# ACKNOWLEDGMENTS

We wish to thank both authors and publishers for permission to quote copyrighted material in figures, tables, and problems, as listed below. We should also like to thank our colleagues for permission to reproduce material prior to publication.

*FIGURES*

2.4 F. A. Cotton, *Chemical Applications of Group Theory,* Wiley-Interscience Publishers, Inc., New York, 1963, p. 247.
2.5 F. R. Jensen and C. H. Bushweller, *J. Amer. Chem. Soc., 91,* 3223 (1969), p. 3224.
3.5 R. W. Fessenden and R. H. Schuler, *J. Chem. Phys., 39,* 2147 (1963), p. 2156.
3.9 S. Sobajima, *J. Phys. Soc. Jap., 23,* 1070 (1967), p. 1072.
3.12 J. A. Pople and A. A. Bothner-By, *J. Chem. Phys., 42,* 1339 (1965), p. 1341.
3.13 M. Barfield and D. M. Grant, *J. Amer. Chem. Soc., 85,* 1899 (1963), p. 1901.
4.3 C. E. Johnson, Jr. and F. A. Bovey, *J. Chem. Phys., 29,* 1012 (1958), p. 1013.
4.4 J. F. M. Oth and J. M. Gilles, *Tetrahedron Lett.,* 6265 (1968), p. 6269.
4.5 J. R. Didry and J. R. Guy, *Compt. Rend. Ac. Sci., 256,* 3042 (1963), p. 3044.
5.2 H. R. Ward, R. G. Lawler, and R. A. Cooper, *J. Amer. Chem. Soc., 91,* 746 (1969), p. 747.
6.4 A. Streitwieser, Jr., *Molecular Orbital Theory for Organic Chemists,* John Wiley and Sons, Inc., New York, 1961, p. 209.
6.8 A. D. Baker, D. P. May, and D. W. Turner, *J. Chem. Soc. (B),* 22 (1968), p. 25.
8.2 E. D. Becker, *J. Chem. Ed., 42,* 591 (1965), p. 591.
8.3 E. Derouane, *Ind. Chim. Belge, 34,* 103 (1969), p. 110.
8.4 R. Freeman, *Mol. Phys., 5,* 499 (1962), p. 505.
9.11 R. Fraisse-Jullien, C. Frejaville, and P. Laszlo, unpublished results.
9.12 A. Kumar and B. D. Nageswara Rao, *Mol. Phys., 15,* 377 (1968), p. 384.
9.16 P. Diehl, R. G. Jones, and H. J. Bernstein, *Can. J. Chem., 43,* 81 (1965), p. 83.
9.18 R. Freeman and W. A. Anderson, *J. Chem. Phys., 37,* 2053 (1962), p. 2062.
9.20 Courtesy R. Evans, Princeton University, Princeton, N. J.
9.21 Courtesy A. Speert, Princeton University, Princeton, N. J.
9.23 P. E. Butler and K. Griesbaum, *J. Org. Chem. 33,* 1956 (1968), p. 1957.
10.2 J. D. Baldeschwieler, *Science, 159,* 263 (1968), p. 266.
11.5 M. Saunders, "Measurement of Rates of Fast Reactions Using Magnetic Resonance," in *Magnetic Resonance in Biological Systems,* A. Ehrenberg, B. G. Malmström, and T. Vanngard (eds.), Pergamon Press, Oxford, 1967, p. 85.
11.6 M. Saunders, "Measurement of Rates of Fast Reactions Using Magnetic Resonance." in *Magnetic Resonance in Biological Systems,* A. Ehrenberg, B. G. Malmström, and T. Vanngard (eds.), Pergamon Press, Oxford, 1967, p. 85.
11.7 G. Schröder, J. F. M. Oth, and R. Merenyi, *Angew. Chem. Intl. Ed., 4,* 752 (1965), p. 759.
11.9 M. Rabinovitz and A. Pines, *J. Amer. Chem. Soc., 91,* 1585 (1969), p. 1588.
12.2 Courtesy D. K. Dalling and D. M. Grant, University of Utah, Salt Lake City, Utah.
12.3 Courtesy D. K. Dalling and D. M. Grant, University of Utah, Salt Lake City, Utah.
12.4 F. J. Weigert, M. Jautelat, and J. D. Roberts, *Proc. Nat. Acad. Sci. U. S., 60,* 1152 (1968), p. 1154.
12.5 Courtesy D. M. Grant, E. Allred, and D. K. Dalling, University of Utah, Salt Lake City, Utah.
12.8 J. D. Roberts, *Chem. Brit.,* 529 (1966), p. 530.
12.9 J. D. Roberts, Chem. Brit., 529 (1966), p. 530.
12.10 J. B. Hendrickson, *J. Amer. Chem. Soc., 89,* 7047 (1967), p. 7054.
12.11 J. B. Hendrickson, *J. Amer. Chem. Soc., 89,* 7043 (1967), p. 7044.
12.12 J. D. Roberts *Chem. Brit., 529* (1966), p. 535.
12.13 J. B. Hendrickson, *J. Amer. Chem. Soc., 89,* 7043 (1967), p. 7044.
13.1 Courtesy A. Speert, Princeton University, Princeton, N. J.
15.3 W. Reusch and C. Djerassi, *Tetrahedron, 23,* 2893 (1967), p. 2894.

| | | |
|---|---|---|
| | 15.4 | H. Audier *et al., Bull. Soc. Chim. Fr.,* 2292 (1964), p. 2296. |
| | 15.5 | Courtesy W. von Philipsborn, University of Zurich, Switzerland. |
| | 15.6 | Courtesy W. von Philipsborn, University of Zurich, Switzerland. |
| | 15.7 | Courtesy W. von Philipsborn, University of Zurich, Switzerland. |
| | 16.1 | S. J. Brois, *J. Amer. Chem. Soc., 90,* 506 (1968), p. 507. |
| | 16.2 | G. A. Yousif and J. D. Roberts, *J. Amer. Chem. Soc., 90,* 6428 (1968), p. 6429. |
| | 16.3 | J. E. Anderson and J. M. Lehn, *J. Amer. Chem. Soc., 89,* 81 (1967), p. 83. |
| | 16.4 | J. S. Gutowsky and T. H. Siddall, III, *J. Amer. Chem. Soc., 89,* 4300 (1967), p. 4302. |
| | 17.1 | C. C. McDonald and W. D. Phillips, *J. Amer. Chem. Soc., 89,* 6332 (1967), p. 6333. |
| | 17.2 | J. T. Gerig, *J. Amer. Chem. Soc., 90,* 2681 (1968), p. 2683. |
| *TABLES* | 3.4 | T. Yonezawa *et al., Bull. Chem. Soc. Jap., 38,* 1224 (1965); K. B. Wiberg, *Tetrahedron, 24,* 1083 (1968); G. Berthier, H. Faucher, and D. Gagnaire, *Bull. Soc. Chim. Fr.,* 1872 (1968). |
| | 3.6 | C. S. Foote, *Tetrahedron Lett., 579* (1963), p. 581. |
| | 4.4 | H. J. Dauben, Jr., J. D. Wilson, and J. L. Laity, *J. Amer. Chem. Soc., 90,* 811 (1968), p. 812. |
| | 4.6B | P. Bucci, *J. Amer. Chem. Soc., 90,* 252 (1968). |
| | 4.6C | B. P. Dailey and J. N. Shoolery, *J. Amer. Chem. Soc., 77,* 3977 (1955), p. 3980. |
| | 4.6D | A. E. Matter *et al., Tetrahedron, 25,* 691 (1969), pp. 693–694. |
| | 4.6E | L. M. Jackman and S. Sternhell, *Applications of Nuclear Magnetic Resonance Spectroscopy in Organic Chemistry,* Pergamon Press, Oxford, 1969, 2nd ed., p. 202. |
| | 11.1 | E. L. Muetterties, *Inorg. Chem., 4,* 769 (1965), p. 770. |
| | 11.2 | M. Saunders, "Measurement of Rates of Fast Reactions Using Magnetic Resonance," in *Magnetic Resonance in Biological Systems,* A. Ehrenberg, B. G. Malmström, and T. Vanngard (eds.), Pergamon Press, Oxford, 1967, p. 89. |
| | 14.2 | F. Weigert and J. D. Roberts, *J. Amer. Chem. Soc., 90,* 3543 (1968), p. 3547. |
| | 14.3 | J. W. Emsley, J. Feeney, and L. H. Sutcliffe (eds.), *High Resolution Nuclear Magnetic Resonance Spectroscopy,* Pergamon Press, Oxford, 1966, vol. 2, pp. 1017–1020. |
| | 14.4 | M. A. Jensen, *J. Organometal. Chem., 11,* 423 (1968), p. 424. |
| | 15.3 | L. P. Kuhn, P. von R. Schleyer, W. F. Baitinger, and L. Eberson, *J. Amer. Chem. Soc., 86,* 650 (1964), p. 654. |
| | 15.7 | S. Bank, W. D. Closson, and L. T. Hodgins, *Tetrahedron, 24,* 381 (1968), p. 382. |
| | 15.8 | G. B. Savitsky, K. Namikawa, and G. Zweitel, *J. Phys. Chem., 69,* 3105 (1965), p. 3107. |
| | 15.11 | W. McFarlane, *J. Chem. Soc.* (B), 28 (1969), p. 28 |
| *PROBLEMS* | 7.1 | H. H. Jaffé and M. Orchin, *Theory and Applications of Ultraviolet Spectroscopy,* John Wiley and Sons, Inc., New York, 1962, p. 205. |
| | 9.2 | U. Scheidegger, Varian A. G., Zurich, Switzerland. |
| | 9.3 | U. Scheidegger, Varian A. G., Zurich, Switzerland. |
| | 9.4 | U. Scheidegger, Varian A. G., Zurich, Switzerland. |
| | 13.7 | U. Scheidegger, Varian A. G., Zurich, Switzerland. |
| | 16.4 | T. Miyadera and Y. Kishida, *Tetrahedron, 25,* 209 (1969), p. 211. |
| SAMPLE PROBLEMS | 9.25 | R. Freeman and W. A. Anderson, *J. Chem. Phys., 37,* 2053 (1962), p. 2063. |
| | 11.3 | G. M. Whitesides and H. L. Mitchell, *J. Amer. Chem. Soc., 91,* 5384 (1969), p. 5385. |

# I

# SPECTROSCOPY

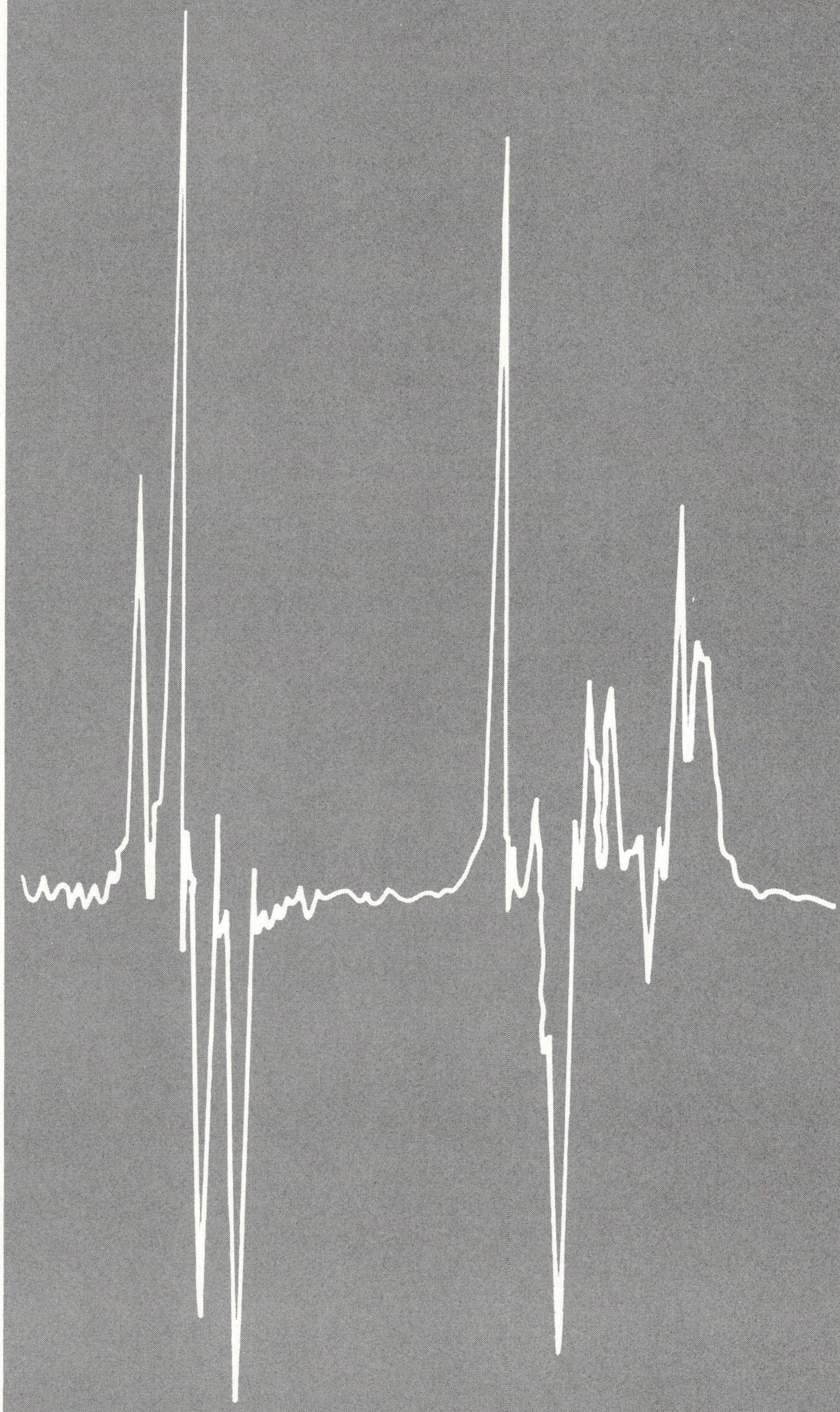

# CHAPTER 1 BASIC CONCEPTS OF SPECTROSCOPY

## 1.1 *Introduction*

The subject of spectroscopy is the interaction of electromagnetic radiation with matter. The subject of organic spectroscopy is the interaction of electromagnetic radiation with assemblies of organic molecules. Absorption spectroscopy is the most widespread; energy quanta from electromagnetic waves are absorbed by the sample submitted to the radiation. The amount of energy absorbed depends upon the wavelength of the incoming beam since, depending upon the value of the electromagnetic frequency $\nu$, transitions between different kinds of energy states are induced in the molecules of the sample. Spectroscopy consists in the recording of the absorption changes as the frequency $\nu$ is varied over a given range within the electromagnetic spectrum. The plot of absorption energy versus frequency usually shows a number of absorption bands whose characteristic position, intensity, width, and shape yield precious information to the chemist.

Energy occurs in discrete quanta; quantum mechanics tells us that the energy states of molecules exist only as discrete sets. Consider, for instance, the chloroform molecule $CHCl_3$; let us focus our attention on the hydrogen atom or, more specifically, the hydrogen nucleus or proton. If chloroform molecules are submitted to a magnetic field $\mathbf{H}_0$ (under suitable conditions, to be specified in Section 1.8), their protons will belong to either of two energy states depending upon whether their spin vectors are oriented parallel or antiparallel with respect to the direction of the applied magnetic field $\mathbf{H}_0$. All of the nuclei oriented parallel to the field (denoted by $\alpha$) will have the same energy, while all the protons oriented antiparallel to the field (denoted by $\beta$) have a different energy, as represented schematically in Figure 1.1.

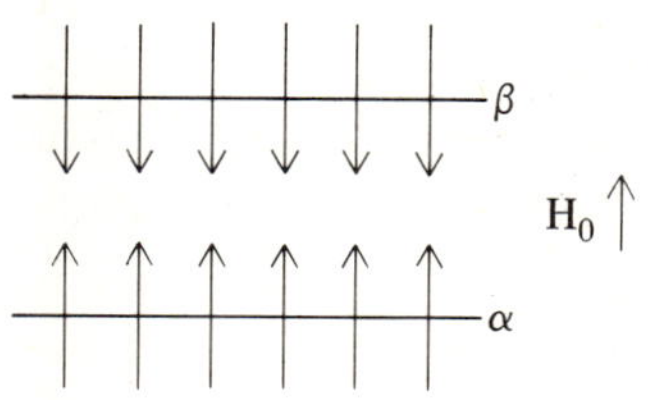


*FIGURE 1.1*
*Alignment of nuclear spins parallel (α) and antiparallel (β) to an external magnetic field $H_0$.*

This is the first fundamental concept, that of energy states. In a large assembly of molecules (chloroform molecules in this instance) an external magnetic field $H_0$ is applied and a clear-cut situation arises; all of the molecules are found to belong to either of two and only two energy states. There are a number of relevant questions about the phenomenon just described. What is nuclear spin? How does the ordering of the nuclear spins (into the $\alpha$ and $\beta$ states) brought about by an external magnetic field $H_0$ occur? What is the probability of transitions occurring between these ordered states? What is the degree of ordering and the relative energies of these ordered states? What happens when the external magnetic field is removed? Let us examine these questions and, in the process, gain some insight into the nature of spectroscopy in general.

## 1.2 *Nuclear Spin*

Like so many characteristics of elementary particles, spin is best defined by its properties since other descriptions would be misleading. For instance, an analogy has often been made with the spinning of a top. Unfortunately, if the nuclear particles were really spinning, the velocity of the periphery would have to be greater than the speed of light, thus violating relativity. In the words of Dirac, "The spin does not correspond very closely to anything in classical mechanics, so the method of classical analogy is not suitable for studying it."[1] The way out, of course, is to consider that spin is

[1] P. A. M. Dirac, *Principles of Quantum Mechanics*, 4th ed., The Clarendon Press, Oxford, 1959, p. 143.

quantized; that is, for a given particle, the component of the spin vector $I$ (or angular momentum vector $\hbar I = hI/2\pi$) in any one direction can only take up the $2I + 1$ values $l = +I, I - 1, \ldots, -I$. Not only is there spin due to the orbital angular momentum of a particle traveling in orbit, there is also an intrinsic spin for the motion of the particle about its own axis. Electrons, protons, and neutrons all have intrinsic spins of $\frac{1}{2}$ (in units of $h/2\pi$), so that only two directions in space can be assumed whose projections on an arbitrary axis can have the two values $+\frac{1}{2}$ and $-\frac{1}{2}$. It turns out that spin and orbital angular momentum are strongly coupled and the total angular momentum can then have the values of $l + \frac{1}{2}$ and $l - \frac{1}{2}$. For instance, when $l = 0$, there are only two states with the spin angular momentum equal to $+\frac{1}{2}$ and $-\frac{1}{2}$: These are denoted as the $1s_{1/2}$ states. When $l = 1$, the total angular momentum can assume the values $+\frac{3}{2}$, $+\frac{1}{2}$, $-\frac{1}{2}$, and $-\frac{3}{2}$; the values $+\frac{1}{2}$ and $-\frac{1}{2}$ occur twice since $l$ is taking the three possible values $+1$, 0, and $-1$. Therefore, these energy states will be denoted $1p_{3/2}$ and $1p_{1/2}$.[2] The $d$ and $f$ states are defined in a similar manner, and the student is already familiar with these terms from what he knows about electrons. It turns out that nucleons (protons and neutrons) behave in some way analogous with electrons. As a filled shell of 2, 8, 18, 32, and so on, electrons corresponds to a stable configuration, it turns out that nuclear "shells" containing 2, 8, 20, 28, and so on, protons or neutrons are especially stable. Of course the nature of the forces acting upon protons and neutrons within the nucleus is very different from that of the forces acting upon electrons. This is why atomic-shell numbers and nuclear-shell numbers are not the same. Nevertheless the analogy with the periodic classification has sufficient merit that a number of nuclear properties, such as the resultant spin quantum number for the nucleus itself, can be predicted from consideration of the shell model. One such prediction is that a closed-shell nucleus will have an angular momentum of zero, and hence a spin number $I = 0$. That is the case for ${}^{4}_{2}\mathrm{He}$ or for ${}^{16}_{8}\mathrm{O}$, which have closed shells of two protons and two neutrons, and of eight protons and eight neutrons, respectively. More generally, one of the regularities in nuclear properties from which the shell model of the nucleus could be developed is the following: In a nucleus with an even number of neutrons and an odd number of protons the neutrons couple their spins to a zero total spin vector, and thus do not influence the total nuclear angular momentum. The same statement holds when the words "neutron" and "proton" are interchanged. Nuclei having one more or one less neutron than a closed shell of protons and of neutrons should have a nuclear spin $I$ depending only upon the level of the last nucleon. For instance, ${}^{17}_{8}\mathrm{O}$, with eight protons and nine neutrons, has one more neutron than the closed shell. This closed shell can be written as $1s_{1/2}1p_{3/2}1p_{1/2}$. The ordering of energy levels is as written, the $1p_{3/2}$ level being lower in energy than the $1p_{1/2}$ level. The extra neutron must go into the next energy level, which is the $1d_{5/2}$ level; this nuclear species indeed has a spin number $I = \frac{5}{2}$.[3] Based upon these and other regularities, rules have been developed for the determination of nuclear spin numbers.

*Rule A.* Nuclei with *even* mass numbers and *even* nuclear charges are spinless. Examples:

${}^{12}_{6}\mathrm{C}$ ${}^{16}_{8}\mathrm{O}$ ${}^{18}_{8}\mathrm{O}$ ${}^{32}_{16}\mathrm{S}$

---

[2] L. Pauling, *General Chemistry*, 3d ed., W. H. Freeman and Co., San Francisco, Calif., 1970.
[3] M. G. Mayer, *Science*, *145*, 999 (1964).

*Rule B.* Nuclei with *odd* mass numbers have a spin which is an *odd* integer multiple of $\frac{1}{2}$. Examples:

| Nuclei | $^{1}H$ | $^{11}B$ | $^{13}C$ | $^{15}N$ | $^{17}O$ | $^{19}F$ | $^{29}Si$ | $^{31}P$ | $e^-$ |
|---|---|---|---|---|---|---|---|---|---|
| Spin number $I$ | $\frac{1}{2}$ | $\frac{3}{2}$ | $\frac{1}{2}$ | $\frac{1}{2}$ | $\frac{5}{2}$ | $\frac{1}{2}$ | $\frac{1}{2}$ | $\frac{1}{2}$ | $\frac{1}{2}$ |

*Rule C.* Nuclei with *even* mass numbers and *odd* nuclear charges have a spin which is an integer. Examples:

| Nuclei | $^{2}H(D)$ | $^{14}N$ |
|---|---|---|
| Spin Number $I$ | 1 | 1 |

The spin quantum number in turn specifies the number of energy states of the nuclear spin, since according to quantum mechanics the number of possible energy states is given by $2I + 1$, where $I$ is the spin quantum number. Since for protons $I$ is equal to $\frac{1}{2}$ only two energy states can exist for their nuclear spins in a magnetic field. For other nuclei this might of course be different, depending upon the value of the spin quantum number $I$. For spinless nuclei no orientation can occur in an external magnetic field, since there are no spin vectors to orient.

## 1.3 *Relative Energies of the α and β States*

The energy of interaction between one nucleus and the magnetic field $\mathbf{H}_0$ is simply the scalar product of the nuclear magnetic moment $\boldsymbol{\mu}_N$ by the magnetic field:

$$E = -\boldsymbol{\mu}_N \mathbf{H}_0$$

The magnetic moment $\boldsymbol{\mu}$ arises from the nucleus having the property of spin and also being a charged particle. It is as if a little magnet were attached along the direction of the spin vector. The magnetic moment $\boldsymbol{\mu}$ is equal to $\gamma_N \hbar \mathbf{I}$, where $\hbar$ = Planck's constant/$2\pi$, $\mathbf{I}$ is the spin vector already mentioned, and the proportionality constant $\gamma_N$ is a number, characteristic of each nuclear species, called the gyromagnetic ratio.

Going back to the hydrogen nuclei in the $CHCl_3$ molecules, their energy of interaction with $\mathbf{H}_0$ is given by

$$E = -\gamma_H \hbar \mathbf{I} \mathbf{H}_0$$

where $I$ can take *only* the two values $+\frac{1}{2}$ or $-\frac{1}{2}$.

In Figure 1.2 the magnitude of the splitting between the two energy states α and β, referred to the energy of the system in the absence of a magnetic field (dotted line), is seen to be proportional to the magnitude of the externally applied magnetic field $\mathbf{H}_0$. It is evident from this expression of the energy that the spin number $I$ specifies the number of different orientations of the nuclear spins with respect to the field $\mathbf{H}_0$, while the gyromagnetic ratio $\gamma$ specifies the exact energy for each orientation in a given magnetic field $\mathbf{H}_0$.

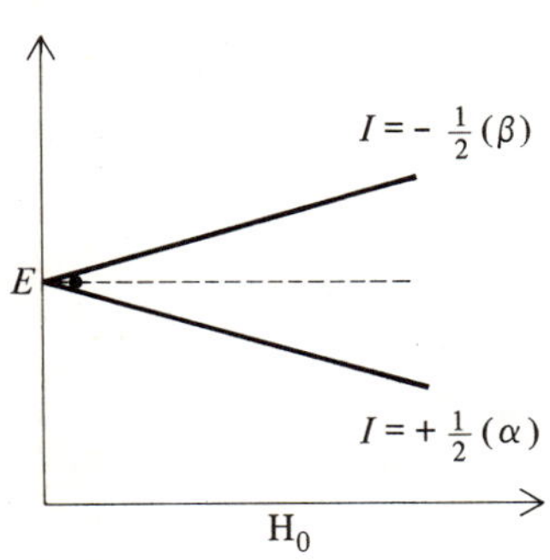


*FIGURE 1.2*
*The relative energies of the (α) and β states as a function of the applied magnetic field* $\mathbf{H}_0$.

## 1.4 *Number of Nuclei in the Two Energy States and the Difference in Population Between the α and β States*

The difference $E_\beta - E_\alpha$ between the energies of the protons $H_\alpha$ and $H_\beta$ is given by

$$E_\beta = -\gamma_H \hbar(-\tfrac{1}{2})H_0 = \tfrac{1}{2}\gamma_H \hbar H_0$$
$$E_\alpha = -\gamma_H \hbar(+\tfrac{1}{2})H_0 = -\tfrac{1}{2}\gamma_H \hbar H_0$$
$$E_\beta - E_\alpha = \gamma_H \hbar H_0$$

Thermodynamics then gives the ratio $N_\alpha/N_\beta$ of the populations as $N_\alpha/N_\beta = \exp(\gamma_H \hbar H_0/KT)$, where $K$ is Boltzmann's constant.

At ordinary temperatures and with the magnetic fields available on Earth, $\gamma\hbar H_0$ is much smaller than $KT$, so that $N_\alpha/N_\beta \sim 1 + \gamma\hbar H_0/KT$. Hence there are

$$N_\alpha = \frac{N}{2}\left(1 + \frac{\gamma\hbar H_0}{KT}\right) \text{ nuclei in state } \alpha$$

$$N_\beta = \frac{N}{2}\left(1 - \frac{\gamma\hbar H_0}{KT}\right) \text{ nuclei in state } \beta$$

For the chloroform protons in a magnetic field $H_0$ of 10,000 Oe, $\gamma\hbar H_0/KT \sim 10^{-5}$ and the two populations $N_\alpha$ and $N_\beta$ differ by only 10 out of 1 million! (The obvious consequence is that if we want to increase the difference between the two populations we shall have to decrease the temperature or, more practically, increase the magnetic field $H_0$ as much as possible.)

Since there are two different, *unequally populated*, energy states spectroscopy becomes possible. That is, it should be possible to induce *transitions* between the two states and use a quantum $h\nu$ of electromagnetic radiation in order to bring one chloroform proton from state $\alpha$ up to the excited state $\beta$.

Let us next find out how the experiment should be done in practice. Setting $h\nu$ equal to the difference in energy,

$$h\nu = \Delta E = \gamma_H \hbar H_0$$

we see that

$$\nu = \gamma_H \frac{H_0}{2\pi}$$

Using the value of $\gamma_H = 2.67530 \times 10^4$ rad/sec Oe, it is found that a frequency $\nu$ of $42.6 \times 10^6$ Hz (42.6 MHz) will be needed for the experiment if the magnetic field $H_0$ used is 10,000 Oe, whereas in a field of 14,000 Oe, protons resonate at 60 MHz or in a field of 23,500 Oe and 51,600 Oe protons resonate at 100 MHz and 220 MHz, respectively. Hence the term 60 MHz or 100 MHz spectrometer. From all of the above it is obvious that unless a frequency of exactly 42.6 MHz is used (with a magnetic field of 10,000 Oe) no chloroform protons will change their spin states; however, as soon as that frequency is reached the protons in the $\alpha$ state will abandon it for the $\beta$ state.

## 1.5 *The Saturation Problem*

We have just seen that the populations of the $\alpha$ state and of the $\beta$ state have different values. These are given by the Boltzmann distribution if and only if these two energy states are in thermal equilibrium. Spectroscopy, in this case the upward transition taking nuclei from the $\alpha$ into the $\beta$ state, has the effect of disturbing the normal distribution of particles in the $\alpha$ and the $\beta$ state from thermal equilibrium. Therefore one witnesses in such an experiment two antagonistic tendencies: the spectroscopic absorption of energy which tends to erase population differences, and thermal collisions which tend to restore the system to its Boltzmann-statistics behavior. At this point it should be emphasized that spectroscopy is feasible only if the two energy levels differ in population; otherwise the probability for an upward transition (absorption) is exactly matched by the probability for a downward transition (emission), the net result being zero change. One is thus led to the conclusion that the spectroscopic experiment can be maintained only if there is *some* mechanism for particles in the excited state to decay to the lower or ground state *faster* than the spectroscopic transition itself brings them into the upper state. This mechanism of decay is called "relaxation." The hydrogen nuclei in the excited state relax into the lower energy state, primarily through thermal collisions. Other molecules in the medium also possess magnetic moments and as they rapidly tumble about, these magnetic moments create small fluctuating magnetic fields having a Fourier component right at the absorption frequency $\nu$,[4] which then induce in a more or less random manner the transition from the upper to the lower state (why not the reverse? Because the net result would then be in violation of the second principle of thermodynamics). This particular relaxation mechanism is called *spin-lattice* relaxation, a particular spin reverting to its initial state by interactions with the neighboring molecules in the lattice. There is also a *spin-spin* relaxation mechanism involving nuclear moments in the same molecule. In spectroscopies other than magnetic resonance, relaxation will occur by different processes such as fluorescence, phosphorescence, and radiationless (thermal) decay (in the case of electronic transitions). The important feature is that the excited state has, by definition, higher energy and must decay back to the ground state in the absence of external perturbation. It can do so spontaneously; the probability for such spontaneous emission is clearly proportional to the number of particles present in the excited state. (This probability is therefore very small in the case of the magnetic resonance experiment.) Alternatively, emission can be *induced*, as in spin-lattice relaxation, and the probability of such induced emission is proportional both to the number of particles in the upper state and to the amount of energy contained in the perturbing radiation (the fluctuating magnetic field in the case of the magnetic resonance experiment). The rate of decay is governed by the particular relaxation process available to the excited state. In all cases involving *only* two energy states the rate law must be an exponential and in practice *most* relaxation processes are found to involve exponential returns to the thermal equilibrium distribution of particles.

## 1.6 *Lineshape*

It is unreal to expect that the energy absorbed by protons in the chloroform molecules will be exactly at the resonance frequency $\nu$. Experimentally, a spread of frequencies is observed with the maximum absorption occurring at $\nu$. This is due to the energy levels themselves not being sharply defined, since the uncertainty principle states that the product of the uncertainty on the measurement of energy $\Delta E$ by the uncertainty on the measurement of time $\Delta t$ *must* be of the order of $\hbar$:

$$\Delta E \, \Delta t \sim \hbar$$

Hence

$$\Delta \nu \, \Delta t \sim \frac{1}{2\pi}$$

The linewidth $\Delta\nu$ is proportional to the inverse of the average time spent in the excited state, which is by definition short compared with the lifetime of the ground state. Hence a sharp absorption line means that the excited state has a long lifetime and broad absorption implies shortlived excited states. In practice, the lifetime of the excited state is most often controlled by the available relaxation processes, which make particles revert to the ground state. If these are due to thermal collisions, as in the spin-lattice relaxation mode for nuclei, then $\Delta t$ becomes the mean time between collisions: The linewidth reflects what is called the correlation time in solution. A more detailed discussion of relaxation will be provided in Chapter 5. Suffice it to say here that there are other factors leading to line broadenings, apart from relaxation

[4] Any ordinary function of time can be expressed as a sum of simple harmonic functions: $f(t) = a_0 + a_1 \cos \omega t + b_1 \sin \omega t + \cdots + a_n \cos n\omega t + b_n \sin n\omega t$. Such a breakdown is called a Fourier decomposition. Thus, there will be a component $a_k \cos k\omega t + b_k \sin k\omega t$ such that $\nu$ (the absorption frequency) is identical to $k\omega/2\pi$.

processes. One of these is the coupling between neighboring magnetic moments in the case of the magnetic resonance experiment; this will also be considered in Chapter 3 on spin-spin coupling. The shape of absorption lines is usually Gaussian or Lorentzian, with the latter predominating. A Gaussian lineshape is a function of the type

$$f(\nu) = \frac{1}{b} \exp\left[-\frac{(\nu - \nu_0)^2}{b^2}\right]$$

in which $\nu$ is the frequency and $\nu_0$ the frequency of the absorption maximum. A Lorentzian lineshape is a function of the type

$$f(\nu) = \frac{ab^2}{(\nu - \nu_0)^2 + b^2}$$

In each of these expressions, $a$ and $b$ are constants which give the exact lineshape. For instance, the width at half-height of a *Gaussian* line is $1.18b$.[5]

Another factor affecting lineshape is the overlapping of absorption bands whenever a complex absorption pattern results from the superposition of individual bands merged into a common envelope. This problem can be tackled with computers, provided that the total number of components and the lineshape for each individual component (e.g., Gaussian or Lorentzian) is known. Analog computers, such as the DuPont Curve Resolver, are best suited for this purpose, but digital computers can also be used.[6]

## 1.7 *Probability of Transitions*

The probability that a given transition will occur can be either zero or finite according to the corresponding selection rules given by quantum mechanics (an example is given at the end of Chapter 2). These are symmetry or parity rules, and are equivalent to the notion that, depending upon the type of spectroscopy considered, transitions will be either allowed or forbidden between states having or not having the same parity. For nuclear magnetic resonance the appropriate selection rule states that the total spin number must change by one unit. Because of energy quantization, there are selection rules for every type of spectroscopy; as discussed in later sections (see Chapter 2).

What happens in the nuclear magnetic resonance experiment when the magnetic field $H_0$ is removed? There is an exponential return to thermal equilibrium. This can be followed by the exponential decay of the nuclear magnetic resonance line and measured by monitoring the intensity of the signal as a function of time. This represents a means for measuring the rate of relaxation in the nuclear magnetic resonance experiment.

In order to introduce some of the most important spectroscopic concepts we have taken the example of the magnetic resonance experiment. In summary, spectroscopy deals with quantized energy states in molecules. When thermal equilibrium is present the energy differences are reflected in differences in populations of the lower and of the higher states. The populations are then given by a Boltzmann distribution. Absorption of energy by the molecules amounts to increasing the populations of the higher states by depletion of that of the lower states. This absorption is conditional upon the transition between two particular states being allowed by a corresponding selection rule. The absorption line is usually Lorentzian and its width is generally dependent upon the frequency of intermolecular collisions, which tend to bring the system back into

[5] R. S. Seshadri and R. N. Jones, *Spectrochim. Acta, 19*, 1013 (1963).
[6] H. Stone, *J. Opt. Soc. Amer., 52*, 998 (1962).

thermal equilibrium. The frequency at which maximum absorption occurs corresponds to the energy quantum separating the lower from the upper state. Relaxation processes render spectroscopy feasible. To a crude approximation, one can think of spectroscopy as an intramolecular process, while relaxation phenomena are mainly intermolecular in nature.

## 1.8 *The Nuclear Magnetic Resonance Spectrometer*

These principles are embodied in the design of nmr spectrometers. The directing magnetic field $H_0$ is produced by a magnet—permanent or electromagnet. The sample is placed in the gap between the two poles of the magnet, and wrapped in a coil which serves to submit it to the scanning magnetic field $H_1$, in a direction perpendicular to that of $H_0$. The absorption of the $H_1$ radiation is detected, amplified, and displayed on an oscilloscope or recorder. In order to vary the applied magnetic field $H_0$, small Helmholtz coils are disposed next to the magnet. By varying with the sweep generator (Figure 1.3) the electric current in these coils, $H_0$ can be modified by very small amounts, and one can thus sweep through a spectrum.

In general, a spectrometer consists of a source of monochromatic radiation, a conduct through which it passes before it hits the sample, a detector which measures the amount of the original radiation which has been absorbed by the sample, and a recorder.

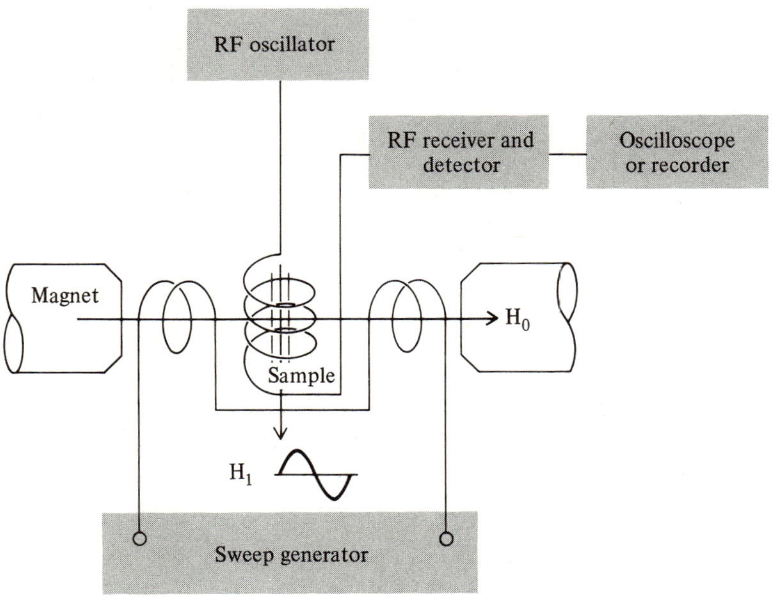


*FIGURE 1.3*
*Simplified diagram of an nmr spectrometer.*

## 1.9 *PROBLEMS*

SAMPLE PROBLEM

Consider the hypothetical vinyl cation, $H_2C{=}\overset{+}{C}H$, a molecule which has been postulated as a reaction intermediate. How many proton magnetic resonance (pmr) lines can, in principle, be observed for this three-spin system?

ANSWER

15 lines.

The three proton spins can adopt one of the following configurations:

| | | | | | | | | | | | |
|---|---|---|---|---|---|---|---|---|---|---|---|
| ↑↑↑ | or | ↑↓↑ | ↓↑↑ | ↑↑↓ | or | ↓↓↑ | ↓↑↓ | ↑↓↓ | or | ↓↓↓ |
| $\alpha\alpha\alpha$ | | $\alpha\beta\alpha$ | $\beta\alpha\alpha$ | $\alpha\alpha\beta$ | | $\beta\beta\alpha$ | $\beta\alpha\beta$ | $\alpha\beta\beta$ | | $\beta\beta\beta$ |

Let us order their energies according to the following diagram:

| $E$ | | | Spin number |
|---|---|---|---|
| | $\beta\beta\beta$ | | $-\frac{3}{2}$ |
| $\beta\beta\alpha$ | $\beta\alpha\beta$ | $\alpha\beta\beta$ | $-\frac{1}{2}$ |
| $\alpha\beta\alpha$ | $\beta\alpha\alpha$ | $\alpha\alpha\beta$ | $+\frac{1}{2}$ |
| | $\alpha\alpha\alpha$ | | $+\frac{3}{2}$ |

The allowed transitions, corresponding to a change of $\pm 1$ in the spin number will be

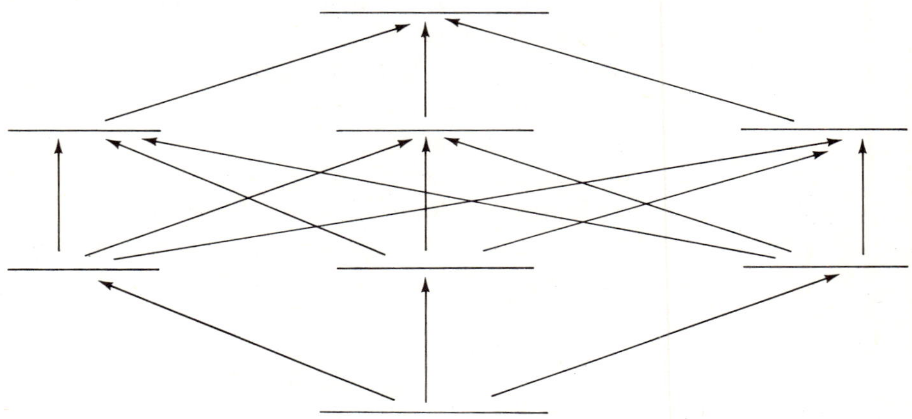

Another representation, very useful for visualizing the transitions between energy states, is that of a cube where each corner corresponds to an energy state and the transitions are given by the 12 edges and the 3 diagonals of the cube:

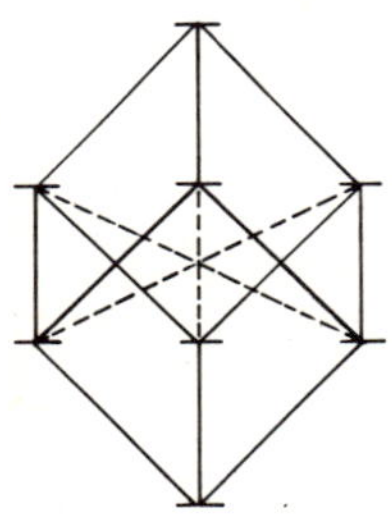

In reality, many vinyl derivatives, such as vinyl chloride and acrylonitrile, give rise to such multiline proton transitions (Figure 1.4). The chlorine or nitrogen spins do not interfere because of fast quadrupolar relaxation, a phenomenon to be discussed in Chapter 5.

1. Account for the spin quantum numbers of carbon-13, nitrogen-14, and deuterium.
2. The half-life for nuclei in the excited state as controlled by spin-lattice relaxation processes can be of the order of hours in solids below 4°$K$, while it is much shorter at higher temperatures or in liquids. Why?
3. What is the significance of the term $\Delta\nu$ in the expression of the Lorentzian line-shape,

$$E(\nu) \propto \frac{1}{(\nu - \nu_0)^2 + (\Delta\nu)^2}$$

(Hint: Establish its relationship to the width at half-height.)

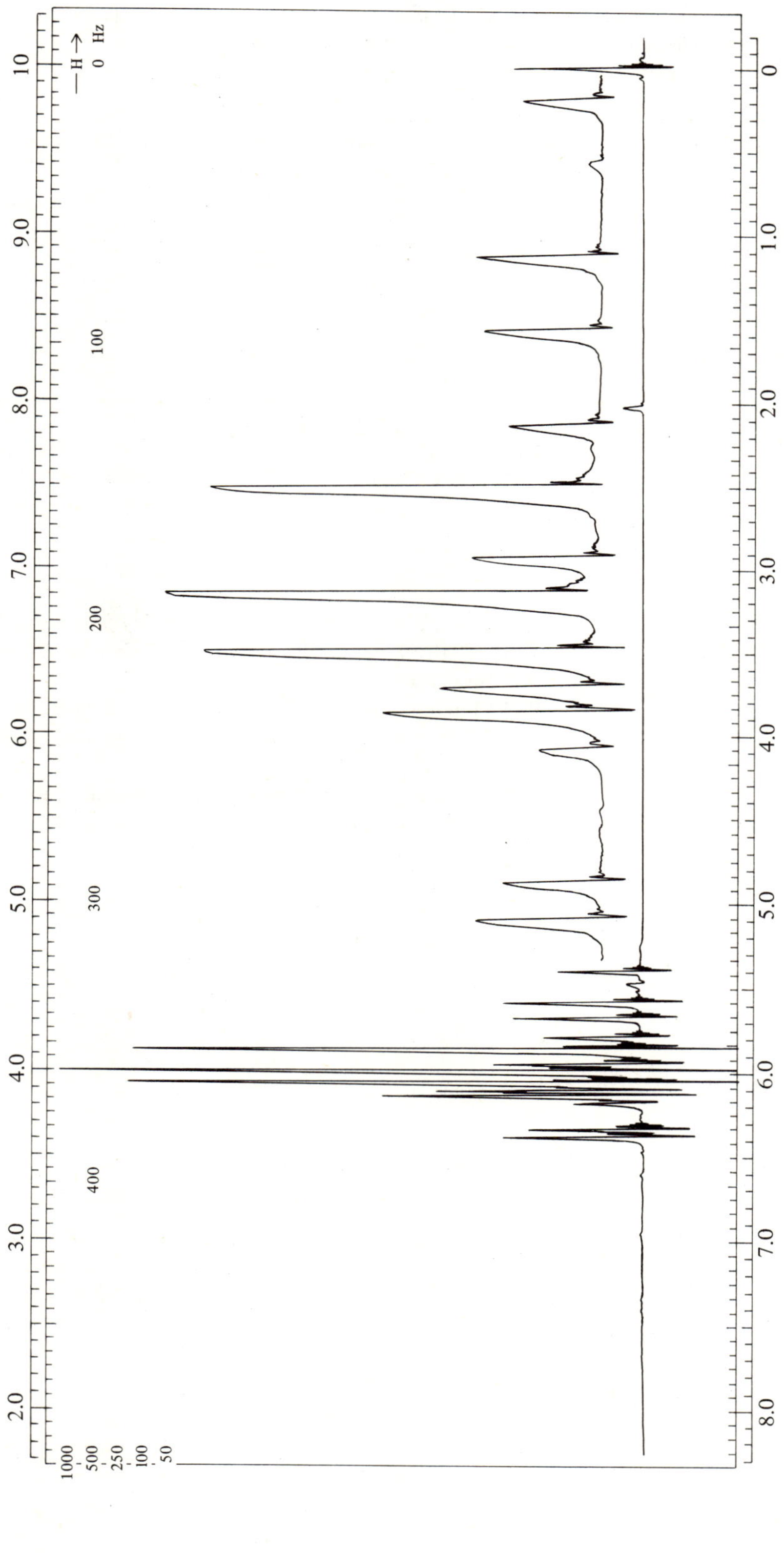


*FIGURE 1.4*
*Spectrum of acrylonitrile (60 MHz). The upper trace is an expansion of the spectrum.*

## 1.10
### *APPENDIX I*
### *The electromagnetic spectrum*

| Spectral region | x-Rays | Far uv, uv | Visible | Near ir, ir, Far ir | Microwave | TV | Radiofrequency esr, nmr | Audio |
|---|---|---|---|---|---|---|---|---|
| Molecular energy | electronic (inner shells) | electron (valence shell) | | vibrational | rotational | | spin states | |
| Wavelength ($\mu$) | $10^{-4}$ $10^{-3}$ | $10^{-2}$ $10^{-1}$ | | 1 10 | $10^{2}$ $10^{3}$ $10^{4}$ | $10^{5}$ | $10^{6}$ $10^{7}$ $10^{8}$ $10^{9}$ | $10^{10}$ $10^{11}$ |
| Frequency (Hz) | $10^{18}$ $10^{17}$ | $10^{16}$ $10^{15}$ | | $10^{14}$ $10^{13}$ | $10^{12}$ $10^{11}$ $10^{10}$ | $10^{9}$ | $10^{8}$ $10^{7}$ $10^{6}$ $10^{5}$ | $10^{4}$ $10^{3}$ |
| Energy (cal/mole) | $10^{8}$ $10^{7}$ | $10^{6}$ $10^{5}$ | | $10^{4}$ $10^{3}$ | $10^{2}$ 10 1 | $10^{-1}$ | $10^{-2}$ $10^{-3}$ $10^{-4}$ $10^{-5}$ | $10^{-6}$ $10^{-7}$ |

### *APPENDIX II*
### *Conversion factors between energy units*

| | kcal/mole | eV | Hz | $cm^{-1}$ |
|---|---|---|---|---|
| 1 $cm^{-1}$ | $2.86 \times 10^{-3}$ | $1.24 \times 10^{-4}$ | $3.00 \times 10^{10}$ | 1 |
| 1 Hz | $9.53 \times 10^{-14}$ | $4.13 \times 10^{-15}$ | 1 | $3.33 \times 10^{-11}$ |
| 1 kcal/mole | 1 | $4.34 \times 10^{-2}$ | $1.05 \times 10^{13}$ | $3.50 \times 10^{2}$ |
| 1 eV | 23.1 | 1 | $2.42 \times 10^{14}$ | $8.07 \times 10^{3}$ |

## 1.11
### *SUGGESTED READINGS*


D. H. Whiffen, *Spectroscopy*, John Wiley and Sons, Inc., New York, 1966.

T. L. Cottrell, *Dynamic Aspects of Molecular Energy States*, John Wiley and Sons, Inc., New York, 1965.

R. P. Feynman, R. B. Leighton, and M. Sands, *The Feynman Lectures on Physics*, 4th ed., Addison-Wesley Publishing Company, Inc., Reading, Mass., 1966. Sections II–34, 35 and III–10, 12.

R. McWeeny, *Spins in Chemistry*, Academic Press, New York, 1970, Lecture 1.

M. G., Mayer, "The Shell Model," *Science 145*, 999 (1964).

# CHAPTER 2 TYPES OF SPECTROSCOPY

Chapter 1 described a typical spectroscopic experiment in which electromagnetic radiation is used to induce transitions between energy states. We happened to choose nuclear energy states, and found that their separation, $\gamma_H \hbar H_0$, is of the order of $10^{-3}$ cal for protons in a magnetic field $H_0$ of 10,000 Oe, a separation much smaller than $KT$.

Let us define the various types of spectroscopies used by organic chemists by the nature of the energy states between which transitions are induced.

## 2.1 *Electron Spin Resonance*

Electron spin resonance (esr) is in principle identical to nuclear magnetic resonance (nmr). The only difference is that due to the negative magnetic moment of the electron, the lower energy state has the electron spin antiparallel to the field; the electron, like the proton, has a spin quantum number of $\frac{1}{2}$. The energy difference between the two spin states is

$$\Delta E = h\nu = g\beta H_0 \qquad \text{with } g \sim 2 \text{ and } \beta = \frac{e\hbar}{2mc}$$

$\Delta E$ is of the order of 1 cal, still much smaller than $KT$. Hence the two levels are nearly equally populated and saturation problems are almost as important as in nmr. It is possible to get absorption by varying either the microwave frequency or the magnetic field $H_0$, but since it is experimentally difficult to vary a microwave frequency over a large range, most esr spectrometers operate at a fixed frequency, close to 9000 MHz, and the magnetic field $H$ is varied.

The esr spectrometer is very similar to the nmr spectrometer. The source of monochromatic radiation, in the microwave region, is a klystron. Because of the wavelengths utilized, microwave components are introduced: The sample is now inserted in a resonant cavity connected by rectangular waveguides to the microwave oscillator and to the detector—a crystal or a bolometer.

## 2.2 *Nuclear Quadrupole Resonance*

Nuclei with spin numbers $I$ greater than or equal to 1 have electrical quadrupole moments $Q$, due to the nonspherical symmetry of the electrical charges of the nucleus:

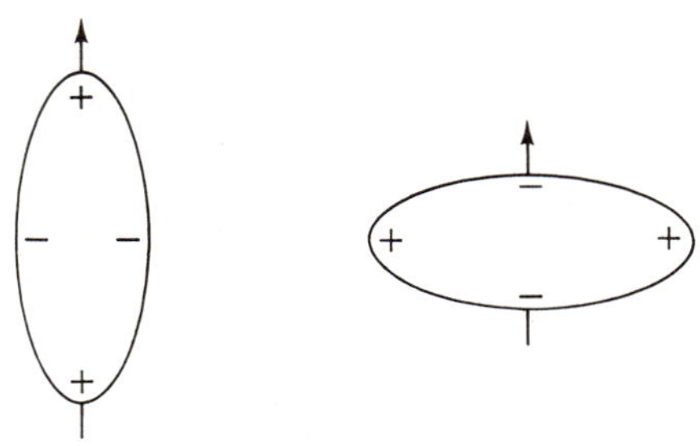

$Q > 0$ $Q < 0$

When placed in an electrostatic gradient, as provided in a polar molecule, these quadrupolar nuclei display additional splitting of the energy states. If the electrostatic field $E$ is axially symmetric (as, for instance, the C—Cl bond in $CH_3Cl$) the values of the various quadrupolar energy states are given by

$$E_m = eQq \frac{[3m^2 - I(I+1)]}{4I(2I-1)}$$

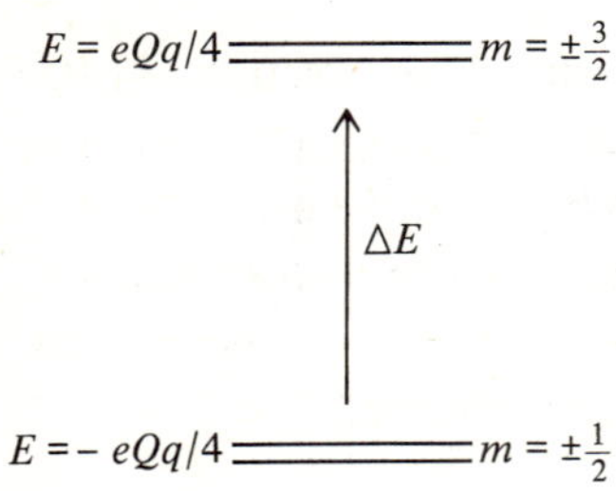


*FIGURE 2.1*
*Quadrupolar energy levels for $^{35}Cl$ in an axially symmetrical molecule.*

Where $e$ is the electron charge, $q = \delta E/\delta Z$, the gradient of the electric field $E$ along the $Z$ axis, which is chosen to coincide with the C—Cl bond, and the nuclear magnetic quantum number $m$ is the projection of the spin vector $I$. (See Section 2.8 for the derivation and explanation of the above equation.) With $I = \frac{3}{2}$ for $^{35}Cl$, $m$ can take $2I + 1$ values: $+\frac{3}{2}$, $+\frac{1}{2}$, $-\frac{1}{2}$, and $-\frac{3}{2}$. Hence

$$E_{3/2} = eQq\,\frac{(27/4) - (15/4)}{6 \times 2} = \frac{eQq}{4} = E_{-3/2}$$

$$E_{1/2} = eQq\,\frac{(3/4) - (15/4)}{6 \times 2} = \frac{-eQq}{4} = E_{-1/2}$$

The selection rule for nuclear quadrupole resonance (nqr) transitions (as for nmr transitions) is $\Delta m = \pm 1$ so that we shall be able to induce a transition between the two pairs of degenerate levels (Figure 2.1) provided that

$$h\nu = \frac{eqQ}{2}$$

Almost all nqr transitions are in the range of 1–2000 MHz. $^{35}Cl$ signals are usually found in the range of 20–50 MHz, and since the quantity $eQ$ is a nuclear constant, it is obvious that quadrupole resonance is sensitive to the polarity of chlorine-containing bonds. For example, methyl chloride resonates at 34,029 MHz, ethyl chloride at 33,080 MHz, and propyl chloride at 32,968 MHz. One important limitation has to be emphasized at this point. In the nuclear magnetic resonance experiment the splitting between the energy states is induced by a large magnetic field $H_0$ as produced in an electromagnet. As a consequence, all phases of matter (gas, liquid, or solid) can be submitted to the nmr experiment. For nuclear quadrupole resonance, however, the splitting is due to the interaction of the nuclear quadrupole moment with a molecular gradient in the absence of any external perturbation. Therefore, nqr measurements have to be made in the solid phase at low temperature, since in a liquid or a gas the direction of a molecular bond fluctuates more rapidly with respect to the laboratory referentials than the nucleus can reorient itself with respect to the bond, and nqr disappears.

The nqr spectrometer consists of an oscillator, the source for the radiofrequency (RF) radiation applied to the solid sample in a Dewar, usually at liquid nitrogen temperature, a detector, and a recorder. In essence, it is a magnetic resonance spectrometer without the magnet. As indicated earlier, quantization of the quadrupolar energy levels is effected by the electric gradients $q$ available in molecules, rather than by an externally applied magnetic field.

In summary, resonance spectroscopies are characterized by splittings between energy states of the order of $10^{-3}$–1 cal, with the frequency of the corresponding transitions in the RF range of $10^{7}$–$10^{10}$ Hz.

## 2.3 *Microwave Spectroscopy*

At energies slightly above those of the spin states one finds rotational states with energy differences of $10^{-2}$–1 cal/mole. These are employed in microwave spectroscopy to determine molecular moments of inertia, which yield extremely useful structural information for small molecules in the gas phase.

Microwave spectroscopy makes use of the quantization of rotational energy states. For a diatomic molecule AB, the total kinetic energy can be expressed as

$$\frac{1}{2}(m_A r_A^2 + m_B r_B^2)\omega^2 = \frac{1}{2} I\omega^2 = \frac{p^2}{2I}$$

where $\omega$ is the angular velocity around the center of mass, $r_A$ and $r_B$ are the distance of atoms A and B from the center of mass, $I$ is the moment of inertia, and $p = I\omega$ is the angular momentum.

In general, the rotational energy can be expressed as

$$E_{rot} = \frac{p_x^2}{2I_{xx}} + \frac{p_y^2}{2I_{yy}} + \frac{p_z^2}{2I_{zz}}$$

since all the cross terms of the form $\sum m_i y_i x_i$ vanish.

For linear molecules, a special kind of symmetrical top, this rotational energy is quantized according to[1]

$$E = \frac{J(J+1)\hbar^2}{2I}$$

since $p^2 = J(J+1)\hbar^2$. Transitions are allowed for $\Delta J = \pm 1$. (See Section 2.7 for the derivation of this selection rule.) Geometries are calculated from moments of inertia derived with the help of isotopic substitution (see the sample problem at the end of the chapter).

A microwave spectrometer is such that the wavelength of the radiation is comparable to the dimensions of the instrument. Therefore, rectangular guide lines are used to conduct the radiation through the various parts of the spectrometer, shown schematically in Figure 2.2. Another important feature is the use of Stark modulation in order to increase the normally low sensitivity of the spectrometer: An electrode is

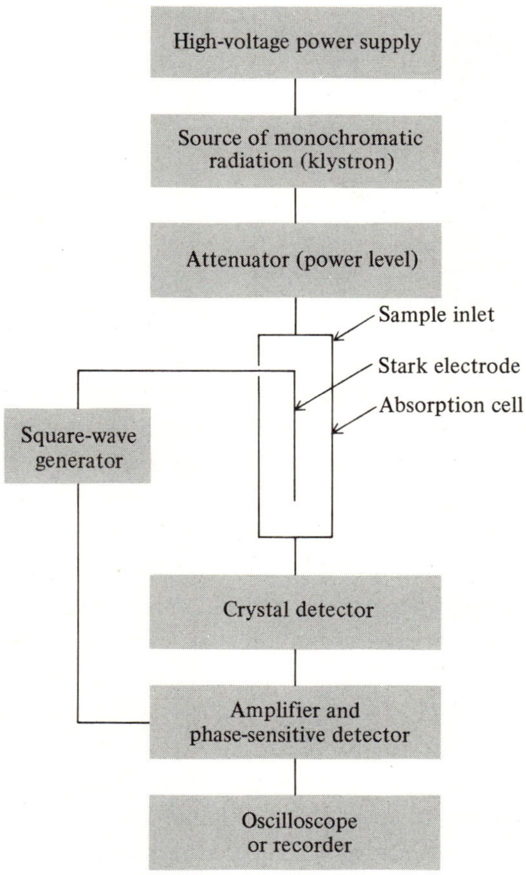


*FIGURE 2.2*
*Simplified block diagram of a microwave spectrometer.*

[1] J. E. Wollrab, *Rotational Spectra and Molecular Structure,* Academic Press, New York, 1967.

inserted in the sample tube in order to submit the sample to a rapidly varying electric field, in the form of a square wave. Phase-sensitive detection and amplification are applied after this modulation of the microwave signal by an electric field, which leads to heightened sensitivity.

## 2.4 *Optical Spectroscopy*

Any molecule possesses three types of internal energy: rotational, vibrational and electronic energies. In Section 2.3 we discussed briefly rotational levels and microwave spectroscopy. Let us now deal with the two remaining energy levels. The so-called optical methods, infrared-Raman and ultraviolet-visible spectroscopy, work with energy levels separated by about 1–10 kcal/mole and about $10^2$ kcal/mole respectively. We will examine each of these in turn.

### *2.4.1 Ultraviolet and Visible Spectroscopy*

In the ultraviolet and visible ranges, transitions occur between the different electronic energy states. Electronic spectra of atoms are usually sharp lines since atoms have relatively few, widely separated, electronic states. In molecules, on the other hand, each electronic state has associated with it a number of vibrational levels, each of which in turn has a series of closely spaced rotational levels. Consequently, there are a number of electronic transitions possible from the ground state to a higher electronic (excited) state (Figure 2.3), resulting in a broad absorption band with perhaps some fine structure due to vibrational interaction.

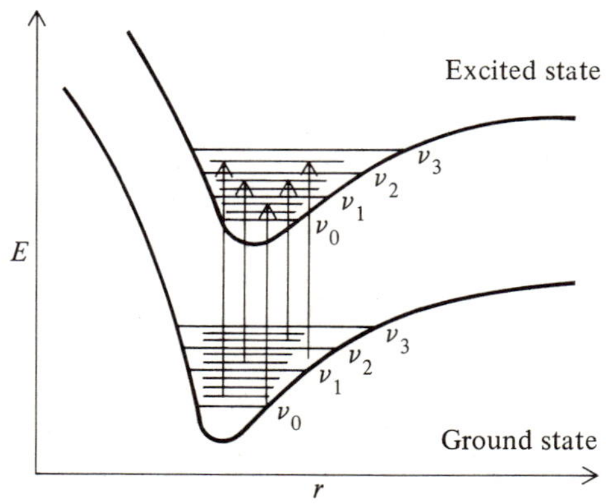


*FIGURE 2.3*
*Electronic transitions from a given rotational and vibrational level in the ground electronic state to a particular rotational and vibrational level in the excited state.*

TYPES OF TRANSITION

The types of excitation depend upon the molecular structure of the compound involved. For simple hydrocarbons, such as methane and ethane, it is possible to promote an electron from the bonding $\sigma$ orbital to the antibonding $\sigma^*$ orbital. This excitation is commonly referred to as a $\sigma \rightarrow \sigma^*$ transition and occurs in the far uv range since the energies involved are rather large. For example, for methane this absorption occurs around 125 nm and for ethane at 135 nm. Molecules containing heteroatoms (such as halides, ethers, amines, etc.) that possess nonbonding lone-pair electrons can undergo excitation that promotes these *n* electrons into the $\sigma^*$ orbital, referred to as $n \rightarrow \sigma^*$ transition. Since this transition requires less energy than the $\sigma \rightarrow \sigma^*$ excitation, it occurs at higher wavelength, usually just below 200 nm, and is sometimes referred to as end absorption.

Molecules possessing $\pi$ bonds, such as ethylene, can promote an electron from the bonding $\pi$ orbital into the antibonding $\pi^*$ orbital or from the bonding $\sigma$ orbital into the antibonding $\pi^*$ orbital, corresponding to $\pi \rightarrow \pi^*$ and $\sigma \rightarrow \pi^*$ transitions, respectively (see Sections 6.1 and 6.2). Compounds containing both $\pi$ bonds and nonbonding lone pairs (e.g., formaldehyde) can undergo an additional transition corresponding to promotion of a lone-pair electron into the antibonding $\pi^*$ orbital, referred to as an $n \rightarrow \pi^*$ transition (see Section 7.1).

PROBABILITY OF TRANSITIONS

Whereas the position of the absorption band depends upon the energy difference between the ground state and excited state of the particular transition, the intensity of absorption depends upon the probability of transition. The probability of transition can be ascertained from selection rules. There are two basic rules and they "prohibit" certain transitions from occurring because of quantum mechanical and symmetry consideration:

1. spin-forbidden transitions;
2. symmetry- and parity-forbidden transitions.[2]

The first rule concerns the multiplicity of spin energy states. In the ground state of molecules the two electrons in any molecular orbital must have their spins paired in accordance with the Pauli exclusion principle (no two electrons may have all four of their quantum numbers identical). However, when a molecule is excited and one of the two electrons is promoted from an occupied orbital into an empty orbital this restriction no longer applies, since the two electrons are in different orbitals. Therefore, the electrons either can be paired or have parallel spins. Since the total spin quantum number $J$ is given by $J = |2S| + 1$, where $S$ is the *sum* of the spins of the individual electrons, paired electrons correspond to $J = 1$ ($S = +\frac{1}{2} - \frac{1}{2} = 0$), the so-called singlet state, and parallel electrons to $J = 3$ ($S = +\frac{1}{2} + \frac{1}{2} = 1$ or $S = -\frac{1}{2} - \frac{1}{2} = -1$), the so-called triplet state. The first selection rule then states that transition between states of different multiplicity is forbidden. That is, electronic transitions can occur only from the singlet ground state to the singlet excited state (and not to a triplet excited state), or from a triplet ground state to a triplet excited state.

The second rule concerns the symmetry of the energy states and can be briefly explained as follows: If two orbitals are orthogonal and do not overlap, as for instance the nonbonding and antibonding $\pi^*$ orbitals in carbonyls (Section 7.1), the probability of a transition of an electron from one to the other is zero. In fact such transitions do occur, but with low intensity and hence low probability, indicating that there is vibrational interaction which distorts the molecule slightly and allows the transition to take place.

### *2.4.2 Infrared and Raman Spectroscopy*

All molecules at any temperature, above and including absolute zero, are undergoing vibrational motions; that is, motions where the bond length and bond angles are periodically displaced from their equilibrium values *without* producing any net translational or rotational movement. Although at first this might seem to indicate a random complex motion, closer examination shows that this is not the case.

An atom has three degrees of freedom, corresponding to movement along any of the three coordinate axes. A molecule with $n$ atoms therefore has $3n$ degrees of freedom. However, only $3n - 6$ of these are genuine vibrations ($3n - 5$ in the case of linear molecules), since the remaining six correspond to the motions of the entire molecule, either as movement along or rotations about any of the three coordinate axes. These $3n - 6$ internal vibrations are known as the normal modes of vibration and may in turn give rise to the so-called fundamental absorption bands either in the ir or the Raman. Whether these fundamental absorptions occur in the ir or Raman is again determined by selection rules derived from symmetry considerations and group

[2] H. H. Jaffé and M. Orchin, *Theory and Applications of Ultraviolet Spectroscopy*, John Wiley and Sons, Inc., New York, 1962, p. 118.

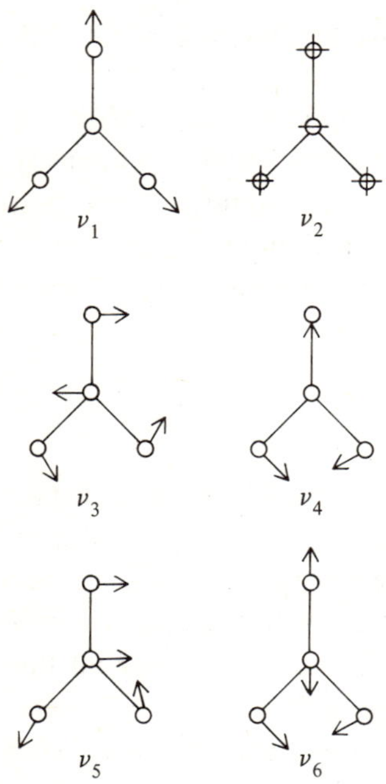


*FIGURE 2.4*
*The six normal modes of vibration of the carbonate ion.*

theory[3] which are beyond the scope of this text. They may be simply stated as follows: for ir activity, a change in the dipole moment relative to the ground state is required; for Raman activity, a change in polarizability relative to the ground state is required upon excitation.

As an example, Figure 2.4 shows the six [3(4) – 6] fundamental vibrations of the planar carbonate ion $CO_3^{2-}$. Of these six fundamental vibrations, $\nu_1$ is Raman active only, $\nu_2$ is ir active only, and $\nu_3$, $\nu_4$, $\nu_5$, and $\nu_6$ are both ir and Raman active.

In some cases absorptions that are active in the ir are inactive in Raman and vice versa. Among simple organic molecules fully and symmetrically substituted ethylenic double bonds, $R_2C{=}CR_2$, and acetylenic triple bonds, $RC{\equiv}CR$, for instance, are ir inactive but Raman active.

Absorption bands do not arise only from the normal modes of vibrations. Both the ir and Raman spectra contain significant absorption not due to the fundamental stretching or bending modes. These additional bands are commonly referred to as overtones, combinations, coupled, and Fermi resonance bands.[4]

Overtones are integral multiples of the fundamental absorption frequency and they decrease in intensity from first to second to higher overtones. Combination bands are sums or differences of two or more fundamental frequencies and are usually low in intensity. Coupling bands arise from the interaction of two absorption frequencies in the same region of the molecule. The resultant band is found at a different position than expected for each of the two individual components. Fermi resonance refers to the interaction or coupling of an overtone or combination band with a nearby fundamental vibration, resulting in enhanced intensity of the overtone or combination band or sometimes a splitting of the bands. Fermi resonance is a consequence of the anharmonicity of vibrations.

The instrumentation of uv-visible and is spectroscopies is basically the same. It consists of four major areas: (1) a radiation source (hydrogen discharge tube for uv, tungsten incandescent lamp for visible, and a Nernst filament or Globar for infrared; (2) an optical area that divides the source into a sample and reference beam as well as decomposes the radiation into the desired wavelength either by means of a prism or a grating; (3) a sampling area that allows the radiation to pass through the desired compounds as well as a reference sample consisting of the pure solvent (if any); and (4) a detector area consisting of photomultiplier tubes for uv-visible and a thermocouple or bolometer for infrared.

## 2.5 *Summary*

This rapid survey of the various spectroscopies is meant to show their basic unity. What the organic chemist does is to select the one most appropriate to answer a given question, almost as one dials a radio station to suit one's mood.

Spectroscopy can be divided into two broad classes: resonance spectroscopies dealing with nuclear, electronic, or quadrupolar resonance phenomena with small energy differences between the ground and excited states; and optical spectroscopies concerned with vibrational and electronic transitions and large energy separations. An important consideration results from this distinction. As a consequence of the

---

[3] F. A. Cotton, *Chemical Applications of Group Theory*, Wiley-Interscience Publishers, Inc., New York, 1963.
[4] P. Yates, G. D. Abrahams, and L. L. Williams, *Tetrahedron Lett.*, 4341 (1969).

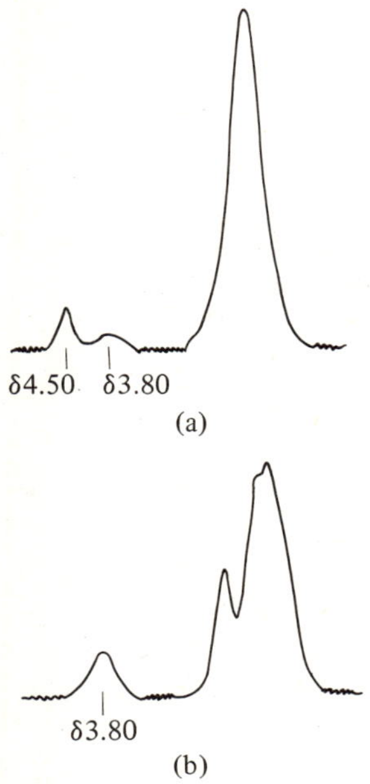


*FIGURE 2.5*
*The complete nmr spectra (60 MHz) of (a) axial-isomer-enriched chlorocyclohexane at −150°, and (b) 100% pure equatorial chlorocyclohexane at −150°.*

rather large energy differences between ground and excited states in vibrational and electronic excitation, the duration of an optical transition is very short, typically of the order of $10^{-13}$ sec, faster than random correlation times in liquids (of the order of $10^{-10}$–$10^{-12}$ sec) and much faster than rates of ordinary chemical reactions. Therefore, the ir spectrum of chlorocyclohexane, for instance, shows distinct absorptions for the equatorial chlorine and the axial chlorine conformer. By contrast, the resonance spectroscopies are much slower and therefore the room-temperature proton nmr spectrum of chlorocyclohexane consists of the weighted average of the spectra for the two conformational isomers. In this particular instance, lowering the temperature is sufficient to decrease the interconversion rate so that distinct spectra for the individual conformers result (Figure 2.5).

## 2.6 *PROBLEMS*

SAMPLE PROBLEM

Calculate the interatomic distances in hydrogen cyanide, assuming as a first approximation that $r_{C-D} = r_{C-H}$, and knowing that the spacings $\Delta\nu$ of $J = 0 \rightarrow J = 1$ lines in the microwave spectrum of HCN are 88,632 MHz, and 72,416 MHz for DCN.

ANSWER

For HCN:

$$E = h\,\Delta\nu = 5.87 \times 10^{-16} \text{ erg}$$

$$E = \frac{\hbar^2}{2I}$$

for the $J = 0 \rightarrow J = 1$ allowed transition.

$$I = \frac{1.11 \times 10^{-54}}{5.87 \times 10^{-16}} = 0.188 \times 10^{-38}$$

$$= \frac{m_1 m_2 r_{12}^2 + m_1 m_3 r_{13}^2 + m_2 m_3 r_{23}^2}{(m_1 + m_2 + m_3)}$$

$$= \frac{12r_{12}^2 + 14r_{13}^2 + 168r_{23}^2}{27 \times 6.02 \times 10^{+23}} = \frac{12r_{12}^2 + 14(r_{12} + r_{23})^2 + 168r_{23}^2}{27 \times 6.02 \times 10^{+23}}$$

$$\rightarrow 12r_{C-H}^2 + 14(r_{C-H} + r_{C-N})^2 + 168r_{C-N}^2 = 30.6 \times 10^{-15}$$

For DCN:

$$24r_{C-H}^2 + 28(r_{C-H} + r_{C-N})^2 + 168r_{C-N}^2 = 39.0 \times 10^{-15}$$

Solving for $r_{C-H}$ and $r_{C-N}$, one gets:

$r_{C-H} = 1.07$ Å
$r_{C-N} = 1.15$ Å

1. The nqr frequencies are 32,069 MHz for isopropyl chloride, and 34,063 MHz for cyclopropyl chloride. How do you interpret this difference, knowing that both $^{35}Cl$ and $^{37}Cl$ have negative quadrupole moments?
2. The quadrupolar coupling constants $eqQ$ for a number of molecules are

| | |
|---|---|
| $CH_3CH_2Cl$ | −66.0 MHz |
| $CH_2{=}CHCl$ | −67.2 MHz |
| $HC{\equiv}CCl$ | −79.7 MHz |

   How do you interpret these variations? ($Q$ is negative for the chlorine nuclei.)
3. The nqr frequency is 34,622 MHz for chlorobenzene. For the dichlorobenzenes, it is 35,729 MHz (ortho), 34,936 MHz (meta), and 34,779 MHz (para). What is the simplest explanation for this trend?
4. How many fundamental vibrational and bending modes does $CO_2$ have? Indicate these normal modes and predict their ir and Raman activity.
5. Why does not the triple bond in acetylene, $HC{\equiv}CH$, or dimethylacetylene, $CH_3C{\equiv}CCH_3$, absorb in the ir?
6. Would you expect the singlet or triplet excited state of a molecule to be lower in energy? Why?

## 2.7 *APPENDIX I* *Derivation of a typical selection rule*

Let us indicate how selection rules are obtained.[5] The probability for absorption of radiation from an energy state $i$ to an energy state $j$ is

$$\beta_{lm} = \frac{2\pi}{3\hbar^2}\mu_{lm}^2$$

where

$$\boldsymbol{\mu}_{lm} = \int \psi_l \boldsymbol{\mu} \psi_m \, d\tau$$

$\psi_l$ and $\psi_m$ are the wave functions for the two states, and the integration is to be taken over all coordinates. For instance, let us consider the transition rules for molecular rotations. Their quantized energy is

$$E = J(J+1)\frac{\hbar^2}{2I}$$

where $I$ is the moment of inertia. A linear diatomic molecule of dipole moment $\mu$ with unnormalized wave functions will have, for example,

| | |
|---|---|
| $\psi(J=0) = 1$ | (spherical symmetry) |
| $\psi(J=1) = \cos\theta$ | (axial symmetry) |
| $\psi(J=2) = 3\cos^2\theta - 1$ | (quadrupole moment symmetry) |

[5] D. H. Whiffen, *Spectroscopy*, John Wiley and Sons, Inc., New York, 1966, pp. 17–19.

The vibrational function is $\psi_v^2$, normalized so that $\int \psi_v^2\, d\tau = 1$, and the dipole moment function is $\mu \cos \theta$.

$$\mu_{0,1} = \int \psi_v \psi(J=0)(\mu \cos \theta)\psi_v \psi(J=1)\, d\tau$$

$$d\tau = dr \sin \theta\, d\theta$$

$$\mu_{0,1} = \int \psi_v^2 \mu\, dr \int_0^\pi \psi(J=0) \cos \theta\, \psi(J=1) \sin \theta\, d\theta$$

$$= \mu \int_0^\pi \cos^2 \theta \sin \theta\, d\theta = \mu \left[ -\frac{1}{3} \cos^3 \theta \right]_0^\pi = \frac{2\mu}{3} \neq 0$$

$$\mu_{0,2} = \mu \int_0^\pi (3 \cos^2 \theta - 1) \cos \theta \sin \theta\, d\theta = 0$$

Hence, the selection rule is $\Delta J = \pm 1$ for linear molecules.

## 2.8 *APPENDIX II* *Derivation of the nuclear quadrupole resonance equation*

Consider a quadrupolar chlorine nucleus, such as in the $Cl_2$ molecule. Due to the axial symmetry of the molecule the electrostatic field created at one of the chlorine nuclei by the electronic distribution will also possess axial symmetry. If $q$ is the gradient of the electrostatic field,

$$\mathbf{q} = \nabla \mathbf{E}$$

it has components $q_{xx}$, $q_{yy}$, and $q_{zz}$ along the three axes of a referential triad having its origin at one of the chlorine nuclei and its $Z$ axis along the chlorine-chlorine bond:

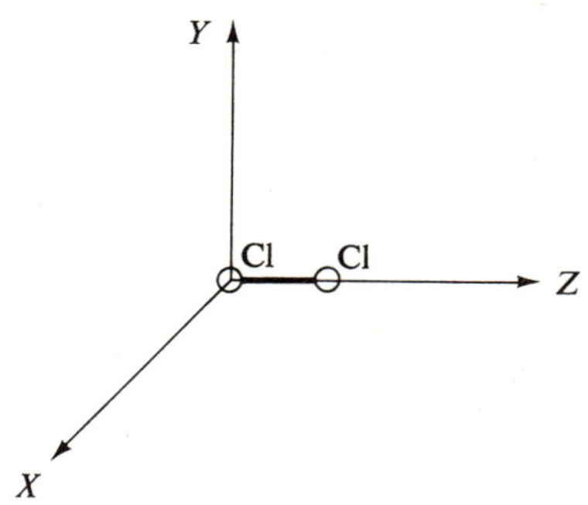


Let us define another referential system ($x'$, $y'$, $z'$) originating at the same chlorine nucleus and such that $z'$ is the direction of the spin vector (i.e., the axis of revolution for the spinning of the nucleus). We represent the revolving nucleus as an ellipsoid for reasons which will be clear from what follows:

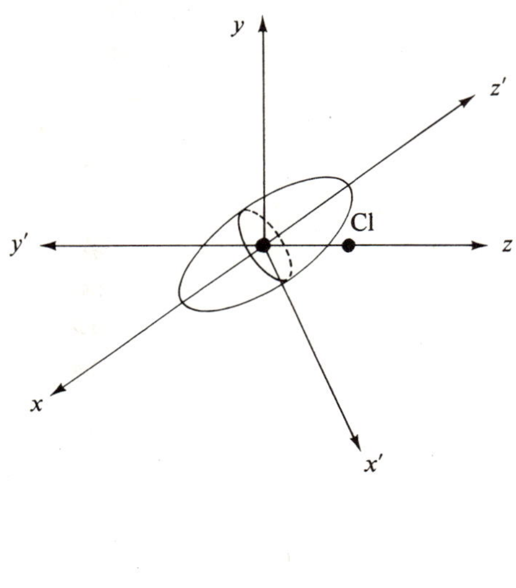

If $dV'$ is a volume element within the nucleus of sides $dx'$, $dy'$, $dz'$, then the following integral expresses the electrostatic energy $W$ due to the charge distribution in the nucleus (with a density $\rho_N$) interacting with the electrostatic gradient $q$:

$$W = \int \rho_N \, q \, dV'$$

which can be expanded as a power series:

$$W = q(0) \int \rho_N \, dV' + q_{x'}(0) \int \rho_N x' \, dV' + q_{y'}(0) \int \rho_N y' \, dV' + q_{z'}(0) \int \rho_N z' \, dV'$$
$$+ \frac{1}{2}\left[ q_{x'x'}(0) \int \rho_N (x')^2 \, dV' + q_{y'y'}(0) \int \rho_N (y')^2 \, dV' + q_{z'z'}(0) \int \rho_N (z')^2 \, dV' \right] + \ldots \quad (1)$$

In this power expansion the first term is a constant; the three terms of the type $q_{x'}(0) \int \rho_N x' \, dV'$ are the dipolar terms ($x'$, $y'$, $z'$ can be considered as the coordinates of a vector, and dipole moments are vectorial in nature). It is an experimental fact that the charge distribution within a nucleus does not lead to a net dipole moment for the nucleus. Whether one deals with a prolate or an oblate ellipsoid,

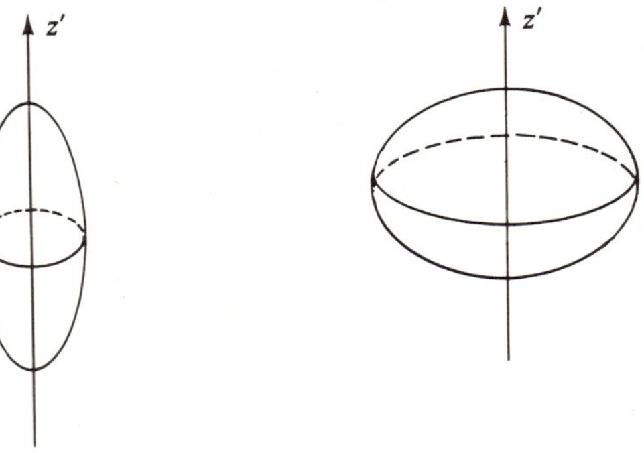


there is no dipole moment perpendicular to the axis of the nucleus, since it is an axis of symmetry, and there is also no dipole moment along the symmetry axis (this would correspond to a pear-shaped nucleus, something which is not observed). Conversely, one can account for the properties of the nucleus by considering only the next higher terms, the quadrupolar terms, in the power-series expansion and neglecting the following higher terms. As a dipole corresponds to a vector, a quadrupole corresponds to a tensor. A tensor is a rather similar mathematical species to a vector. In Cartesian coordinates, a vector **V** has as coordinates $V_x$, $V_y$, and $V_z$. A tensor **T** has nine such coordinates, $T_{xx}$, $T_{yy}$, $T_{zz}$, $T_{xy}$, $T_{yx}$, and so on. Certain tensors are such that all the coordinates vanish except for the three diagonal terms $T_{xx}$, $T_{yy}$, and $T_{zz}$; they can then be represented by an ellipsoid whose three principal axes correspond to these values. Such is the case for the distribution of charge within the nucleus. Returning to equation (1), it is seen that each of the integrals of the type $\int \rho_N (x')^2 \, dV'$ is the principal component of a tensor which we shall refer to as the quadrupole moment tensor and label as $eQ$, where $e$ is the electronic charge introduced here to provide a convenient unit system:

$$W = \frac{1}{2}\left[ q_{x'x'}(0) \int \rho_N (z')^2 \, dV' + q_{y'y'}(0) \int \rho_N (y')^2 \, dV' + q_{z'z'}(0) \int \rho_N (z')^2 \, dV' \right]$$
$$= \frac{1}{2}\left[ q_{x'x'}(0) \cdot eQ_{x'x'} + q_{y'y'}(0) \cdot eQ_{y'y'} + q_{z'z'}(0) \cdot eQ_{z'z'} \right] \quad (2)$$

The Laplace equation describes the *electronic* distribution in space:

$$\nabla^2 V = 4\pi\rho_e$$

where $\rho_e$ is the electronic density. Since $\rho_e = 0$ at the nucleus

$$\nabla^2 V = 0 \tag{3}$$

which is equivalent to

$$q_{x'x'} + q_{y'y'} + q_{z'z'} = 0 \tag{4}$$

since $\mathbf{q} = \nabla\mathbf{E} = \nabla^2 V$. Furthermore, the cylindrical symmetry of the nucleus makes

$$q_{x'x'} = -\tfrac{1}{2}q_{z'z'} \tag{5}$$

Hence

$$W = \tfrac{1}{2}eq_{z'z'}(Q_{z'z'} - Q_{x'x'}) \tag{6}$$

$$= \tfrac{1}{2}q_{z'z'}\left[\int\rho_N(z')^2\,dV' - \int\rho_N(x')^2\,dV'\right] \tag{7}$$

Let us here effect a change from Cartesian to cylindrical coordinates:

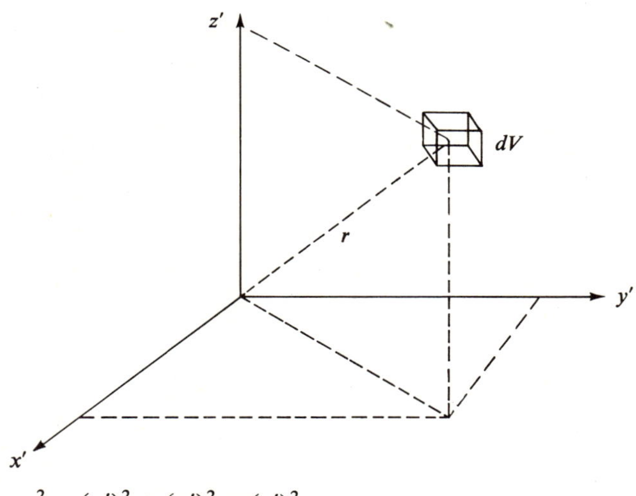


$$r^2 = (x')^2 + (y')^2 + (z')^2$$

$$= 2(x')^2 + (z')^2 \qquad \text{(symmetry)}$$

$$(x')^2 = \tfrac{1}{2}[r^2 - (z')^2] \tag{8}$$

so that equation (7) becomes

$$W = \tfrac{1}{2}\,q_{z'z'}\left[\int\rho_N(z')^2\,dV' - \frac{1}{2}\int\rho_N r^2\,dV' + \frac{1}{2}\int\rho_N(z')^2\,dV'\right]$$

$$= \tfrac{1}{2}\,q_{z'z'}\left[\frac{3}{2}\int\rho_N(z')^2\,dV' - \frac{1}{2}\int\rho_N r^2\,dV'\right]$$

$$= \tfrac{1}{4}\,q_{z'z'}\int\rho_N\left[3(z')^2 - r^2\right]dV' \tag{9}$$

Let us now change all of our coordinates to the *molecular* referentials $x$, $y$, $z$. If $\theta$ is the angle of the $Ox$ and $Ox'$ axes, then

$$q_{z'z'} = q_{xx} \sin^2 \theta + q_{zz} \cos^2 \theta$$

$$q_{z'z'} = q_{zz}(\cos^2 \theta - \tfrac{1}{2} \sin^2 \theta) \tag{10}$$

by using (5) again.

$$q_{z'z'} = q_{zz}(\tfrac{3}{2} \cos^2 \theta - \tfrac{1}{2}) \tag{11}$$

and the *classical* electrostatic energy $W$ is given by

$$W = \tfrac{1}{4} e q_{zz} Q^* (\tfrac{3}{2} \cos^2 \theta - \tfrac{1}{2}) \tag{12}$$

where Q* means that it is expressed in the former referential system ($x'$, $y'$, $z'$). In order to jump from this expression of the classical energy $W$ to that of the Hamiltonian $\mathscr{H}$ whose expectation values will be the quantized states for the system, it suffices to remark that

$$\cos \theta = \frac{I_z}{I} \tag{13}$$

in which $I_z$ is the projection of the spin vector **I** along the $Z$ axis.

$$\mathscr{H} = \tfrac{1}{8} e q_{zz} Q^* \left( \frac{3I_z^2 - I^2}{I^2} \right) \tag{14}$$

[The transformation from (13) to (14) was rendered possible by the application of a powerful theorem of quantum mechanics, the so-called Wigner–Eckert theorem.]

In the expression of (14), $I_z^2$ and $I^2$ are operators, and the corresponding expectation values are, again according to quantum mechanics:

| Operators | Expectation values |
|---|---|
| $I_z^2$ | $I_z^2 = m^2$, by definition |
| $I^2$ | $I(I+1)$ |

Hence, one can write the energy as

$$E_m = \tfrac{1}{8} e q_{zz} Q^* \frac{3m^2 - I(I+1)}{I(I+1)} \tag{15}$$

Now, it only remains to change $Q^*$ to the molecular reference system. $Q$ is defined for the aligned state $m = I$ of the nucleus with respect to the axis of alignment $z$:

$$E_{m=I} = \tfrac{1}{4} q_{zz} \int \rho_N (3z^2 - r^2)\, dV$$

by rewriting equation (9) in the new coordinate system:

$$E_{m=I} = \tfrac{1}{4} q_{zz} e Q \tag{16}$$

$$= \tfrac{1}{8} e q_{zz} Q^* \frac{3I^2 - I(I+1)}{I(I+1)}$$

$$= \tfrac{1}{8} e q_{zz} Q^* \frac{(2I - 1)}{(2I + 1)} \tag{17}$$

Setting equations (16) and (17) equal:

$$Q^* = 2Q \frac{I+1}{2I-1} \tag{18}$$

and therefore the final expression quoted in the text is obtained:

$$E_m = eq_{zz}Q \frac{3m^2 - I(I+1)}{4I(2I-1)} \tag{19}$$

## 2.9 *SUGGESTED READINGS*


A. Carrington, "Electron Paramagnetic Resonance," *Chem. Brit.*, *4*, 301 (1968).
R. S. Drago, "Nuclear Quadrupole Resonance," *Anal. Chem.*, *38*, 31A (1966).
A. Carrington, "Electron Resonance of Gaseous Free Radicals," *Chem. Brit.*, *6*, 71 (1970).
E. B. Wilson, Jr., "Microwave Spectroscopy in Chemistry," *Science*, *162*, 59 (1968).
R. S. Berry, "The Decay of Molecular Excited States," *Rec. Chem. Prog.*, *31*, 9 (1970).

# CHAPTER 3 SPIN-SPIN SPLITTING OF ENERGY LEVELS AND COUPLING CONSTANTS

## 3.1 *The Coupling Phenomenon*

Nuclear or electronic spins do not exist in isolation; they are part of molecular systems containing other particles with magnetic moments. Coupling to these additional magnetic moments must be considered. We shall see how this phenomenon can give insight into molecular geometry, stereochemistry, electronic densities, and other interesting facets of molecular structure.

Consider the simplest case of two particles having the property of spin and hence magnetic moments. These two particles are in the same molecule. By way of example, let us choose one nucleus and one electron. If the nucleus were a proton, we would be dealing with the hydrogen radical H•. One can visualize these magnetic moments as two tiny bar magnets. They influence each other across the molecule and across space. Hund's rule states that the parallel arrangement of the two electron spin vectors is to be preferred over their antiparallel orientation. The energy difference between these two arrangements, parallel or antiparallel to each other, is a measure of the interaction between the two spins and hence of their *coupling.* Let us restate these qualitative statements in terms both more quantitative and more illustrative. The energy for the system consisting of one electronic and one nuclear spin submitted to the action of a magnetic field $H_0$ is

$$E = g_e\beta_e H_0 m_S - g_N\beta_N H_0 m_I + a g_e\beta_e m_S m_I \quad (1)$$

(We are talking loosely about the energy, when this is in fact a Hamiltonian operator.) In equation (1) the first term is the energy of interaction between the electronic spin and the directing field $H_0$ and $m_S$ is the value for the projection of the spin vector $S$ along the direction of the field $H_0$. Similarly, the second term is the energy of interaction of the nuclear spin whose projection along $H_0$ is $m_N$ (the constant $g_N\beta_N$ is identical to $\gamma_N h/2\pi$). The negative sign comes from the fact that the magnetic moments of the electron and the nucleus have opposite signs. The last term describes the coupling interaction between the $I$ and the $S$ spins; the product $g_e\beta_e$ is a constant, chosen so that the *coupling constant a* can be expressed in convenient units, to be specified below.

This information is sufficient to construct the energy diagram step by step as shown in Figure 3.1. In this diagram the splitting of the energy levels for the electronic spin is much greater than that for the nuclear spins because of the larger magnetic moment of the electrons. Over each energy state is written the energy increment by which it differs from the energy state from which it originates (dotted lines). For example, the lower electronic state corresponds to the $\beta_e$ orientation of the electronic spin (antiparallel to the field, $m_S = -\frac{1}{2}$); it is split into two nuclear energy levels, the lower of which corresponds to the $\alpha_N$ ($m_N = \frac{1}{2}$) orientation of the nuclear spin. Finally, the coupling as expressed by the third term in equation (1) will lower or increase the energy of each of the four possible states according to the sign of the product $m_S m_I$.

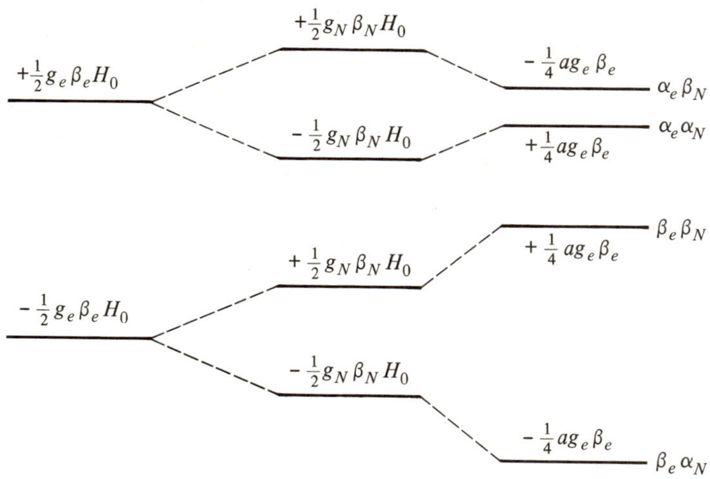


*FIGURE 3.1*
*Energy-level diagram for the interaction between an electronic and a nuclear spin.*

For instance, the $\beta_e\beta_N$ state is raised in energy by $\frac{1}{4}ag_e\beta_e$, whereas the $\alpha_e\beta_N$ state is lowered in energy by the same amount.

What are the possible transitions between these four energy states? At this point a significant simplification occurs because *only one spin flips at a time*. That is, one can either conduct an esr experiment for which the selection rule is

$$\Delta m_S = \pm 1 \qquad \text{and} \qquad \Delta m_N = 0$$

*or* an nmr experiment defined by the complementary selection rule

$$\Delta m_S = 0 \qquad \text{and} \qquad \Delta m_N = \pm 1$$

Accordingly, the diagram of Figure 3.1 can be seen to involve four distinct transitions, two esr and two nmr transitions. [Note: the energy separation between the two nmr transitions $\Delta E = ag_e\beta_e$ is the same as the energy separation between the two esr transitions (Figure 3.2).]

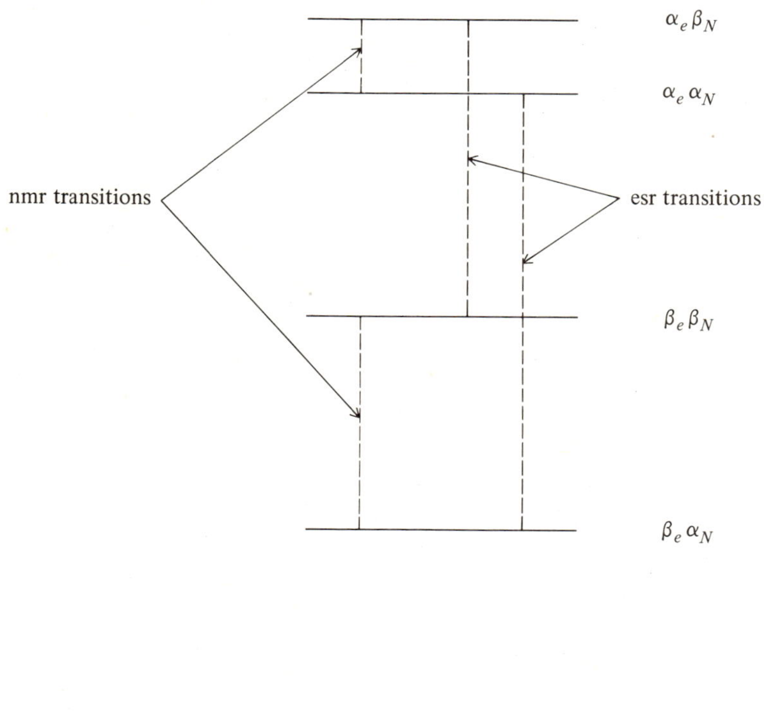


*FIGURE 3.2*
*Transitions (nmr and esr) for a system consisting of one electronic and one nuclear spin.*

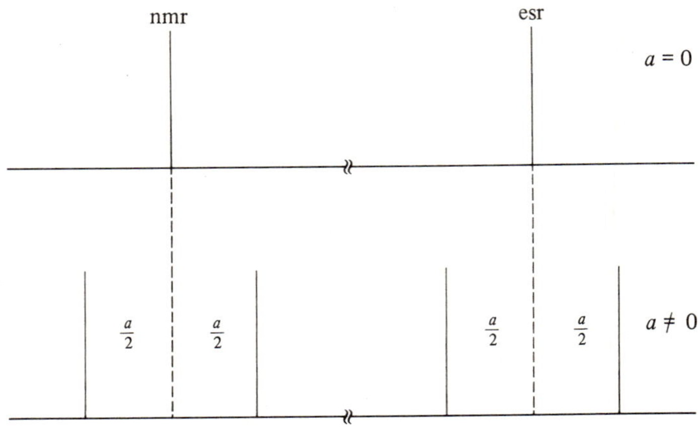


*FIGURE 3.3*
*Appearance of the nmr and esr spectra in the absence (top) and presence (bottom) of nuclear-electronic coupling.*

In the corresponding spectra (Figure 3.3), each of the two nmr lines or esr lines is displaced by $\pm a/2$ (in relative units) from the value of the resonance frequency. A single line is observed in each spectrum in the absence of coupling, but a doublet of lines separated by $a$ Hz is observed in each spectrum when coupling is present.

### *3.1.1 Nuclear Magnetic Resonance of Radicals*

The two doublets can be observed effectively in practice only if the particle responsible for the splitting of the resonance line maintains its spin state for a time long with respect to $(\Delta E)^{-1} = a^{-1}$. As indicated earlier, the lifetimes of the excited spin states are determined by the available relaxation processes. In organic free radicals the coupling constants $a$ of the unpaired electronic spin with the nuclei in the molecule are generally in excess of 0.3 MHz (Note: These coupling constants are generally referred to in the literature as "hyperfine coupling constants"; they are expressed in gauss or oersteds; and correspond to the same number in the cgs system. One gauss is approximately 2.8 MHz. The selection of a unit of magnetic field to express energies, a somewhat unfortunate usage, originates simply in the resonance condition $h\nu = g_e\beta_e H_0$.) Nuclear relaxation times $T_N$ are generally greater than a millisecond, hence much larger than the reciprocal of $a$ (ca. $3 \times 10^{-6}$ sec). The doublet splitting is therefore observable in the esr spectrum. The converse, however, is not necessarily true. Electronic relaxation times being of the order of $10^{-5}$ sec (the relaxation processes available to free radicals are more efficient because of the greater magnetic moment of the electron) it is generally impossible to see this doublet splitting in the nmr spectrum of a free radical. Nevertheless the coupling of electronic and nuclear spins manifests itself in the nmr spectrum as a shift of the resonance line, usually a *paramagnetic* shift which lowers the value of the applied field $H_0$ necessary for resonance to occur at a given frequency. This can be seen from the above expression of the energy [equation (1)]: when fast exchange interconverts the $\alpha_e$ and $\beta_e$ levels their relative populations are again given by the Boltzmann statistic. If one applies equation (1) not only to an individual system of one electron and one nucleus, but also to the real assembly of all such systems, the last term then becomes

$$\frac{-ag_e\beta_e g_e\beta_e H_0}{4kT} \tag{2}$$

since the average value of $m_S m_I$ is $(\frac{1}{4})[-g_e\beta_e H_0/kT]$ (as was seen earlier the energy $g_e\beta_e H_0$ is much smaller than $kT$). The negative sign indicates that the extra shift of the magnetic resonance is paramagnetic. It is proportional both to the coupling constant $a$ and to the magnitude of the applied field $H_0$ and inversely proportional to $kT$.

Replacing the constants with their values the final expression of the hyperfine constant $a$ as a function of the paramagnetic shift $\delta_p$ is

$$a(\text{Oe}) = -0.04255 \frac{\delta_p(\text{Hz}) \times T(^\circ\text{K})}{\text{H}_0(\text{Oe}) \times (g_e)^2} \tag{3}$$

With $T = 300^\circ$K, a magnetic field $H_0 = 23{,}500$ Oe, and $g_e$ (the so-called Landé factor) being very close to $-2$ for most organic free radicals, $a$ is seen to be approximately $1.35 \times 10^{-4}\, \delta_p$ Hz. A chemical shift of 100 Hz can easily be measured accurately, and is thus seen to correspond to coupling with the unpaired spin of the order of 0.1 Oe. This phenomenon, the so-called "contact shift," is very valuable for the determination of the magnitude of *small* coupling constants in organic radicals. It is actually superior to the alternative esr method in this respect, since such small splittings are difficult to measure in spectra. Furthermore, this technique is capable of yielding not only the magnitude but also the sign of the coupling constant.

*3.1.2 Coupling to Several Equivalent Nuclei*

Let us now consider the behavior of an electron (with a spin of ½) in an organic molecule such as the methyl radical, $CH_3\cdot$, in the presence of an external magnetic field $H_0$. One expects a priori that there should be further splitting of the two possible magnetic energy levels of the electron depending on the orientations of the three proton nuclear spins: parallel or antiparallel with respect to $H_0$, as shown in Figure 3.4.

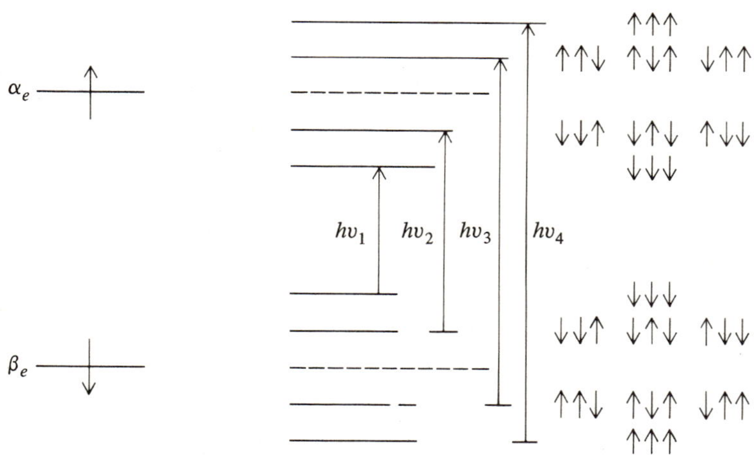


*FIGURE 3.4*
*Interaction of electronic and nuclear spins in the esr of the* $CH_3\cdot$ *radical.*

This energy diagram assumes only that the $\beta_e$ and $\alpha_N$ energy states are lower in energy (relative to the corresponding $\alpha_e$ and $\beta_N$ states) and that electronic and nuclear spins follow Hund's rule and are preferably paired. It is then apparent from the diagram that instead of a single resonance frequency $\nu$ there are four different transitions possible, with frequencies $\nu_1$, $\nu_2$, $\nu_3$, and $\nu_4$. The nuclear spins do not flip during the process of the electronic spin undergoing resonance (selection rules $\Delta m_S = \pm 1$, $\Delta m_N = 0$).

Indeed, a quartet of lines is observed, with equal spacings and binomial intensities,[1] in the esr of methyl radicals (Figure 3.5).

[1] They are proportional to the coefficients in the expansion of the Newton binomial $(1 + 1)^n$. (See page 113.)

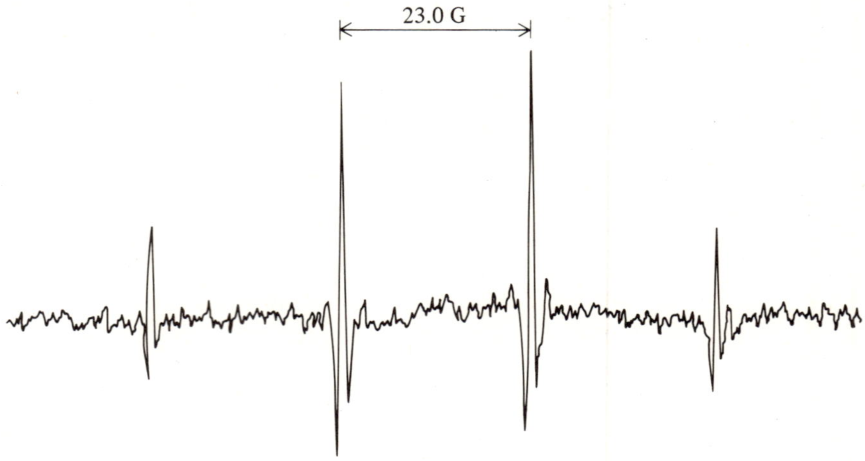


FIGURE 3.5
*Electron spin resonance of* $CH_3\cdot$ *radical.*

### 3.1.3 Nuclear–Nuclear Coupling

In a similar manner two nuclear spins can interact, as illustrated in Figure 3.6 for the HF molecule. In the interactions depicted in Figure 3.6 the $^1H$ transitions obey the selection rules $\Delta m_H = 1$ and $\Delta m_F = 0$ while the rules for the $^{19}F$ transitions are $\Delta m_H = 0$ and $\Delta m_F = 1$.

These two diagrams must be interrelated since both involve the same energy term representing the coupling between the $^{19}F$ and the $^1H$ nuclear moments. The separation between the two fluorine transitions is identical to that separating the two hydrogen transitions, since

$$(\beta_F\beta_H - \alpha_F\beta_H) - (\beta_F\alpha_H - \alpha_F\alpha_H) \equiv (\beta_H\beta_F - \alpha_H\beta_F) - (\beta_H\alpha_F - \alpha_H\alpha_F) \tag{4}$$

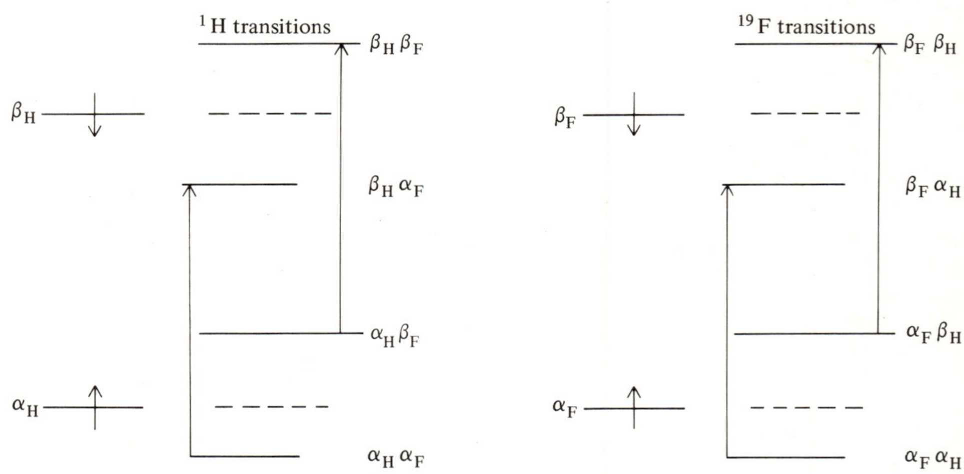


FIGURE 3.6
*Interaction of hydrogen and fluorine nuclear spins in the HF molecule.*

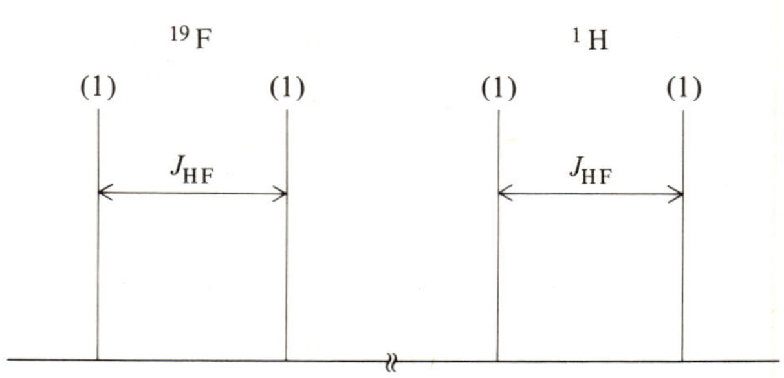


FIGURE 3.7
*The* $^{19}F$ *and* $^1H$ *spectra of the HF molecule.*

Consequently there is an equal splitting in the corresponding spectrum, as shown in Figure 3.7.

A summary and comparison of the properties of spin-spin interactions in esr and nmr is given in Table 3.1.

*TABLE 3.1*

*Comparison of esr and nmr Coupling Constants*

| Interaction | esr | nmr |
|---|---|---|
| Nomenclature | Hyperfine constants | coupling constants |
| Notation | $a_N$ | $D_{ij}$ or $J_{ij}$ |
| Magnitude and unit | 0–50 Oe | 0–$10^3$ Hz |
| Fundamental properties | independent of the magnitude of the applied field $H_0$ | Independent of magnitude of the applied field $H_0$<br>Reciprocal, *i.e.*, $J_{ij} = J_{ji}$ |

### *3.1.4 Origin of the Electron Spin Resonance Coupling Phenomenon*

In esr the coupling constant $a$ can be divided into a direction-independent term and a direction-dependent term. The direction-dependent term originates in the coupling through space of the magnetic moments and is the quantum mechanical equivalent for the coupling of two magnetic dipoles:

$$a = \left(\frac{\mu_e}{S}\right)\left(\frac{\mu_N}{I}\right)\left\langle\frac{3\cos^2\theta - 1}{r^3}\right\rangle S_z I_z \tag{5}$$

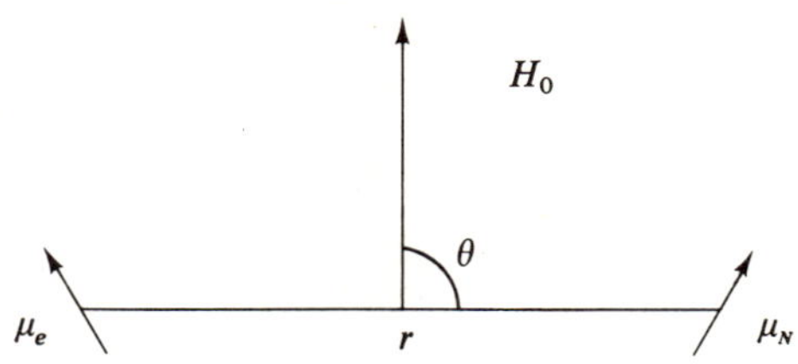


In this expression $r$ is the radius vector and $\theta$ is the angle made between $r$ and the direction of the applied magnetic field $H_0$; the bracket, $\langle\ \rangle$, indicates that the average must be taken over all the possible orientations of the electronic ($\mu_e$) and nuclear ($\mu_N$) magnetic moments. The direction-dependent term $(3\cos^2\theta - 1)$ vanishes in liquids due to fast tumbling: $\int_0^\pi (3\cos^2\theta - 1)\sin\theta\, d\theta = 0$. It is possible, however, for the electron-nucleus distance $r$ to become zero as well and therefore, although the $(3\cos^2\theta - 1)$ term vanishes, the whole average can remain finite. It can be shown[2] that the coupling constant $a$ assumes a finite value corresponding to the so-called Fermi contact term:

$$a = -\frac{8\pi}{3} g_N \mu_N g_e \mu_e \rho_N \tag{6}$$

where $g_N$ and $g_e$ designate the $g$ factors for the nucleus and the electron, and $\mu_N$ and $\mu_e$ are the corresponding magnetic moments. $\rho_N$ is the *unpaired electron density at the nucleus*, that is, the value at the nucleus of the difference between the average numbers of $\alpha$ and $\beta$ electrons. The probability $\rho_N$ is then simply given by $|\psi(0)|^2$, the square of the amplitude of the corresponding wave function in which the coordinate $r$ has gone to zero. There will be a finite value of $a$ if and only if the unpaired electron is in an orbital devoid of a node at the nucleus: an $s$ orbital or a hybrid having some $s$ character. An example is the hydrogen radical H• in which $a_H = 507$ Oe. Likewise, it can be calculated that the hyperfine constant *to carbon* (to the isotope of carbon having a spin, i.e.,

[2] A. Carrington and A. D. McLachlan, *Introduction to Magnetic Resonance*, Harper & Row, New York, 1967.

carbon-13) of an electron in a 2*s* orbital would be 1190 Oe. However, when the experimental esr spectrum of the methyl radical is measured, $a$ is found to be $a_C = 40$ Oe. This measurement indicates that the predominant electronic structure of methyl radicals $CH_3\cdot$ is an $sp^2$ hybrid on carbon with the remaining $p_z$ orbital to accommodate the unpaired spin. The alternative picture of a pyramidal structure, in which the $p_x$, $p_y$, and $p_z$ atomic orbitals on carbon are used for bonding to the three hydrogens and the 2*s* orbital contains the lone electron, accounts for less than 4 percent of the total structure. This evidence suggests strongly that the methyl radical has trigonal hybridization and hence is planar. A paradox has now become evident. How can there possibly be coupling to the three protons in the methyl radical (as shown in Figure 3.5), when these protons are in the nodal plane of the $p_z$ orbital containing the unpaired spin?

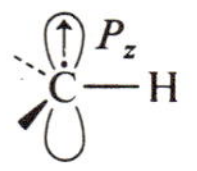

The explanation is $\sigma$-$\pi$ interaction placing part of the unpaired electron in a 1*s* hydrogen orbital. The *equivalent* picture is that of exchange polarization of the $\sigma$ core (Figure 3.8), amounting to the

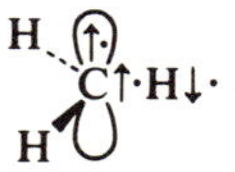

resonance form for the methyl radical being preferred over the

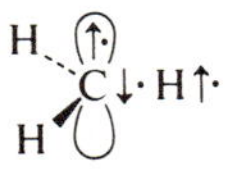

resonance form.

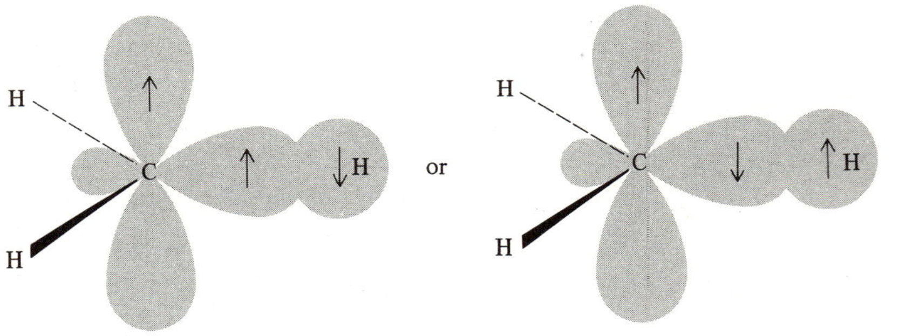


*FIGURE 3.8*
*Exchange polarization of the $\sigma$ core in the methyl radical.*

The *magnitude* of the interaction is given by the McConnell relation:

$$a_H = Q_{CH}\rho_C \tag{7}$$

where $\rho_C$ is the electron density in the $p_z$ orbital and $Q_{CH}$, the proportionality constant, is in the range of $-22$ to $-28$ Oe.

In accordance with the above mechanism $a_H$ is found to be only 23 Oe in methyl radicals. That very little spin density is generated at the hydrogen 1*s* orbital of $CH_3\cdot$ can be appreciated by comparing this value with that of the hydrogen radical H•, for which spin density at the nucleus $|\psi(0)|^2$ is much higher, corresponding to an experimental $a_H$ value of 507 Oe. A number of values of $a_H$ for simple hydrocarbon radicals are gathered in Table 3.2. It is evident that as a first approximation the McConnell relationship accounts well for the data. Note in particular that different values of Q *must* be used for positively charged and negatively charged radicals.

*TABLE 3.2*

*Representative Values of $a_H$ for Some Organic Radicals*

| Radical | Spin density on H atom | Spin density on C atom | $a_H$ (Oe) |
|---|---|---|---|
| $H\cdot$ | 1 | — | 507 |
| $CH_3\cdot$ | small | 1 | 23.04 |
| $C_6H_6^{\overset{-}{\cdot}}$ | very small | $\frac{1}{6}$ | 3.75 |
| $C_8H_8^{\overset{+}{\cdot}}$ | very small | $\frac{1}{8}$ | 1.5 |
| $C_5H_5\cdot$ | very small | $\frac{1}{5}$ | 5.98 |
| $C_6H_6^{\overset{+}{\cdot}}$ | very small | $\frac{1}{6}$ | 4.44 |
| $C_7H_7\cdot$ | very small | $\frac{1}{7}$ | 3.91 |

Electron spin resonance spectroscopy can also be applied profitably to the study of *diradicals*. These molecules with two unpaired electrons are in a triplet state, since the spin angular momentum $m_S$ can take the three values $+1$, 0, and $-1$. Correspondingly, there should thus be two esr transitions *of the same energy* connecting the $-1$ and the 0, and the 0 and the $+1$ states, respectively. Selection rules forbid the $\Delta m_S = 2$ transition from the lower state with $m_S = -1$ to the higher state with $m_S = +1$. However, the degeneracy of the two allowed esr transitions can be removed by coupling. When the two electronic spins in the molecule are close enough to interact strongly, their interaction is described by an expression closely resembling equation (5). Since it is direction dependent, the spectrum will be defined by the relative orientation of the magnetic field $H_0$ with respect to the molecular plane for the triplet species. Fixed orientations of that plane can be achieved in molecular crystals, which are rotated into various orientations with respect to the field. It then becomes possible not only to observe the two $\Delta m_S = \pm 1$ transitions, with slightly differing frequencies, but also the weaker $\Delta m_S = \pm 2$ transition, which becomes observable because of the presence of the dipolar coupling. It is possible to infer from this information the relative "distance" of the two unpaired spins in their time-averaged configuration. This is an experimental determination of the extent of correlation between the two electronic spins.

## 3.2 *Spin-Spin Interaction in Nuclear Magnetic Resonance*

Spin-spin interaction in nuclear magnetic resonance is of two kinds, direct and indirect coupling of the nuclear moments. The direct coupling constant (the energy of interaction between two magnetic dipoles) is given by equation (8):

$$D_{ij} = \frac{h\gamma_i\gamma_j}{4\pi^2 r_{ij}^3} \cdot \tfrac{1}{2}\langle 3\cos^2\theta - 1\rangle \qquad (8)$$

where $\gamma_i$ and $\gamma_j$ are the gyromagnetic ratios of nuclei $i$ and $j$, respectively, $r_{ij}$ is the internuclear distance between the nuclei, and $\theta$ is the angle between the direction of the applied field $H_0$ and $r_{ij}$.

There is an important difference between esr and nmr. Since in nmr the *internuclear* distance $r$ cannot be zero, direct coupling constants *do* vanish with the $\langle 3\cos^2\theta - 1\rangle$ term. Rapid Brownian motion and the resulting equal probability of all molecular orientations with respect to the external magnetic field in normal solutions averages the $\langle 3\cos^2\theta - 1\rangle$ term to zero. However, if this tumbling is restricted and made anisotropic,[3] and if the molecule is forced to orient with respect to the magnetic

[3] That is, certain directions in space become more favored.

field, then the $\langle 3 \cos^2 \theta - 1 \rangle$ term does not vanish and direct coupling constants will influence the spectrum. The mesophases of liquid crystals are ideally suited for such orientation and have been widely employed recently to obtain these *direct* coupling constants.

### *3.2.1 Liquid Crystals and Direct Coupling Constants*

There exists a state of matter intermediate between the liquid phase and the solid phase. It retains some of the order of crystals while at the same time having in part the degrees of freedom characteristic of liquids. This so-called mesophase can be further subdivided according to its structure. For instance, the nematic mesophases usually consist of elongated rod-like molecules (such as *p*-azoxyanisole)

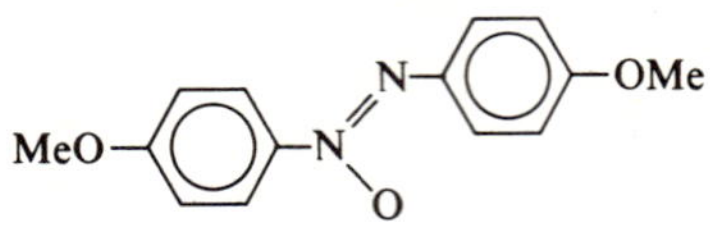


which tend to orient with their long axes parallel to one another. The smectic state, characteristic of detergents, is made of layers of such nematic domains. The cholesteric state, first observed with some derivatives of cholesterol, is a chiral, helically ordered arrangement of molecules, in which the orientation within a nematic domain changes gradually instead of remaining constant.

Upon application of an intense magnetic field, such as that available in a magnetic resonance spectrometer, the liquid crystalline molecules tend to orient with their long axes parallel to the field. If such mesophases are used as solvents, a normal molecule dissolved in them will also tend to orient somewhat with respect to the magnetic field, due to solute-solvent intermolecular forces.

One is thus led to expect that both solvent and solute molecules will have nmr spectra affected by the $D_{ij}$ coupling constants. In practice, the solvent molecules are better ordered so that multiple direct couplings between their protons fragment the spectrum into so many lines that it fades into the background noise. This convenient property leaves apparent the spectrum of the solute, as it is affected by its very partial orientation.

Of course one can do the reverse experiment, in which nematic solutes serve to order slightly the solvent molecules. For instance, one can dissolve 15 percent poly-$\gamma$-benzyl-L-glutamate in methylene chloride, and introduce that sample in the cavity of the spectrometer (Figure 3.9), in which case the spectrum of the polybenzyl glutamate vanishes, leaving only that of the less-ordered solvent (and eventually solute) molecules.

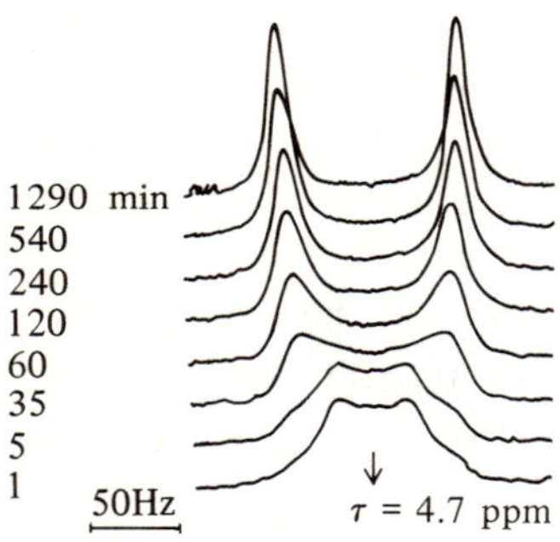


*FIGURE 3.9*
*Variation of the doublet of methylene chloride in the liquid crystalline state with the residence time of the sample in the magnetic field.*

### *3.2.2 Indirect Coupling Constants*

The more familiar indirect coupling constants arise from interaction of the magnetic moments modulated by the intervening valence electrons in the molecule. The mechanism for the coupling phenomenon is a polarization of the electronic distribution by each magnetic moment which transmits information about its spin state—$\beta$ or $\alpha$—to the other magnetic moment. One can visualize this phenomenon using the picture shown in Figure 3.10. Assuming the $\alpha$ orientation for the $\mu_1$ moment, the electronic spins in an adjacent chemical bond will tend to be polarized accordingly (see Section 3.1.2). This perturbation is transmitted from bond to bond until it affects the $\mu_2$ moment, which will then have either an enhanced or a reduced "preference" for the $\alpha$ spin state. The very same argument, which would, incidentally, call for the sign of coupling constants to alternate according to whether there is an odd or even

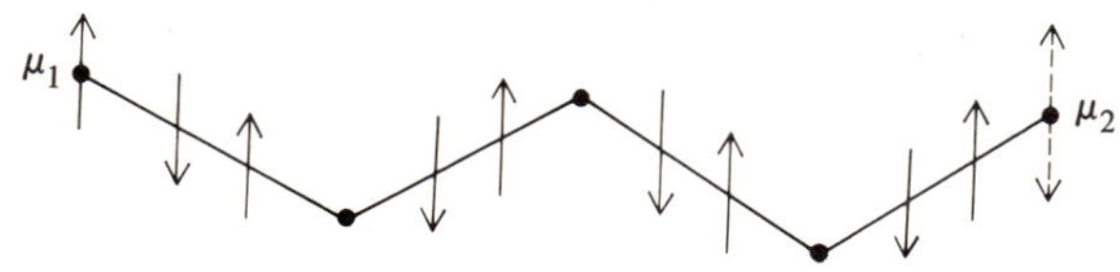


*FIGURE 3.10*
*Indirect coupling as effected through valence electrons (depending upon the magnitude and sign of the coupling, the α or the β orientation of $\mu_2$ will be favored).*

number of intervening bonds, can be applied to the second magnetic moment. The magnitude of the interaction must then be of the form

$$E = hJ_{1,2}\mathbf{I}_1\mathbf{I}_2 \tag{9}$$

That is, the energy of the coupling interaction is the scalar product of the spin vectors multiplied by a number $J_{1,2} = J_{2,1}$ (the coupling constant). It is clear that, given the mechanism of the interaction, it must be independent of the *direction* of the magnetic moments with respect to the applied field $H_0$. Coupling constants are a molecular property (just as are dipole moments) as well as an intensive property. These indirect coupling constants are very useful because they can be related to the stereochemistry of the molecule as indicated in the following sections.

### *3.2.3 One-Bond Coupling Constants*

For organic molecules, the most important one-bond coupling constants are those between two $^{13}$C carbons ($^1J_{^{13}C-^{13}C}$) or between one $^{13}$C carbon and the attached $^1$H proton ($^1J_{^{13}C-H}$). Since $^{13}$C exists with approximately 1.1 percent natural abundance, these coupling constants are, at least in principle, readily available.

It has been found empirically that they are extremely useful indices for hybridization at the carbon atoms. Table 3.3 compares experimental values with *s* character on carbon. One can visualize the origin of this correlation by noting that indirect coupling of the nuclear spins is effected by the intervening electrons; since *p* orbitals have a node at the nucleus, but not *s* orbitals, it is apparent, at least as a first approximation, that coupling constants should be related to the *s* character of the hybrid orbital.

*TABLE 3.3*
*$J_{^{13}C}$ Coupling Constants and Percent s Character*

| Bond type | *s* Character on carbon (percent) | Indirect coupling constants (Hz) | |
|---|---|---|---|
| | | $J_{^{13}C-C}$ | $J_{^{13}C-H}$ |
| C—C | 25 | ~30 | ~125 |
| C=C | 33 | ~60 | ~159 |
| C≡C | 50 | ~140 | ~250 |

From Table 3.3 one can derive the approximate relationships between hybridization on unsubstituted hydrocarbons and these $^1J$ couplings:

$$^1J_{C-H} \sim 500 \times (s\text{ character}) \tag{10}$$

$$^1J_{C-C} \sim 575 \times (s\text{ character}) \times (s\text{ character}) \tag{11}$$

Inductive factors also affect these coupling constants. The same naive picture continues to account for the observations. A substituent on carbon more electronegative than hydrogen will draw *p* electrons into the C—X bond, thus increasing the *s* character in the C—H bond. If $^1J_{C-H}$ is considered, its value is indeed increased by such a substitution. The effect of several substituents has been found to be additive, and has also been explained in such rehybridization terms.

Let us first recall the hybridization picture of methane.[4] Out of three 2*p* atomic

[4] C. A. Coulson, *Valence*, 2nd ed., Oxford University Press, 1961, p. 198.

orbitals ($p_x$, $p_y$, $p_z$) and 2*s* atomic orbital on carbon, four hybrid atomic orbitals are generated:

$\frac{1}{2}(s + \sqrt{3}\, p_a)$

$\frac{1}{2}(s + \sqrt{3}\, p_b)$

$\frac{1}{2}(s + \sqrt{3}\, p_c)$

$\frac{1}{2}(s + \sqrt{3}\, p_d)$

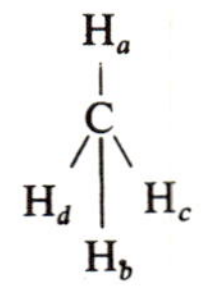


with $\sqrt{3}\, p_a = p_x + p_y + p_z$, and so on.

If attention is now focused upon an individual C—H bond of methane, one can calculate the bond orders $p_{\text{C—H}}$ either between the hybridized 2*s* orbital on carbon and the 1*s* orbital on hydrogen, or between the hybridized 2*p* orbitals on carbon and the 1*s* orbital on hydrogen. The bond order $p_{rs}$ (see Chapter 6) between two atoms *r* and *s* in a closed-shell molecule is defined as

$$p_{rs} = 2 \sum_{j}^{\text{occ.}} c_{jr}\, c_{js} \tag{12}$$

where $c_{jr}$ and $c_{js}$ are the atomic orbital coefficients in the LCAO expansion for atoms *r* and *s* in the *j*th molecular orbital. Hence:

$$p_{\text{C—H}}(2s - 1s) = \frac{1}{\sqrt{2}}\,\frac{1}{\sqrt{2}} = 0.50 \tag{13}$$

$$p_{\text{C—H}}(2p - 1s) = \frac{\sqrt{3}}{\sqrt{2}}\,\frac{1}{\sqrt{2}} = 0.866 \tag{14}$$

Let us further define the bond index $W_{\text{C—H}}$ as the square of the corresponding bond order:

$$W_{\text{C—H}} = (p_{\text{C—H}})^2 \tag{15}$$

$$W_{\text{C—H}}(2s - 1s) = 0.25 \tag{16}$$

$$W_{\text{C—H}}(2p - 1s) = 0.75 \tag{17}$$

Notice that these bond indices, as defined, are related directly to the $sp^3$ hybridization of the methane molecule since they measure the percent *s* character or *p* character in the C—H bond. The sum is of course equal to unity, the bond order for the C—H bond. For hydrocarbons other than methane (for which the values of $W_{\text{C—H}}$ are symmetry determined) an approximate calculation must be performed to obtain values of $W_{\text{C—H}}$. Methods have recently become available for calculating the wave functions and the energies of hydrocarbons, and we have listed in Table 3.4 values of $W_{\text{C—H}}$ (2*s* – 1*s*) from the extended Hückel theory (EHT), complete neglect of differential overlap (CNDO), and self-consistent field (SCF) methods; these values themselves disagree, because of the large errors inherent in these calculations, but their relative order is the same whatever the method of calculation. They are to be compared with the magnitudes of the ${}^1J_{\text{C—H}}$ coupling constants, listed in the fourth column. The correlation is very satisfactory; it is seen that there is a direct relationship between

*TABLE 3.4*

*Bond Indices and Carbon-Hydrogen Coupling Constants*

| | $W_{1s-2s}$ | | | |
|---|---|---|---|---|
| Compound | CNDO | EHT | SCF | $^1J_{^{13}C-^1H}$ |
| Methane | 0.250 | 0.152 | 0.238 | 125 |
| Ethane | | 0.176 | | 126 |
| Propane ($CH_2$) | 0.220 | | | 123 |
| $(CH_3)_2CH^+$ | | | | 168 |
| $(C_6H_5)_2CH_2$ | | | | 126 |
| $(C_6H_5)_2CH^+$ | | | | 164 |
| Cyclohexane | | 0.206 | | 124 |
| Cyclopentane | | 0.209 | | 128 |
| Cyclobutane | | | | |
| eq. | 0.237 | 0.240 | | 134 |
| ax. | 0.265 | | | |
| Cyclopropane | 0.288 | 0.270 | | 161 |
| Bicyclobutane | | 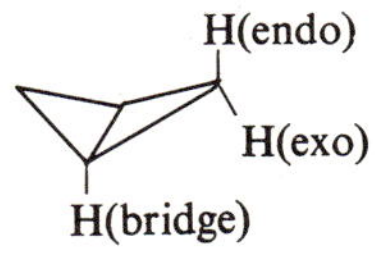  | | |
| exo | 0.283 | | | 152 |
| endo | 0.301 | | | 170 |
| bridge | 0.366 | | | 203 |
| Ethylene | 0.333 | 0.248 | 0.291 | 157 |
| Acetylene | 0.500 | 0.377 | 0.403 | 250 |
| Cyclopropene | | | 0.354 | 220 |
| $C_3H_3^+$ | | | 0.367 | 265 |
| Cyclobutadiene | | | 0.309 | 190 |

hybridization as measured by the percent $s$ character ($W_{2s-1s}$) and $^1J_{C-H}$. This proportionality relationship is often written as

$$^1J_{C-H} \sim 500\, W_{C-H}(2s - 1s) \tag{10}$$

### *3.2.4 Calculation of Nuclear Spin Coupling Constants*

These correlation equations (10) and (11) have a theoretical basis. The coupling between two nuclear spins can be considered as a distortion of the intervening electron cloud. There are methods available to calculate such perturbations of the electronic structure. The importance of such perturbation theory is considerable but beyond the scope of this book. Suffice it to state its guiding principle: because of the perturbation on the system, the wave functions for the ground state get mixed with contributions from all of the excited states of the molecule in proportion to their distance from the ground state.

For instance, the dominant term for the one-bond $^{13}C{-}^1H$ coupling constants under discussion—the so-called Fermi contact term—is essentially the product of the values of the spin densities in the $s$ orbitals at carbon and at hydrogen by the mutual polarizability of the $2s$(C) and $1s$(H) orbitals, as expressed by the corresponding bond index $W_{C-H}(2s - 1s)$ which was just defined. The coupling constant can be written:

$$^1J_{C-H} = \left(\frac{1}{h}\right)(2\beta\hbar)^2\gamma_C\gamma_H\pi_{C-H}(2s - 1s)\langle 2s_C|\delta|2s_C\rangle\langle 1s_H|\delta|1s_H\rangle \tag{18}$$

In this expression, $h$ is Planck's constant, $\hbar = h/2\pi$, $\beta$ is the Bohr magneton, and $\gamma_C$ and $\gamma_H$ are the gyromagnetic ratios for the carbon and hydrogen nuclei, respectively. The mutual polarizability of the two interacting $s$ orbitals $\pi_{C-H}$, is given by

$$\pi_{C-H}(2s - 1s) = -4\sum_i^{\text{occ.}}\sum_j^{\text{unocc.}}\frac{1}{E_j - E_i}c_{iC}c_{iH}c_{jC}c_{jH} \tag{19}$$

where $c_{iC}$, $c_{iH}$, and so on, are atomic orbital coefficients in the LCAO expansion. An approximation often made, the so-called closure approximation, is to remove the

$1/(E_j - E_i)$ terms from the summation, and replace them with $1/\Delta E$, in which $\Delta E$ represents a *mean* excitation energy. A simplified expression would thus be

$$\pi_{C-H}(2s - 1s) = -\frac{4}{\Delta E} \sum_i^{\text{occ.}} \sum_j^{\text{unocc.}} c_{iC}\, c_{iH}\, c_{jC}\, c_{jH} \tag{20}$$

The sum is taken over all the occupied $i$ molecular orbitals, and also over the manifold of the unoccupied $j$ molecular orbitals (the excited states of the system). One notices the similarity between this expression of the polarizability and that for the bond index, the essential difference being the inclusion of all the "bond orders" between occupied and unoccupied molecular orbitals. As for the last two factors, in equation (18) the integrals $\langle 2s_C|\delta|2s_C\rangle$ and $\langle 1s_H|\delta|1s_H\rangle$ are simply the values of the spin densities $\rho_N$ at the carbon and hydrogen nuclei for the valence $s$ orbitals. As an approximation, one can reduce (18) to (21) by effecting the assimilation of the polarizability $\pi_{C-H}$ with the bond index $W_{C-H}$:

$$^1J_{C-H} = \text{constant} \times \frac{1}{\Delta E} \times W_{C-H}(2s - 1s) \times \rho_{2s_C} \times \rho_{1s_H} \tag{21}$$

Hence, if $1/\Delta E$ can also be assumed not to vary markedly throughout a hydrocarbon family, $^1J_{C-H}$ is indeed predicted to be proportional to the bond index $W_{C-H}$ between the 2$s$ orbital on carbon and the 1$s$ orbital on hydrogen. The reason why 2$p$ atomic orbitals on carbon do not contribute to the $^1J_{C-H}$ coupling constants is of course that they have a node at the nucleus; therefore the $\langle 2p_C|\delta|2p_C\rangle$ spin density is zero. The concept of $^1J_{C-H}$ coupling constants as an excellent measure of hybridization at carbon has been applied extensively. Only a few of its most important applications will be presented here. One can extract from Table 3.5 the data on a number of hydrocarbons differing in $s$ character, and hence in the "acidity" of their C—H bonds. The correlation is excellent in a qualitative fashion, and $^1J_{C-H}$ coupling constants can be used as a semiempirical estimate of the acidity of C—H bonds in hydrocarbons. In another application, Foote[5] has correlated the $^{13}$C—H coupling constants with the magnitude of the endocyclic interatomic angle $\theta$ in cycloalkanes (Table 3.6). A plot of $\theta$ versus $^1J_{C-H}$ is close to being linear: Its slight curvature and its slope have been accounted for correctly in the treatment proposed by Mislow.[6]

A third application of the relationship between $^{13}$C—H coupling constants and hybridization is to carbonium ions; Table 3.4 will convince the reader that carbonium

*TABLE 3.5*

*Hybridization as Measured by $^1J_{C-H}$ and Acidity of C—H Bonds*[a]

| Hydrocarbon | $^1J_{C-H}$ | $pK_a$ |
|---|---|---|
| Acetylene | 250 | 25 |
| Ethylene | 157 | 42 |
| Benzene | 159 | 41 |
| Cyclopropane | 161 | 44 |
| Cyclobutane | 134 | 47 |
| Methane | 125 | 47 |
| Ethane | 126 | 48 |
| Propane ($C_2$) | 123 | 49 |
| Cyclopentane | 128 | 49 |
| Cyclohexane | 124 | 49 |

[a] For a more extensive compilation, see R. C. Fort, Jr., and P. von R. Schleyer, *Advances in Alicyclic Chemistry, vol. 1*, Academic Press, New York, 1966, p. 358.

[5] C. S. Foote, *Tetrahedron Lett.*, 579 (1963).
[6] K. Mislow, *Tetrahedron Lett.*, 1415 (1964).

*TABLE 3.6*

*Correlation of $^1J_{C-H}$ with Endocyclic Angle $\theta$ for Cycloalkanes*

| Ring size $n$ | $^1J_{C-H}$ (Hz) | $\theta$ (deg) |
|---|---|---|
| 3 | 161 | 60 |
| 4 | 134 | 89.3 |
| 5 | 128 | 103.3 |
| 6 | 124 | 111 |
| 7 | 123 | 112 |
| 8 | 122 | 112 |
| 9 | 121 | 115 |
| 10 | 118 | 116 |
| 12 | 123 | 112 |

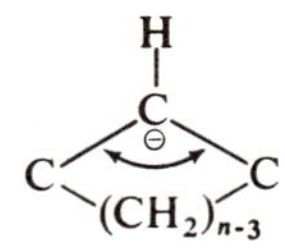


ions such as the isopropyl or the diphenylmethyl cation have trigonal hybridization (compare their values of $^1J_{C-H}$ with those for the corresponding hydrocarbons propane and diphenylmethane, respectively). Likewise, the cyclopropenyl cation $C_3H_3^+$ has gained in *s* character by comparison with its parent hydrocarbon, cyclopropene. Notice that the $^1J_{C-H}$ coupling constants in these charged species have greater magnitudes than would have been predicted on the basis of the corresponding bond index alone. Superimposed upon the change due to hybridization, there is the effect of increased nuclear charge at the carbon nucleus, which has been shown[7] to increase the $^{13}C-H$ coupling constants in the absence of any rehybridization.

*3.2.5 Sign of Indirect Coupling Constants*

Since coupling constants measure the energy of interaction between two nuclear moments, they are expected to have either a positive or negative sign, as is found. By definition, the positive sign corresponds to pairing of the two nuclear spins. For instance, the H—H coupling constant in the hydrogen molecule is found to be positive, which corresponds to the polarization picture in Figure 3.11, where pairing of the electronic spins (Hund's rule) places a slight excess of paired nuclear spins at the nuclei by the correlation of the electronic and of the nuclear spins.

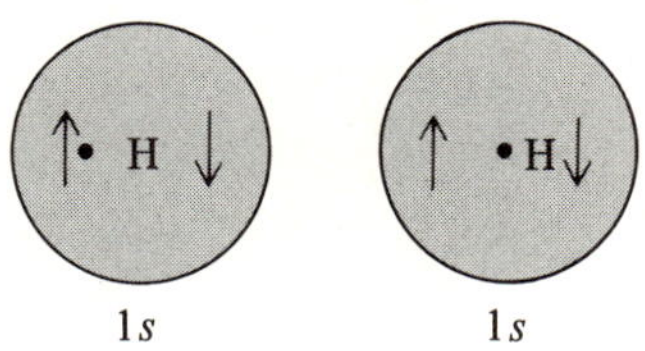


*FIGURE 3.11*
*Spin pairing in the hydrogen molecule.*

In conformity with this picture, the C—H and C—C coupling constants are to be of the same positive sign. Negative coupling constants can also be observed, as in geminal ($^2J_{H-H}$) coupling.

*3.2.6 Geminal Coupling Constants*

The interproton coupling interaction in methane has the value −12.4 Hz. Table 3.7 shows how it is affected upon changes of geometry and upon substitution. It is intriguing that electron withdrawal by electronegative substituents makes the coupling constant more positive, while electron withdrawal by adjacent $\pi$ centers has the converse effect of making the geminal coupling constant more negative. Such effects can be understood on the basis of the following description: Indirect coupling between *protons* is expressed by the predominant term

$$J_{H\text{-}H'} = h(\tfrac{4}{3}\beta\gamma_H)^2 S_h^4(0)\pi_{h,h'} \qquad (22)$$

where $S_h(0)$ is the electron density at the nucleus for a 1*s* hydrogen orbital, which is

[7] D. M. Grant and W. M. Litchman, *J. Amer. Chem. Soc.*, *87*, 3994 (1965); T. L. Brown and J. C. Puckett, *J. Chem. Phys.*, *44*, 2238 (1966).

*TABLE 3.7*

*Geminal Coupling Constants $^2J$ as a Function of Geometry and Substitution*

| Compound | Constant | Compound | Constant | Compound | Constant | Compound | Constant |
|---|---|---|---|---|---|---|---|
| $H_2C\langle^{H}_{H}$ | −12.4 | $CH_4$ | −12.4 | $CH_4$ | −12.4 | $CH_4$ | −12.4 |
| 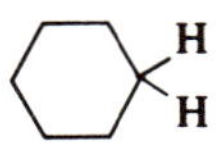 | −13.0 | $CH_3F$ | −9.6 | $SiH_4$ | +2.8 | $CH_3COCH_3$ | −14.9 |
| | | $CH_3Cl$ | −10.8 | $GeH_4$ | +7.7 | $C_6H_5{-}CH_3$ | −14.5 |
|  | −4.3 | $CH_3Br$ | −10.2 | $SnH_4$ | +15.3 | $CH_2(CN)_2$ | −20.4 |
| $H_2C{=}C\langle^{H}_{H}$ | +2.5 | $CH_3I$ | −9.2 | | | | |

virtually a constant for all protons, and $\pi_{h,h'}$, the mutual polarizability of the two 1*s* hydrogen orbitals, is given by

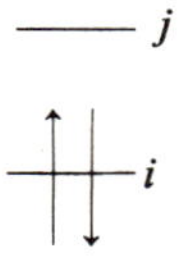


$$\pi_{h,h'} = -4 \sum_{i}^{\text{occ.}} \sum_{j}^{\text{unocc.}} \frac{1}{E_j - E_i} C_{ih} C_{ih'} C_{jh} C_{jh'} \tag{23}$$

where $C_{ih}$, etc., are atomic orbital coefficients in the LCAO expansion.

Let us investigate the *meaning* of these expressions in the case of a $CH_2$ fragment, for which the four molecular orbitals are the following:

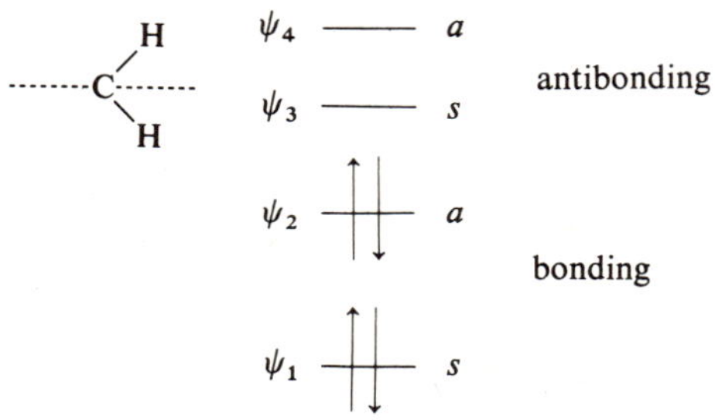


According to these expressions and Figure 3.12, the coupling constant is made up of four different contributions corresponding to the electronic excitations:

$\psi_1 \rightarrow \psi_3$ $\psi_1 \rightarrow \psi_4$
$\psi_2 \rightarrow \psi_3$ $\psi_2 \rightarrow \psi_4$

The product $C_{ih}C_{ih'}C_{jh}C_{jh'}$ is positive for transitions connecting states of the same symmetry, negative otherwise. The corresponding contribution to $^2J_{H-H}$ has the opposite sign, and is weighted by the inverse excitation energy, $\psi_2 \rightarrow \psi_3$ making a more substantial contribution than $\psi_1 \rightarrow \psi_4$.

| Transition | Sign of contribution to $^2J_{H-H'}$ |
|---|---|
| $\psi_2 \rightarrow \psi_3$ | + |
| $\psi_2 \rightarrow \psi_4$ | − |
| $\psi_1 \rightarrow \psi_3$ | − |
| $\psi_1 \rightarrow \psi_4$ | + |

The effect of an electronegative substituent is to remove electron density from the hydrogens in $\psi_1$, thus necessarily making the other symmetrical orbital $\psi_3$ more hydrogen-like, and leaving $\psi_2$ and $\psi_4$ unaffected. The bigger change will be to increase the positive contribution from the $\psi_2 \rightarrow \psi_3$ transition, hence a positive change in the

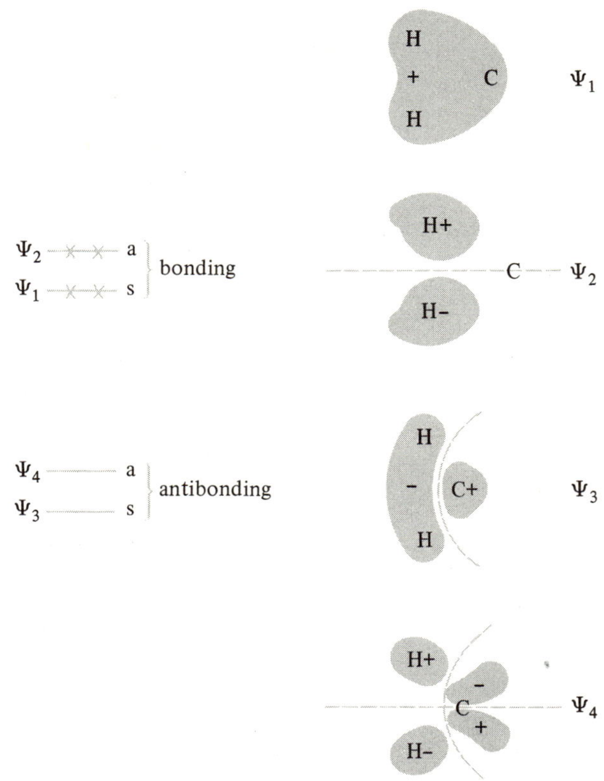


*FIGURE 3.12*
*Molecular orbitals for a* $CH_2$ *fragment.*

coupling constant. [The positive contribution from $\psi_1 \rightarrow \psi_4$ will be reduced, but by a lesser amount since $(E_3 - E_2)^{-1} \gg (E_4 - E_1)^{-1}$.] The effect of an adjacent phenyl group will be to remove electrons from the hydrogen region of $\psi_2$, and by way of compensation to make $\psi_4$ more hydrogen-like, resulting in an overall increase of the negative contribution to the coupling. It is indeed observed that the values of $^2J_{H-H}$ become more and more negative as the number of adjacent $\pi$ centers is increased (Table 3.7), and this contribution varies with the angle between the plane bisecting the $CH_2$ group and that of the $\pi$ center (Figure 3.13).

Returning to Table 3.7, the increasingly positive $^2J_{H-H}$ values in the series $CH_4$,

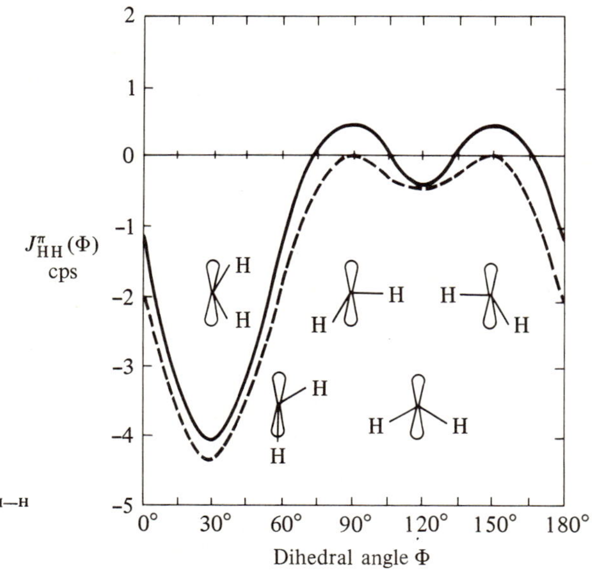


*FIGURE 3.13*
*The* $\pi$ *contribution to* $^2J_{H-H}$ *coupling constants.*

$SiH_4$, $GeH_4$, and $SnH_4$ can also be understood in terms of the above theory, since replacement of the carbon central atom by Si, Ge, and Sn will decrease the separation between bonding and antibonding molecular orbitals and increase that in each group. Hence $\psi_2 \rightarrow \psi_3$ will become more positive, and by a greater amount than that by which $\psi_1 \rightarrow \psi_3$ becomes more negative.

As for the angular dependence, when the $CH_2$ group is part of a ring system, it is observed that the $^2J_{H-H}$ coupling constants become more positive when the HCH angle increases from 109.5° ($CH_4$), to 116° ($C_3H_6$) and to ~120° ($C_2H_4$). These hybridization changes can also be accounted for theoretically, and are summarized in Table 3.8.

*TABLE 3.8*

*Ring Size Effect on Geminal $^2J_{H-H}$ Coupling Constants*

| Ring size $n$ | Range of values[a] for $^2J_{H-H}$ (Hz) |
|---|---|
| 3 | −4 to −6 |
| 4 | −8 to −10 |
| 5 | −9 to −12 |
| 6 | −12.5 to −14 |

[a] In the absence of adjacent π centers or other strongly perturbing substituents.

### *3.2.7 Vicinal Coupling Constants*

Vicinal coupling constants are defined as the interaction between two nuclear spins formally separated by three chemical bonds, such as the H—H coupling in ethane. For that reason, they are denoted $^3J$. One could describe them using a framework similar to that for one-bond and two-bond coupling constants. Let us vary the presentation a little, however, and mention here a point of interest which could have been raised equally well with the other coupling types. The coupling phenomenon is due to the electron spin density present at nucleus 1 when a perturbing magnetic dipole is placed at nucleus 2. This very simple concept has as consequence that $^3J_{H-H}$ ought to be proportional to the hyperfine coupling constant for the corresponding radical. One can thus make a comparison between the nmr coupling constants for a hydrocarbon and the esr coupling constant for the corresponding free radical obtained by abstraction of H · (Table 3.9). There is indeed a striking correspondence between the two series of results. Furthermore, another feature stands out from Table 3.9: It is a

*TABLE 3.9*

*Correlation of Nuclear Magnetic Resonance and Electron Spin Resonance Coupling Constants*

| Hydrocarbon | | $J_{H-H}$ (Hz) | Radical | | $a_H$ (Oe) |
|---|---|---|---|---|---|
| $H_2C{=}CH_2$ | *cis* | 11.7 | $\cdot HC{=}CH_2$ | *cis* | 34.2 |
| | *trans* | 19.0 | | *trans* | 68.5 |
| $H_3C{-}CH{=}CH_2$ | *cis* | 10.0 | $H_3C{-}\dot{C}{=}CH_2$ | *cis* | 32.9 |
| | *trans* | 16.8 | | *trans* | 58.9 |
| $H{-}C{\equiv}C{-}H$ | | 9.1 | $\cdot C{\equiv}C{-}H$ | | 16.1 |

very general observation that *trans* (vicinal) coupling constants are greater in magnitude than *cis* coupling constants. For instance, in the cyclohexane ring system *trans* diaxial coupling constants are large, whereas *gauche* couplings such as between axial-equatorial pairs or between diequatorial protons are small, ca. 12 Hz and 3 Hz, respectively. Both experimentally and theoretically it is possible to obtain values of $^3J_{H-H}$ coupling constants corresponding to the dihedral angle $\theta$ (Figure 3.14) varying between 0 and $2\pi$. When $^3J_{H-H}$ is plotted against $\theta$, the graph obtained is the so-called Karplus curve (Figure 3.15), corresponding to equations (24) and (25):

$$^3J_{H-H} = 10 \cos^2 \theta, \qquad 0 \leq \theta \leq 90° \qquad (24)$$

$$^3J_{H-H} = 16 \cos^2 \theta, \qquad 90 \leq \theta \leq 180° \qquad (25)$$

The geometrical relationship of two protons A and B on two adjacent centers is not uniquely specified by the value of $\theta$. Even if one assumes constant C—H bond lengths, the C—C bond length and the H—C—C in-plane angles are other important parameters. According to theory an increased separation of proton A and the carbon Q in the fragment $H_A—C_P—C_Q—H_B$ is expected to decrease the coupling between the nuclear spins at A and B, since the value assumed by the $q$ hybrid orbital corresponding to the Q carbon at nucleus A would be decreased. For *cis* protons on adjacent $sp^2$ carbons, their coupling constant shows a linear dependence with respect to the C—C bond order $p$:[8]

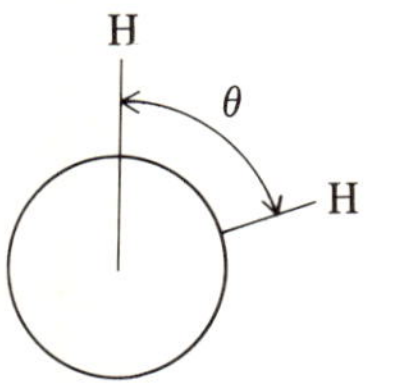


*FIGURE 3.14*
*The dihedral angle between two C—H bonds.*

$$J \sim 11.6p - 0.2 \tag{26}$$

typical experimental values being 11.7 Hz (ethylene), 7.5 Hz (benzene), 8.6 Hz (naphthalene $\alpha\beta$), and 6.7 Hz (naphthalene $\beta\beta$).

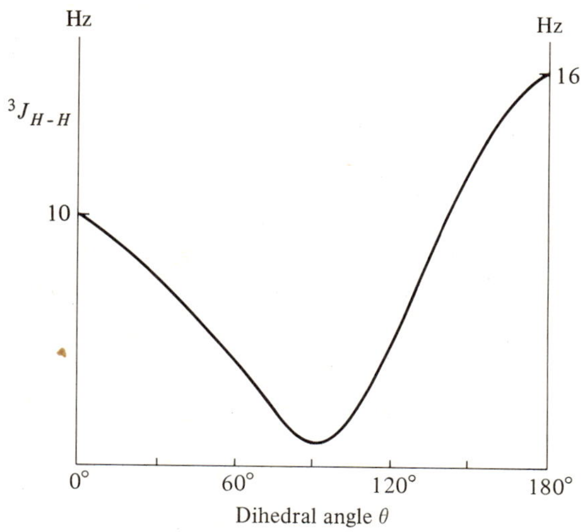


*FIGURE 3.15*
*Vicinal coupling $^3J$ as a function of dihedral angle $\theta$.*

Likewise, the coupling constant of the protons indicated goes from 5.14 Hz in cyclohexadiene to 1.94 Hz in cyclopentadiene:

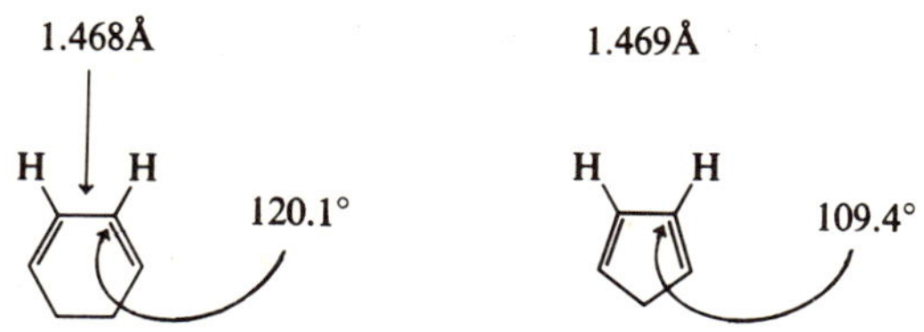


What has been described for protons can be extended to other nuclear spins with $I = \frac{1}{2}$. For the first two rows of the periodic table: $^{13}C$, $^{19}F$, $^{29}Si$, and $^{31}P$ are examples for which there is considerable experimental evidence in support of such parallel behavior, as illustrated in Figure 3.16. When no preferred rotamer is present, the expression derived from experimental data for the coupling constants averaged over the rotation around a single bond are

$$^3J_{H—H} = 14.5 - 0.8 \sum E \tag{27}$$

$$^3J_{H—F} = 53 - 3.4 \sum E \tag{28}$$

where $\sum E$ is the sum of the electronegativities of the substituents in an ethane molecule of the general type $R_1R_2HC—CHR_3R_4$ or $R_1R_2HC—CFR_3R_4$.

[8] N. Jonathan, S. Gordon, and B. P. Dailey, *J. Chem. Phys.*, *36*, 2443 (1962); M. A. Cooper and S. L. Manatt, *J. Amer. Chem. Soc.*, *92*, 1605 (1970); *91*, 6325 (1969).

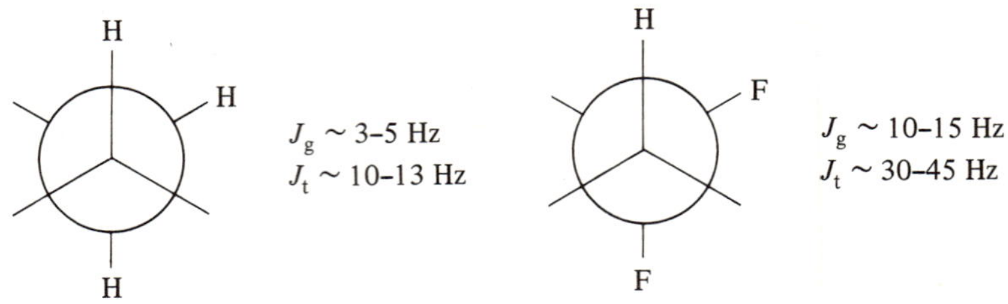


*FIGURE 3.16*
*Comparison of* H—H *and* H—F *coupling constants.*

*3.2.8 Long-Range Coupling* $^4J_{H-H}$

Finally, we turn our attention to the long-range coupling constants. Unless a hydrocarbon molecule incorporates a considerable amount of strain, the only such couplings with a substantial magnitude are those "through 4 bonds," which are of course written as $^4J_{H-H}$. A useful empirical regularity is the so-called "M or W rule"[9] which states that these coupling constants are maximum (ca. +3 Hz) when the coupled protons are coplanar with the three intervening carbons in a straight zigzag path, schematized by the letters *M* or *W*:

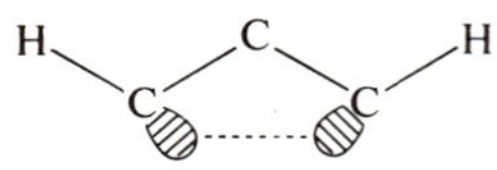


A rationalization has been proposed for the physical mechanism of this coupling in terms of overlap between the back lobes of the two carbon-hydrogen bonds[10]; this is, however, an oversimplification since molecular orbital calculations[11] show that the through-space contribution is the greatest in the axial-equatorial case:

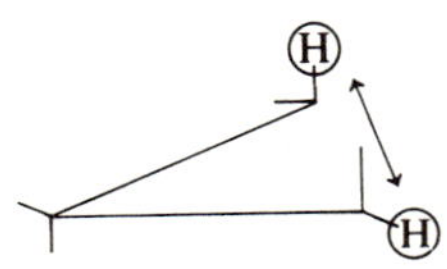


Approximate values of $^4J_{H-H}$ are listed in Table 3.10 for the more important cases.

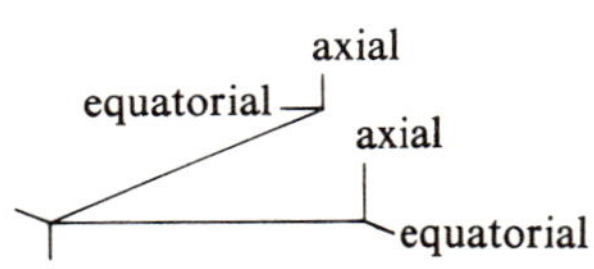


*TABLE 3.10*

*Dependence of* $^4J_{H-H}$ *Coupling Constants on Stereochemistry*

| Axial–axial | Axial–equatorial | Equatorial–equatorial |
|---|---|---|
| −0.9 Hz | −0.4 Hz | +1.8 Hz |

Delocalization of the molecular orbitals will enhance the values of these coupling constants, either through strain of the σ framework or from accumulation of intervening double bonds. In the latter case, hyperconjugative-type mechanisms manifest themselves in that coupling constants between protons with an allylic or a homoallylic relationship will be highest when C—H bonds are colinear with the $p_z$ orbitals[12]:

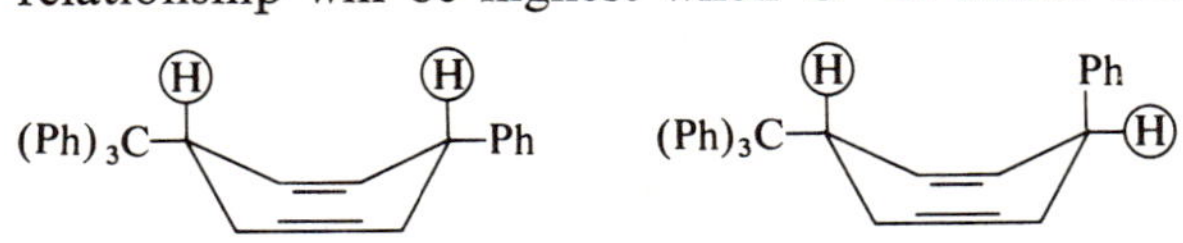


$^5J_{H-H}$ = 11 Hz $^5J_{H-H}$ = 7.5 Hz

[9] A. Rassat, C. W. Jefford, J. M. Lehn, and B. Waegell, *Tetrahedron Lett.*, 233 (1964).
[10] J. Meinwald and A. Lewis, *J. Amer. Chem. Soc.*, *83*, 2679 (1961).
[11] P. R. Carey and R. Ditchfield, *Mol. Phys.*, *15*, 515 (1968); V. F. Bystrov and A. U. Stepanyants, *J. Mol. Spectry.*, *21*, 241 (1966).
[12] D. J. Atkinson and M. J. Perkins, *Tetrahedron Lett.*, 2335 (1969).

## 3.3 *PROBLEMS*

### SAMPLE PROBLEM

In some molecules of the type $R'_2CH{-}CHR_1R_2$, the following values are measured for the vicinal HCCH coupling constants ($\pm 0.2$ Hz):

| | |
|---|---|
| $R_1 = R_2 = H$ | 7.3 Hz |
| $R_1 = CH_3,\ R_2 = H$ | 8.0 Hz |
| $R_1 = R_2 = CH_3$ | 11.2 Hz |

They are temperature independent between $-60°$ and $+40°$. How does one account for these structural variations? Do they correspond to our expectations?

### ANSWER[13]

Let us assume that only staggered conformations are important; the Newman projections are as follows:

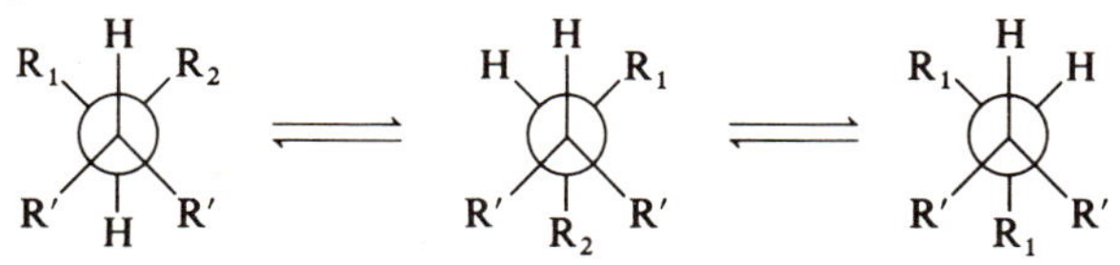

1 2 3

When $R_1 = R_2 = H$, all three rotamers (1), (2), and (3) are equally populated and one must have

$$^3J = \tfrac{1}{3}(J_{trans} + 2J_{gauche}) = 7.3 \text{ Hz} \quad \text{(methyl case)}$$

Let us assume further that additional methyl groups will avoid being staggered between *both* R′ groups. It follows that

$$^3J = \tfrac{1}{2}(J_{trans} + J_{gauche}) = 8.0 \text{ Hz} \quad \text{(ethyl case)}$$
$$^3J = J_{trans} = 11.2 \text{ Hz} \quad \text{(isopropyl case)}$$

These three equations contain only two unknowns ($J_{trans}$ and $J_{gauche}$), and are thus overdetermined. If a single set of solutions can be found, all the assumptions will be justified a posteriori. This is precisely the case: The above equations are solved easily, yielding

$$J_{trans} = 11.2 \pm 0.2 \text{ Hz}$$
$$J_{gauche} = 5.0 \pm 0.4 \text{ Hz}$$

in accord with the Karplus-type dihedral-angle dependency.

1. Which of the two formulations below is appropriate for the hydrocarbon $C_{11}H_{10}$, given the fact that $^1J_{C-H} = 142$ Hz for the bridge hydrogens?

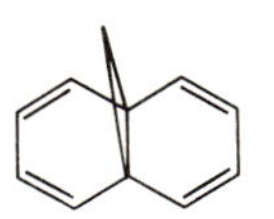

1

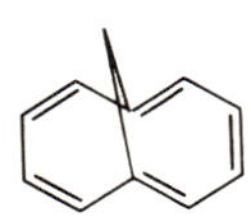

2

2. Indicate the long-range, $^4J$, $^1H{-}^1H$ coupling constants which you would expect to be of importance in the bicyclo[2.2.1]heptane molecule:

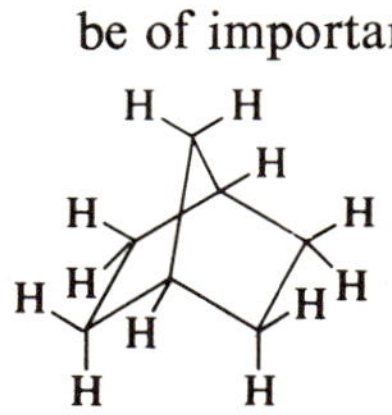

[13] S. Forsén, W. E. Frankle, P. Laszlo, and J. Lubochinsky, *J. Magnetic Resonance, 1*, 291 (1969).

3. The linewidth of the angular methyl of a *trans* decalin, in the nmr spectrum, is greater than that of the corresponding resonance of a *cis* decalin. Explain.

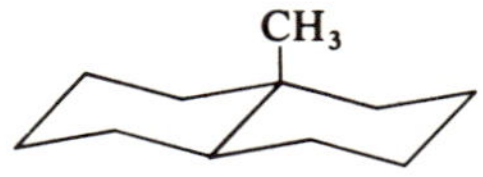

4. The possibility that the sulfur heteroatom is tetravalent in benzothiadiazole—the skeleton of which is shown below—has been explored by determining the coupling constants within the six-membered ring:

   $J_{12} = 8.9$ Hz
   $J_{13} = 1.1$ Hz
   $J_{14} = 0.9$ Hz
   $J_{23} = 6.6$ Hz

   What is your conclusion? Make your argument quantitative by deriving the corresponding bond orders from the vicinal coupling constants above.

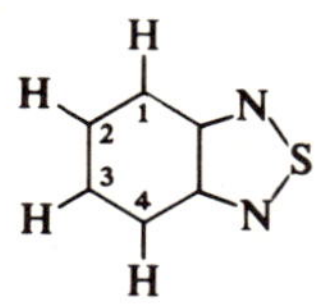


5. Predict the appearance of the esr of this nitroxide free radical:

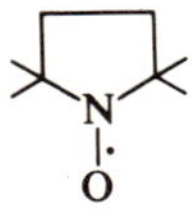


6. The $^1J_{C-H}$ coupling constants for the 9 and 10 positions in the central ring of anthracene are, of course, the same and equal to 156 Hz. When anthracene is reacted with HCl and $AlCl_3$, a salt $C_{14}H_{11} \cdot Al_2Cl_7$ is generated. Among its features, the $^{13}C$—H coupling constants for the 9 and 10 positions are now different, the 9 position has $^1J_{C-H} = 127.5$ Hz whereas for the 10 position $^1J_{C-H} = 162.3$ Hz. Explain this result.

## 3.4 *APPENDIX*

*Summary of interproton coupling constants*

| Type | *J*, typical (Hz) | Range (Hz) |
|---|---|---|
| $H_2$ | 280[a] | |
| $CH_4$ | 12.4[a] | |
|  | ~13 | +5 to −30 |
| 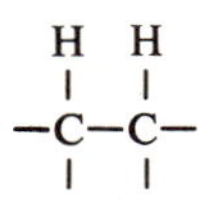 | ~7 | 0–9[b] |
| 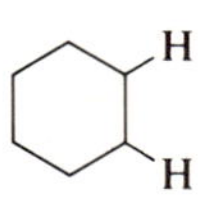 | a, a ~10<br>a, e ~5<br>e, e ~3 | 8–14<br>0–7<br>0–5 |
| 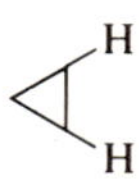 | *cis* ~9<br>*trans* ~6 | 6–12<br>4–8 |

| Type | *J*, typical (Hz) | Range (Hz) |
|---|---|---|
| 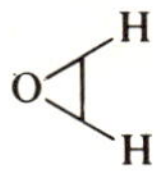 | *cis* ~3<br>*trans* ~4 | 2–4<br>2–5 |
| 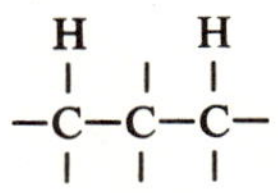 | ~2 | 0–7[c] |
| 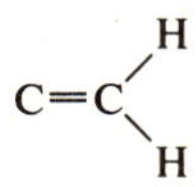 | ~0 | 0–5 |
| 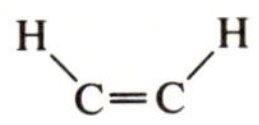 | ~10 | 0–15 |
| 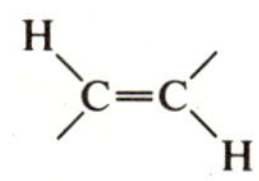 | ~16 | 5–30 |
| 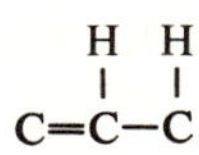 | ~5 | 4–10 |
| 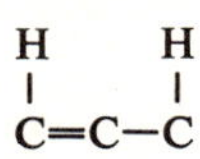 | ~1 | 0.5–3.0 |
| 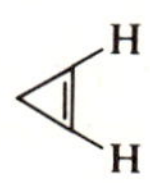 | ~2 | 0–2 |
| 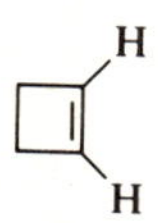 | ~4 | 2–4 |
| 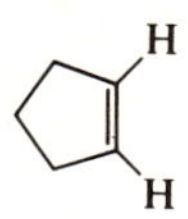 | ~5.5 | 5–7 |
| 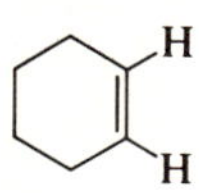 | ~10 | 8–11 |
| 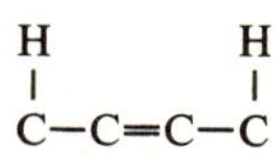 | ~0 | 0–1.5 |
| 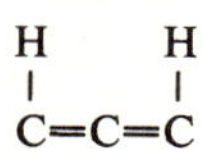 | ~5 | 4–7 |
| 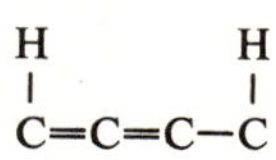 | ~2 | 2–4 |
| 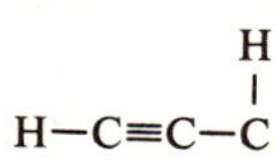 | ~2 | 2–4 |
| 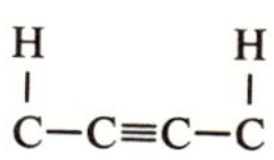 | ~1 | 1–3 |
| 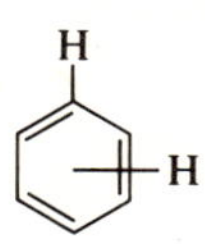 | *ortho* ~8<br>*meta* ~3<br>*para* ~1 | 6–10<br>1–4<br>0–2 |

*(Continued)*

*(Continued)*

| Type | *J*, typical (Hz) | Range (Hz) |
|---|---|---|
| 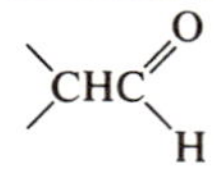 | ~2 | 1–3 |
| 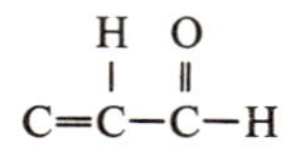 | ~6 | 5–8 |
| 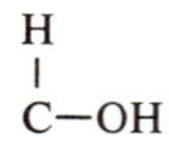 | ~5 | 4–10[d] |

[a] Determined from the partially deuterated compounds.
[b] Strongly dependent on dihedral angle (see text).
[c] Dependent on spatial orientation.
[d] Free rotation and no exchange of O—H protons.

## 3.5 *SUGGESTED READINGS*

C. Trindle, "Bond Description of Delocalization," *J. Amer. Chem. Soc., 91*, 219 (1969).

C. Trindle and O. Sinanoğlu, "Local Orbital and Bond Index Characterization of Hybridization," *J. Amer. Chem. Soc., 91*, 853 (1969).

G. R. Luckhurst, "Liquid Crystals as Solvents in Nuclear Magnetic Resonance," *Quart. Rev.* (London), *22*, 179 (1968).

J. A. Pople and A. A. Bothner-By, "Nuclear Spin Coupling Between Geminal Hydrogen Atoms," *J. Chem. Phys., 42*, 1339 (1965).

T. Yonegawa, I. Morishima, M. Fujii, and H. Kato, "Molecular Orbital Calculations of Nuclear Spin Coupling Constants. I. Preliminary Results for Proton–Proton Coupling Constants," *Bull. Chem. Soc. Jap., 40*, 487 (1967).

M. Barfield and B. Chakrabarti, "Long Range Proton Spin–Spin Coupling," *Chem. Rev., 69*, 757 (1969).

G. E. Maciel, J. W. McIver, Jr., N. S. Ostlund, and J. A. Pople, "Approximate Self-Consistent M.O. Theory of Nuclear Spin Coupling. I, II, III, IV, and V," *J. Amer. Chem. Soc., 92*, 1, 11, 4151, 4497, 4506, 5829 (1970).

W. T. Dixon, "A Molecular Orbital Theory of Proton Spin-Spin Coupling Constants," *J. Chem. Soc.,* (A)1879 (1967).

S. Sternhell, "Long-Range $^1H—^1H$ Spin-Spin Coupling in Nuclear Magnetic Resonance Spectroscopy," *Rev. Pure and Appl. Chem., 14*, 15 (1964).

S. Sternhell, "Correlation of Interproton Spin-Spin Coupling Constants with Structure," *Quart. Rev., 23*, 236 (1969).

W. McFarlane, "Nuclear Spin-Spin Coupling Between Directly Bound Elements," *Quart. Rev.* (London), *23*, 187 (1969).

M. A. Cooper and S. L. Manatt, "The Effect of Steric Compression on Proton-Proton Spin-Spin Coupling Constants. Further Evidence and Mechanistic Considerations," *J. Amer. Chem. Soc., 92*, 4646 (1970).

J. B. Pawliczek and H. Günther, "H—H Kopplungskonstanten und π-elektronenstruktur in ungesättigten 6-ringen," *Tetrahedron, 26*, 1755 (1970).

M. A. Cooper and S. L. Manatt, "A Detailed Evaluation of the Dependence of $^3J$(H—H) on Bond Angle in Alkenes and Cycloalkenes," *Org. Magn. Resonance, 2*, 511 (1970).

# CHAPTER 4 NUCLEAR ENERGY LEVELS IN MOLECULES: CHEMICAL SHIFT

In general, if we consider a transition between two molecular energy states, the quantum $h\nu$ delivered is likely to depend slightly upon the molecular structure. It should be influenced, for instance, by the nature of the substituents and by their distance to the "chromophore" studied. It should also be dependent upon the nature of the medium (gaseous, polar liquid, etc.).

Let us discuss again the nmr experiment. Energy is absorbed by the proton nuclei of the chloroform molecules at a certain frequency $\nu$ when a magnetic field $H_0$ is provided in the gap of the electromagnet. Actually, the effective field $H_{eff}$ at the hydrogen nucleus is smaller than $H_0$, the magnetic field provided by the instrument. *The electrons in the vicinity of the nucleus tend to shield it from the perturbation by the applied field.* They will circulate so as to produce an electrical current and hence a magnetic field, reacting against the applied directing field $H_0$, so that $H_{eff} = H_0 (1 - \sigma)$, where $\sigma$ is called the shielding constant. This general property of matter is called diamagnetism.

If we replace the chloroform molecules with molecules of methyl chloride there is no reason for $H_{eff}$ at the $CH_3Cl$ protons to be equal to $H_{eff}$ for the $CHCl_3$ protons. They are indeed different. The difference is tiny, something like one part per million! But the resonance frequencies will differ correspondingly, and such small frequency differences can be detected with modern spectrometers (which have resulted from the enormous progress in electronics during World War II). To give an idea of the resolution of an nmr spectrometer, two resonance lines can be resolved easily by standard commercial instruments if they are 1 Hz apart at the 100 MHz resonance frequency for protons (i.e., for a field of 23.5 kOe). The comparison has been made that a telescope with a comparable resolution would separate the two ears of a cat sitting on the moon!

That the resonance frequency depends somewhat upon molecular structure is a very useful property, called the chemical shift. For protons (and, of course, for deuterons since the two species are simply related by the gyromagnetic ratios specifying different resonance frequencies, the molecular environment around the $^1H$ and around the $^2H$ nuclei being virtually identical) the range of chemical shifts is ~30 ppm. For carbon-13, it is ~400 ppm. For fluorine-19, it is ~1000 ppm.

## 4.1 *The Reference Problem*

In order to measure chemical shifts, one would like to measure the difference between the frequency $\nu$ observed for a certain molecule and the frequency $\nu_0$ which would correspond to the applied field $H_0$ were it not for the shielding electrons. It is not possible to perform such an absolute measurement with enough precision. What is done, instead, is to compare the resonance frequency with that of a standard molecular species. This can be accomplished in two ways: with an external reference or with an internal reference. In the case of an external reference, the reference material is isolated from the sample, usually by means of a separate glass container which is inserted into the nmr tube containing the sample. Unless the reference container is exactly spherical, however, corrections must be made for the differences in bulk susceptibility (to magnetization) for the two separate samples. Since it is difficult to obtain perfect spheres and since it is unpleasant to have to make corrections, it is generally more convenient to dissolve directly in the solution studied a small amount of foreign substance, an *internal* reference.

Criteria for the selection of an internal reference are that it should give a sharp line, preferably in a blank region of the spectrum, and be chemically inert and volatile for recovery of the pure sample. To avoid favored orientations with respect to the solvent molecules spherically symmetrical molecules are preferred. For pmr studies the reference compounds most frequently employed are $Si(CH_3)_4$ [TMS], $C(CH_3)_4$, and $C_6H_{12}$; for $^{13}C$ work $CS_2$, $C_6F_6$, or the carbonyl carbon of $O{=}C(OCH_3)_2$; and for $^{19}F$ resonance $CFCl_3$ or one of the Freons.

To measure a chemical shift is then simply to measure the difference in frequency between a given line (or "resonance") and that of the reference compound. For instance, the chloroform resonance is observed ~436 Hz downfield from the TMS resonance at 60 MHz. If the operating frequency were 100 MHz, it would occur at $436 \times 100/60 = 725$ Hz, because of the proportionality between the resonance frequency $\nu$ and the magnetic field $H_0$. It is for this reason that chemical shifts are expressed in units independent of the frequency of the spectrometer used, usually in parts per million from the reference. For instance, the chloroform resonance occurs at 7.25 ppm downfield from the tetramethylsilane reference ($7.25 \text{ ppm} = 436/60 \times 10^6 = 725/100 \times 10^6$). For proton resonances this is denoted by $\delta$. A more old-fashioned scale which is still used sets TMS equal to a value of 10.00 instead of 0.00. In this $\tau$ scale chloroform appears at $2.75\tau$.

This $\tau$ scale was defined rather strictly; it applies only to infinite dilution values in carbon tetrachloride solution. In general, it is preferable to derive the infinite dilution shift from a plot of the chemical shift versus concentration in a certain solvent. For proton studies, however, it is sufficient to use solutions of less than 3 percent $w/v$. For carbon-13 and fluorine-19 the magnetic resonances are more sensitive, as evidenced by the ranges of chemical shifts quoted above, and therefore concentration effects are more noticeable.

## 4.2 *Semiempirical Correlations: Polar Effects*

Theoretically, the concept of chemical shift is extremely difficult to explain. There are very few molecules for which chemical shifts can be calculated so that the result has some resemblance to the experimental value; HF is one such favorable example.

The difficulty can be reduced if smaller demands are set. For instance, the proton resonances for the methyl halides are $CH_3I = 2.16$, $CH_3Br = 2.68$, $CH_3Cl = 3.05$, and $CH_3F = 4.26$ ppm relative to TMS (the methane protons absorb at 0.28). Our naive argument will be as follows. Around protons the electron distributions are almost spherical. Any electrostatic distortion of these $1s$ orbitals drawing electron density away from the nuclei should and does reduce their shielding, as in the above sequence. In other words, it is possible to understand this particular sequence of chemical shifts in terms of a derived concept, that of electronegativity, since it is apparent that the order followed is the same. It is indeed possible to use proton chemical shifts to set or to complete electronegativity scales. It should be clear, however, that the procedure used is far from being a priori and rigorous; on the contrary, it is a good example of the semiempirical correlations so typical of physical methods as used by organic chemists.

Another elegant and important example is the work of Taft[1] on the fluorophenyl

---

[1] R. W. Taft *et al.*, *J. Amer. Chem. Soc.*, *85*, 709, 3146 (1963); *86*, 4199 (1964); *87*, 2489, 4385, 4387 (1965); *89*, 2391, 2397, 5957 (1967).

label. Since $^{19}F$ chemical shifts cover a range of several hundred parts per million (ppm) and can be measured easily to 0.05 ppm, the fluorine nucleus can be used as a sensitive probe for electronic effects. Fluorophenyl compounds were used since the aromatic nucleus offers what is essentially a rigid conducting cavity for transmission of the effects. In Taft's words: "The fluorine atom may be regarded as a distant but sensitive observer removed from the confusion of the battlefield (that is, field of interaction of solvent and substituent) by the rigid benzene ring."

Excluding the *ortho* substituents, which are likely to exert proximity effects, let us look at some data in Table 4.1. Examination of the data in Table 4.1 reveals that the *meta* shifts correlate very well with electronegativity along a line of the periodic table as well as vertically. It is also apparent that the *para* shifts respond to $\pi$ delocalization of charge.

*TABLE 4.1*

*Fluorine Chemical Shifts (ppm from Fluorobenzene) for Substituted Fluorophenyls*

| Substituents: | $CMe_3$ | $NMe_2$ | OMe | F | $SiMe_3$ | $PMe_2$ | SMe | Cl |
|---|---|---|---|---|---|---|---|---|
| *para* | +5.55 | +15.9 | +11.7 | +6.8 | +0.50 | +1.40 | +4.40 | +3.20 |
| *meta* | +0.45 | −0.08 | −1.05 | −3.10 | +0.85 | +0.05 | −0.40 | −2.50 |

Taft therefore defines *inductive* ($\sigma_I$) and *resonance* ($\sigma_R$) parameters by the following equations:

$$\delta^m \text{ (ppm)} = -7.1\,\sigma_I + 0.6$$

$$\delta^p - \delta^m \text{ (ppm)} = -30\,\sigma_R$$

$$\text{Hammett } \sigma = \sigma_I + \sigma_R$$

These substituent constants are very useful: nmr chemical shifts can thus be used to predict reactivities, to evaluate $d_\pi$–$p_\pi$ interactions in thioanisoles, to measure the difference between the resonance abilities of $CH_3$ and $CD_3$, and so on.

A last example of a semiempirical correlation involving chemical shifts is that of the dependence of proton and carbon chemical shifts upon the electron densities at the various carbons of an aromatic molecule. It is observed that, if the proton resonance of benzene is taken as a reference, the cyclopentadienide anion protons resonate 1.7 ppm to higher field, and those of the tropylium cation 1.9 ppm to lower field relative to benzene.

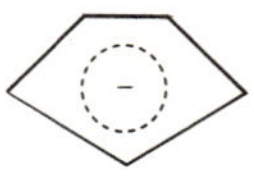

Cyclopentadienide

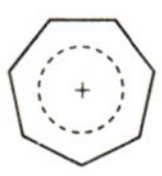

Tropylium

In simple terms, the $C_5$ ion has a higher electron density ($\frac{1}{5}$ or 0.2 electron per carbon, thus shielding the proton nuclei) and the $C_7$ ion has a lower density (by $\frac{1}{7}$ or 0.143 electron per carbon, thus deshielding the proton nuclei) compared to benzene. As a first approximation, after small corrections are applied to compensate for the anisotropy effects (discussed at the end of the section) the chemical shift variation is found to be proportional to the change in the electron densities. The proportionality constant is about 10 ppm per unit of charge for proton resonance. For carbon, it is of the order of 160 ppm per electronic charge. In such a fashion one can account at least partly for the observed chemical shifts of a wide variety of aromatic compounds.

## 4.3 *Components of Chemical Shift*

We have seen what the semiempirical approach can do. Another approach dissects the chemical shift into various components, each corresponding to a simple physical concept. This approach is more a priori than the preceding one, but it is a trifle more complicated to use and a little less satisfactory in its agreement with the experimental results.

### *4.3.1 van der Waals Effects*

If we start again with the crude image of the shielding electrons in a roughly spherical distribution around a given proton, it is obvious that not only should we be able to perturb this protective electron cloud by shifting it away from the nucleus by an electronegative substituent, but also *in a noninductive fashion* by having a neighboring electron cloud repel the electron density around the proton, as illustrated diagrammatically in Figure 4.1. These through space interactions, called London–van der Waals forces or dispersion forces, occur between two molecular systems because of electronic correlation: the electrons moving in system 1 feel the presence of the electronic motion in system 2. As a result, even in molecules with spherical symmetry, at any instant there will be a net dipole moment induced by the electrons in the second system. As a consequence instantaneous dipole moments will exist in both systems and their nonzero time average interaction is the dispersion force. By disturbing the electron distribution around the nuclei these dispersion forces can influence the chemical shift of the nuclei under observation.

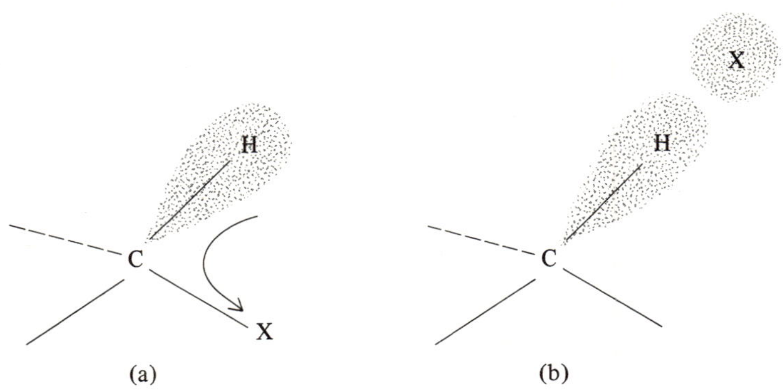


*FIGURE 4.1*
*Perturbation of electron cloud around a proton (a) by inductive withdrawal by an electronegative substituent, or (b) through space London–van der Waals forces.*

This London–van der Waals contribution to the chemical shift was first calculated theoretically for the interaction of two hydrogen atoms by Marshall and Pople.[2] As expected it was found that the shielding of the two hydrogen nuclei by the surrounding electrons decreased as the two atoms approached each other more and more closely. Van der Waals forces depend drastically upon the distance (by the sixth and the twelfth power, the so-called 6-12 potential) and it is thus not at all surprising that the van der Waals contribution to the chemical shift also depends upon distance through an $r^{-6}$ term. The expression used for calculating the van der Waals contribution to chemical shift is

$$\Delta\sigma_w\,(\text{ppm}) = \frac{-60}{r^6}\,Q$$

where $\Delta\sigma_w$ is the effect on the shielding of a proton by another atom at a distance $r$ (in angstroms) and $Q$ is a factor dependent upon the nature of the group interacting with the proton considered. Numerical values of $Q$ for a few atoms are listed in Table 4.2. As seen from the table the value of $Q$ increases with the charge on the atom and

[2] T. W. Marshall and J. A. Pople, *Mol. Phys.*, *3*, 339 (1960).

with its size. It is actually proportional to the product of the polarizability $\alpha$ and the ionization potential $I$. From the expression given it is readily seen that for H···H interactions this van der Waals component will be substantial only for distances below 3 Å, while for H···X interactions detectable effects will be observed whenever $r$ is smaller than 4 Å.

*TABLE 4.2*

*Values of Q*

| Group | H | F | $F^-$ | $Cl^-$ | Br | $Br^-$ | $I^-$ |
|---|---|---|---|---|---|---|---|
| $Q$ value | 1 | 3.1 | 3.6 | 12.5 | 14 | 18 | 30 |

### 4.3.2 Electrostatic Field Effects

The application of an electric field should displace the surrounding electrons to or from the resonating nucleus. The electric fields present in polar molecules are sufficient. Buckingham, Schaefer, and Schneider,[3] as well as Musher[4] have proposed a theory to account for such effects. The equation derived by these authors for calculating the change in chemical shift $\Delta\sigma_E$ as a function of the electrostatic field $E$ is

$$\Delta\sigma_E\,(\text{ppm}) = a \times 10^{-12}E_z + b \times 10^{-18}E^2$$

where $E$ is the electrostatic field, $E_z$ its component along the bond considered, and $a$ and $b$ are constants. For protons $a$ is of the order of $-3$ (a downfield shift results from an $E_z$ component drawing electrons away from the proton in an A—H bond) and $b$ is of the order of $-1$. For fluorine nuclei, $a$ is of the order of $+10$ and $b$ of the order of $-40$. In general, the magnitudes of $a$ and $b$ are much larger for other nuclei than for protons. While $a$ can be either positive or negative, $b$ is always negative.

Let us now consider an example of the application of these concepts to perfluorocyclohexane for which there is a difference of 18.2 ppm between the chemical shifts of the axial and equatorial $^{19}$F nuclei. Assuming the chair form, C—F and C—C bond lengths of 1.32 and 1.55 Å, respectively, and a C—F bond moment of 1.4 D, Emsley[5] calculated a difference between the $E_z$ components at the axial and at the equatorial sites of $-48 \times 10^3$ esu. However, this accounts for only approximately 0.5 ppm of the 18.2 ppm difference. The largest component is from the $E^2$ term; not only is there a difference in $E^2$ originating in the bond dipoles equal to $93 \times 10^9$ esu, but it is also possible to insert the van der Waals contribution by remarking that these instantaneous induced-dipole–induced-dipole interactions will be time averaged to a meanvalue of $E^2$:

$$\langle E^2 \rangle = \frac{3\alpha I}{r^6}$$

hence

$$\langle E^2 \rangle = 0.39 \times 10^{12} \text{ esu}$$

so that the $b \times 10^{-18}$ term is $-40 \times 10^{-18} \times 0.4 \times 10^{12} = -16$ ppm, in good accord with the experimental value.

### 4.3.3 Ring Currents

Let us consider the classical example of benzene. If the resonance frequency is measured at infinite dilution in some inert solvent, say cyclohexane, then the benzene protons absorb approximately 1.5 ppm to lower field than the olefinic protons of 1,3-cyclohexadiene. If the spectrum is instead recorded as a function of concentration, it is then observed that the benzene proton resonance moves downfield with respect

[3] A. D. Buckingham, T. Schaefer, and W. G. Schneider, *J. Chem. Phys.*, *32*, 1227 (1960).
[4] J. I. Musher, *J. Chem. Phys.*, *37*, 34 (1962).
[5] J. W. Emsley, J. Feeney, L. H. Sutcliffe, *High Resolution Nuclear Magnetic Resonance Spectroscopy*, Pergamon Press, Oxford, 1965, vol. I, p. 139.

to an external reference with increasing dilution. These are the experimental facts. They strongly suggest an explanation based upon the $\pi$-electronic delocalization present in the benzene ring. When an external magnetic field $\mathbf{H}_0$ is applied to the system, the delocalized electrons will circulate to produce a ring current counteracting the effect of the applied field (Figure 4.2). The delocalized aromatic electrons are induced by the presence of the strong magnetic field $\mathbf{H}_0$ to circulate, thus generating a current, hence a magnetic field $\mathbf{h}_0$, which will tend in most regions of space to reduce the magnitude of the applied field $\mathbf{H}_0$. At any point in space, the effective magnetic field $\mathbf{H}_{\text{eff}}$ will be given by the vectorial sum $\mathbf{H}_0 + \mathbf{h}_0$. As a result, protons in the plane of the aromatic ring will be exposed to a total magnetic field greater than $\mathbf{H}_0$ and will be deshielded. Protons above the aromatic plane will be shielded. Of course, one must realize that the aromatic rings tumble rapidly in solution, and the symmetry axis of the aromatic ring can assume all the possible orientations with respect to the applied field. However, what counts is the projection of $\mathbf{H}_0$ along the symmetry axis, and summing over all the possible orientations gives a nonzero value. As a consequence it is possible to draw a map of the values of $\mathbf{h}_0$ for all the points of space close to the aromatic ring. This work has been done by Johnson and Bovey.[6] Cylindrical coordinates $z$ and $\rho$ define a point in space with respect to the center of the benzene ring, and lines of equishielding can be calculated to map the magnetic anisotropy of the ring (Figure 4.3). Bovey and Johnson assume that the benzene $\pi$ electrons are contained in two toric[7] volumes, one on each face of the aromatic ring and separated by ~1.28 Å. The analytical expression they obtained corresponding to Figure 4.3 for the component of $\mathbf{h}_0$ normal to the plane of the ring, is

$$\mathbf{h}_0 = \frac{ne^2}{6\pi mc^2 a} \frac{1}{[(1+\rho)^2 + z^2]^{1/2}} \left[K + \frac{1-\rho^2-z^2}{(1-\rho)^2+z^2} E\right]$$

where $ne^2/6\pi mc^2$ is the intensity $I$ of the ring current and $a$, the radius of its circular orbit, is taken as that of the benzene hexagon, 1.397 Å; the quantities $K$ and $E$ are complete elliptic integrals whose values are listed in mathematical tables. This free-electron model has been improved upon subsequently, for instance in the derivation of the magnitude of the ring current in individual rings of polycyclic hydrocarbons.[8] All these treatments have met with considerable success in correlating a large body of experimental data. Recently, however, the validity of the physical picture presented above, the existence of a "ring current,"[9] has been seriously questioned by Musher.[10] He claims that the effect is due only to the planarity of the benzene ring, to its intrinsic anisotropy if you wish, and that it is not necessary to postulate a merry-go-round circulation of electrons.

*FIGURE 4.2*
*Lines of force of the field* $\mathbf{h}_0$.

Not only aromatic hydrocarbons with $(4n+2)$ $\pi$ electrons display large anisotropies of their proton chemical shifts, but also hydrocarbons with $(4n)$ $\pi$ electrons, which of course are not aromatic. In fact, the $4n$ cyclic polyenes *which are planar* are theoretically predicted[11] to produce magnetic anisotropies such that in-plane protons are shielded and out-of-plane protons are deshielded. Note that this is the reverse of

[6] C. E. Johnson, Jr., and F. A. Bovey, *J. Chem. Phys.*, *29*, 1012 (1958).
[7] A torus is the doughnut-shaped volume generated by the rotation of a circle around an outside axis in its plane.
[8] J. D. Memory, G. W. Parker, and J. C. Halsey, *J. Chem. Phys.*, *45*, 3567 (1966).
[9] J. A. Pople, *J. Chem. Phys.*, *24*, 1111 (1956).
[10] J. I. Musher, *J. Chem. Phys.*, *43*, 4081 (1965).
[11] J. A. Pople and K. G. Untch, *J. Amer. Chem. Soc.*, *88*, 4811 (1966).

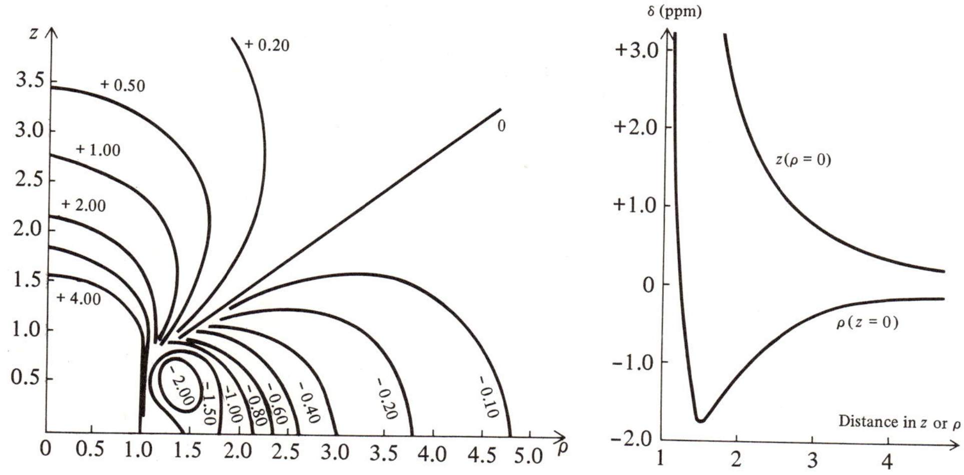


*FIGURE 4.3*
*Anisotropy of the benzene ring according to Johnson and Bovey[6]; equishielding lines (ppm) are depicted for cylindrical coordinates, a point in space being defined by its vertical distance z to the plane of benzene and by its horizontal distance ρ to the axis of the benzene ring. Both z and ρ are expressed in benzene radius units, i.e., 1.397 Å.*

what is observed with $(4n + 2)$ hydrocarbons. Sondheimer[12] has synthesized a number of these fascinating $(4n)$ annulenes, and they indeed behave according to predictions. According to Hückel's rule cyclic conjugated hydrocarbons with $(4n + 2)$ $\pi$ electrons are aromatic, whereas cyclic polyenes with $(4n)$ $\pi$ electrons are antiaromatic. The complementarity of the "aromatic" and of the "antiaromatic" character which is displayed by molecular orbital theory for these two classes of hydrocarbons is nicely extended by the complementarity of the ring currents they generate under the influence of an applied magnetic field. Antiaromatic hydrocarbons support paramagnetic ring currents, whereas aromatic systems have diamagnetic ring currents. Their effects are summarized in Table 4.3. If diamagnetic ring currents for aromatic molecules are easily visualized, it is a little more difficult to form a picture of paramagnetic ring currents. In this respect it helps to point out that antiaromatic hydrocarbons are characterized by contribution of antibonding electrons to the overall electronic structure, which make the electronic distribution more localized. Due to their localization on the various carbon centers these antibonding electrons can be regarded as the origin of the paramagnetic ring currents. All of the available experimental evidence supports this concept of ring currents. Diamagnetic susceptibility is enhanced in aromatic hydrocarbons, and decreased in antiaromatic hydrocarbons.

*TABLE 4.3*

*Effect of Ring Currents on Proton Resonances*

| Ring current | Inner protons[a] | Outer protons[a] |
|---|---|---|
| diamagnetic | shielded | deshielded |
| paramagnetic | deshielded | shielded |

[a] In the nodal plane of the $\pi$-electronic distribution.

For instance, benzenoid aromatics are characterized by in-plane magnetic susceptibilities $\chi_x$ and $\chi_y$ much greater than the out-of-plane component $\chi_z$. The molar susceptibility $\chi_M$ is defined as

$$\chi_M = \tfrac{1}{3}(\chi_x + \chi_y + \chi_z)$$

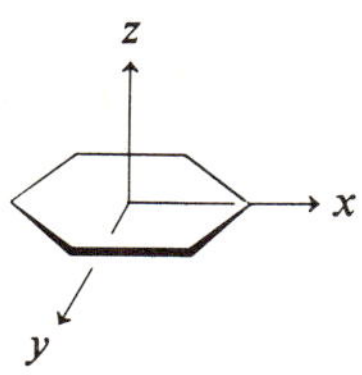


[12] F. Sondheimer *et al.*, *Special Publication No. 21*, The Chemical Society, London, 1967, p. 75.

*TABLE 4.4*

*Diamagnetic Susceptibility Exaltation as a Criterion of Aromaticity*

| Hydrocarbon | $\chi_m$ (calculated) | $\chi_m$ (measured) | Δ |
|---|---|---|---|
| Cyclohexane | 68.1 | 68.1 | 0.0 |
| Cyclohexene | 57.5 | 58.3 | −0.8 |
| 1,3,5-Cycloheptatriene | 59.8 | 51.7 | 8.1 |
| Cyclooctatretraene | 53.9 | 54.8 | −0.9 |
| Benzene | 54.8 | 41.1 | 13.7 |
| Tropylium ion $C_7H_7^+$ | 66.4 | 46.9 | 19.5 |
| Homotropylium ion $C_8H_9^+$ | 78 | 60 | 18 |
| [16]-Annulene | 105 | 109.6 | −4.6 |

It can be calculated as a sum of increments for the various localized bonds. Delocalization, then, is manifested by an exaltation of the experimental diamagnetic susceptibility $\chi_M$ as compared to the calculated value[13] (Table 4.4).

It is seen from Table 4.4 that only the aromatic molecules have positive exaltations: cyclooctatetraene is a normal polyolefin by this token, but cycloheptatriene is very close to being aromatic (correspondingly, its geometry is not very far from being planar). On the other hand, [16]-annulene, a $4n$ system, is antiaromatic and displays a paramagnetic rather than a diamagnetic exaltation of the susceptibility.

An even better confirmation is produced by the direct determination of the magnetic susceptibility tensor elements[14]; $\chi_z - \frac{1}{2}(\chi_x + \chi_y)$ is equal to $-7.4$ in 1,3-cyclohexadiene versus $-59.7$ in benzene. Benzene clearly displays a considerable increase in the diamagnetic susceptibility relative to the diene. A good example of the application of these concepts is the molecule [16]-85-annulene.[15] This $(CH)_{16}$ cyclic hydrocarbon has alternating *cis* and *trans* double bonds:

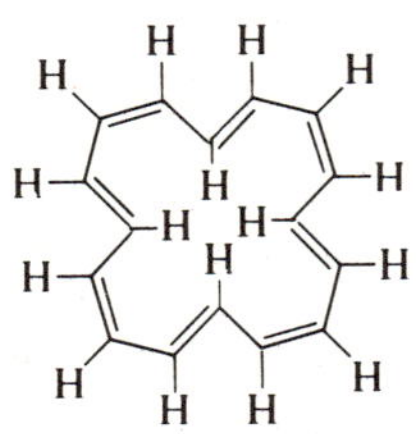

At temperatures above $-30°C$, a single peak is seen at $\delta = 6.7$ for all the protons: fast exchange of the various protons between the different sites occurs due to a combination of valence isomerization interchanging single and double bonds and simultaneous rotation around each of the single bonds, which exchanges the four inner protons with four of the outer protons. Both of these processes are frozen at $-130°C$ and the spectrum then consists of a triplet at $\delta = 10.6$ for the four inner protons and a multiplet centered about $\delta = 5.3$ for the 12 outer protons. The inner protons are deshielded with respect to values normally observed for olefins (ca. 6 ppm) and the outer protons are shielded. As a conclusion from the observation of this *paramagnetic* ring current the molecule can be said to be antiaromatic, in agreement with the prediction formulated on the basis of Hückel's rule ($4n$ with $n = 4$). When this neutral hydrocarbon is reacted with potassium in tetrahydrofuran at 0°C another species is formed with a strikingly different nmr spectrum. That this species is the dianion $C_{16}H_{16}^{=}$ can be confirmed independently by polarography or by following the reac-

[13] H. J. Dauben, Jr., J. D. Wilson, and J. L. Laity, *J. Amer. Chem. Soc.*, *90*, 811 (1968).
[14] J. M. Pochan and W. H. Flygare, *J. Amer. Chem. Soc.*, *91*, 5928 (1969).
[15] J. F. M. Oth and J. M. Gilles, *Tetrahedron Lett.*, 6259 (1968); J. F. M. Oth, G. Anthoine, and J. M. Gilles, *Tetrahedron Lett.*, 6265 (1968).

tion in the cavity of an esr spectrometer until the free radical species (radical anion) disappears. The spectrum of this negatively charged species, expected to be aromatic ($4n + 2$ $\pi$ electrons with $n = 4$), consists of two low-field triplets at $\delta = 8.8$ and 7.4, corresponding to eight hydrogens and four hydrogens, respectively, plus a high-field triplet at $\delta = -8.2$ (upfield from TMS). The aromaticity implied by this finding of a *diamagnetic* ring current is corroborated by other indications: for instance, the molecule is thermally stable for two days at 100°C (Figure 4.4).

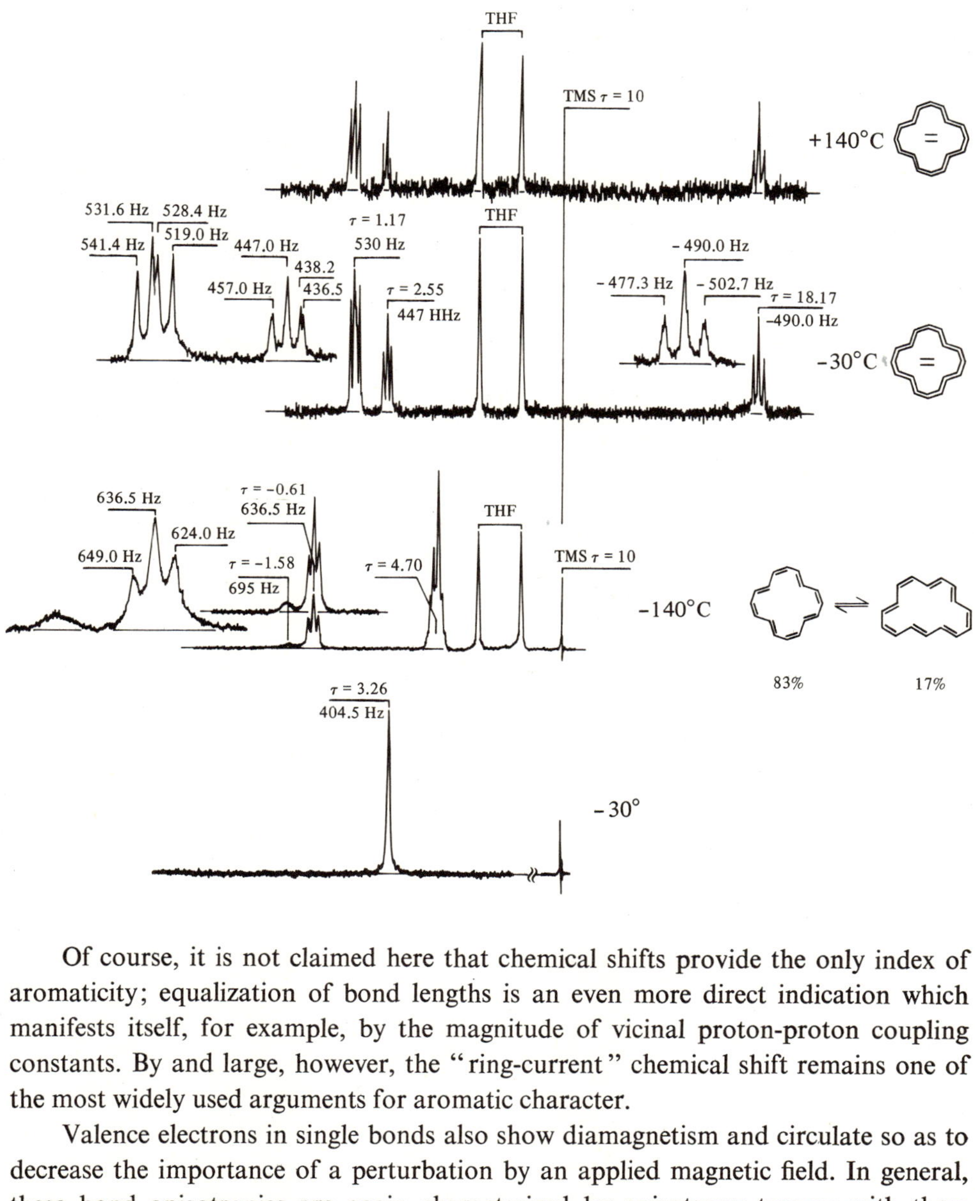


*FIGURE 4.4*
*60 MHz nmr spectra of* [16]-*annulene at* $-30°C$ *and* $-140°C$ *(bottom) and its dianion at* $-30°C$ *and* $+140°C$ *(top).*

Of course, it is not claimed here that chemical shifts provide the only index of aromaticity; equalization of bond lengths is an even more direct indication which manifests itself, for example, by the magnitude of vicinal proton-proton coupling constants. By and large, however, the "ring-current" chemical shift remains one of the most widely used arguments for aromatic character.

#### *4.3.4 Bond Anisotropies*

Valence electrons in single bonds also show diamagnetism and circulate so as to decrease the importance of a perturbation by an applied magnetic field. In general, these bond anisotropies are again characterized by anisotropy tensors with three different magnetic susceptibilities $\chi_x$, $\chi_y$, and $\chi_z$ along the three principal axes. For axially symmetric bonds if the $z$ direction is taken along the bond $\chi_x$ and $\chi_y$ are the same. The problem again is to assess qualitatively and quantitatively the effect of a given bond A—B on the shielding of a neighboring nucleus. This problem is very difficult in the general case and can only be treated accurately with sophisticated quantum mechanical calculations. The result of such an a priori calculation for the

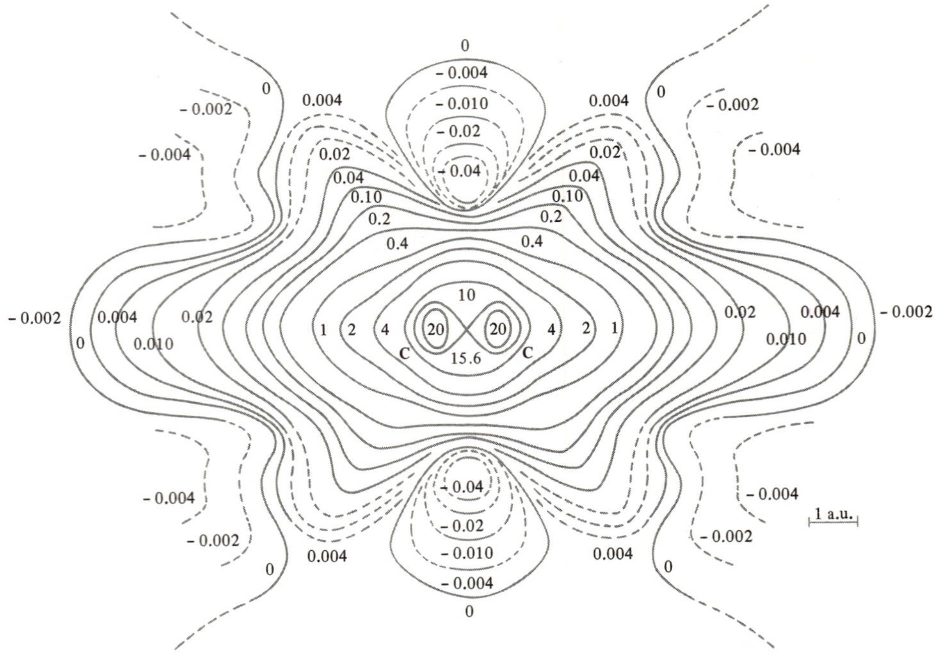


*FIGURE 4.5*
*Equishielding lines (ppm) for the carbon-carbon single bond; distances measured in atomic units (see scale), 1 a.u. ~ 0.53Å*

carbon-carbon single bond[16] is shown in Figure 4.5. In these calculations it is found that there *tends* to be deshielding along the axis of the bond and shielding perpendicular to that axis, *in the immediate neighborhood of the bond*, that is, at distances less than 2–3 Å. Unfortunately, not only have such theoretical calculations been performed only on a limited number of very simple molecules (such as ethane), but their validity is questionable for extrapolation to large complex molecules in which the electronic distribution within a bond may no longer have the same symmetry. However, rather than renouncing all ambition of understanding these bond anisotropy contributions one will often go to a much more crude treatment. The bond under consideration, whether C—H, C—C, C=C, C=O, C≡C, and so on, is assumed to be axially symmetrical and $\Delta\chi$ is defined as $\chi_z - \chi_x = \chi_z - \chi_y$. A magnetic point dipole is situated at the midpoint of a bond or at the center of electric charge and the bond anisotropy is then taken to be simply that from the magnetic dipole. In this approximation, devised by McConnell, the anisotropy part of the shielding is given as

$$\Delta\sigma = \frac{\Delta\chi}{3r^3}(1 - 3\cos^2\theta)$$

where $r$ and $\theta$ are the radius vector and the angular coordinates of a nucleus (for instance a proton) with respect to the point dipole. For example, $\Delta\chi$ has a value of the order of $3 \times 10^{-6}$ cm$^3$/mole for the carbon-carbon single bond.

This approximation has asymptotic validity only; it is very good at distances $r = 4$–$5$ Å, at which the term $\Delta\sigma$ has become extremely small. At closer range, $\Delta\sigma$ as calculated by this method can be far off its true value and should be used with caution. The reader can easily convince himself by examining Figure 4.6, on which the cone corresponding to the equation above (the nodal surface with $\Delta\sigma = 0$) has been superimposed. That these bond anisotropies are important contributors to the chemical shift can be appreciated from examples such as aldehydic protons, which resonate

[16] J. R. Didry and J. R. Guy, *Compt. Rend.*, *256*, 3042 (1963).

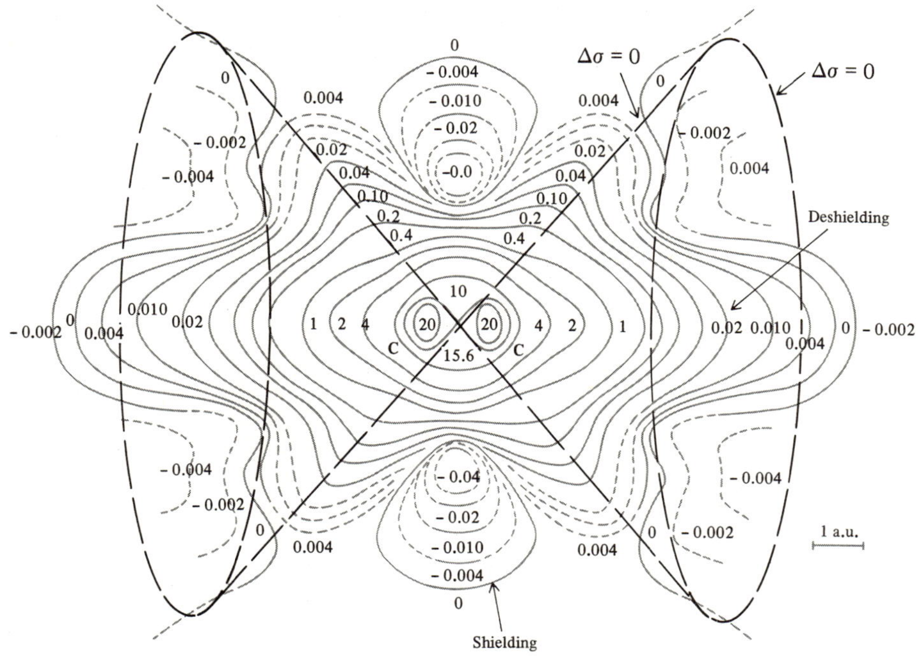


*FIGURE 4.6*
*Comparison for a C—C single bond of the McConnell approximation (dashed lines) with a more exact picture (see Figure 4.5).*

at extremely low fields ($\delta \sim 10$ ppm). Another example is acetylenic protons observed around $\delta = 3$, *upfield* from olefinic protons, when a simple inductive model would predict that proton chemical shifts should follow rather the order: acetylene < ethylene < ethane. Equishielding surfaces have been proposed for the carbonyl and acetylenic functionalities as well. However, these are only idealized representations of a semi-empirical nature which give at best a rough estimate of the magnitude of shielding or deshielding of a particular nucleus and must be used with caution.

To return to the example of perfluorocyclohexane $C_6F_{12}$, discussed earlier, the C—C and C—F bond anisotropies contribute only ~0.5 ppm to the difference in chemical shift between the axial and equatorial fluorines. For such a molecule, the dominant effect is due to the electrostatic fields from the bond dipole moments. For the cyclohexane molecule $C_6H_{12}$, however, the axial protons resonate 0.5 ppm upfield from the equatorial protons, and the effect is largely due to C—C and C—H bond anisotropies.

## 4.4 *The g Factor in Electron Spin Resonance*

Electron spin resonance is very similar to nuclear magnetic resonance. The esr experiment is summarized by the energy diagram in Figure 4.7. The spectrum is always recorded by varying $H_0$ with a fixed frequency $\nu$, typically 10,000 MHz. For the free electron $g = 2.00232$, and $\beta = eh/2\,mc$. In nmr the gyromagnetic ratio $\gamma$ is considered an intrinsic property of the nucleus. Changes in the resonance frequency are interpreted by a change in the screening of the nucleus by the adjoining electrons. In esr the effective magnetic moment of the electron $\mu_e = -g\beta I$ is considered to be different in different situations, so that $g$ appears to vary. Of course, the two viewpoints are equivalent. The value of $g$ varies because of spin-orbit coupling which alters the magnetic moment of the electron. But $g$ varies much less than chemical shifts do; its value stays very close to that of the free electron. Stone[17] has developed a quantitative

[17] A. F. Stone, *Mol. Phys.*, *6*, 509 (1962).

theory of $g$ values. For aromatic hydrocarbon radicals, it is related to the $\pi$-orbital energy of the unpaired electron:

$$\Delta g = A + Bm_0$$

where $m_0$ is the Hückel energy coefficient of the lowest half-occupied orbital and $A$ and $B$ are constants. $g$ is also found to be proportional to $U_{1/2}$, the polarographic half-wave potential.

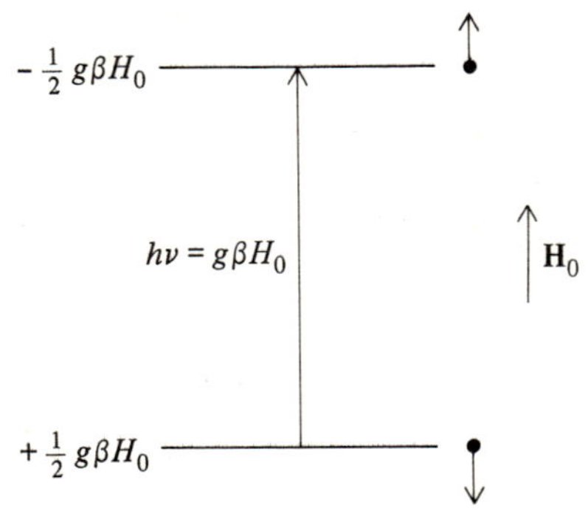


*FIGURE 4.7*
*Energy states for the spin of unpaired electrons.*

## 4.5 *PROBLEMS*

SAMPLE PROBLEM

1. At 60 MHz pure acetone is observed 38.4 Hz to high field of dilute acetone in carbon tetrachloride solution. Knowing that the magnetic field in a cylindrical sample as used here is given by the expression

$$H = H_0\left(1 - \frac{2\pi}{3}\chi_v\right)$$

where $\chi_v$ is the bulk susceptibility of the sample, and the values $\chi_v = -0.460 \times 10^{-6}$ for acetone, $\chi_v = -0.684 \times 10^{-6}$ for carbon tetrachloride, calculate the true solvent shift.

ANSWER

$$\Delta\sigma = \frac{2\pi}{3}\Delta\chi_v \cong 2\Delta\chi_v = 0.224 \times 10^{-6} \times 2$$

$$= 0.45 \text{ ppm} \sim 27.0 \text{ Hz}$$

Therefore only 11.4 Hz of the chemical shift difference is due to the difference between the acetone-acetone and acetone-carbon tetrachloride interactions, a so-called solvent effect.

SAMPLE PROBLEM

2. The carbon-13 chemical shifts of mesitylene referred to internal $CS_2$ are (in parts per million)

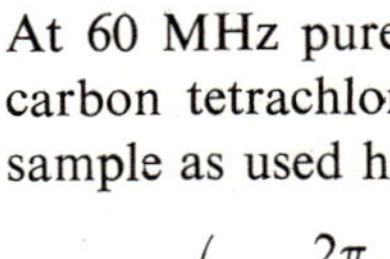


| | |
|---|---|
| 1 | + 66.1 |
| 2 | + 56.4 |
| $CH_3$ | +172.2 |

When mesitylene is protonated in a strongly acidic medium, the mesitylenonium ion is produced:

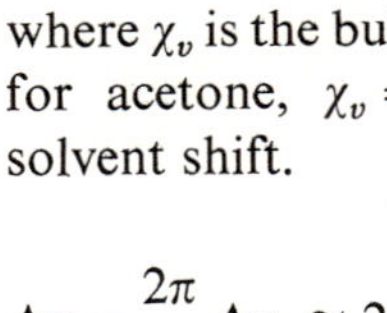


| | |
|---|---|
| 1 | +139.2 |
| 2,4,6 | − 0.5 |
| 3,5 | + 58.3 |
| $CH_3$(2,4,6) | +166.2 |

What can one conclude about the distribution of positive charge in the cation?

ANSWER

Since the change in the chemical shift $\Delta\delta$ is proportional to the amount of charge $\Delta q$:

$\Delta\delta C_{2,4,6} = -0.5 - 56.4 = -56.9 = A\Delta q_{C_{2,4,6}}$

All of the charge is expected to reside in the $\pi$ cloud. Hence:

$2\Delta q_{3,5} + 3\Delta q_{2,4,6} = 1$ (unit of positive charge)

$$\Delta q_{3,5} = \frac{1 - 3\Delta q_{2,4,6}}{2}$$

and

$$\Delta\delta_{C_{3,5}} = -7.8 = A\Delta q_{3,5} = A\left(\frac{1 - 3\Delta q_{2,4,6}}{2}\right)$$

$2 \times 7.8 = 3A\Delta q_{2,4,6} - A$

$15.6 = 3(-56.9) - A$

$A = 170.7 + 15.6 = 186.3$

Therefore,

$$\Delta q_{2,4,6} = \frac{186.3}{56.9} \cong 0.32$$

$\Delta q_{3,5} \cong 0.02$

In excellent agreement with molecular orbital theory and corresponding to the following valence bond resonance forms:

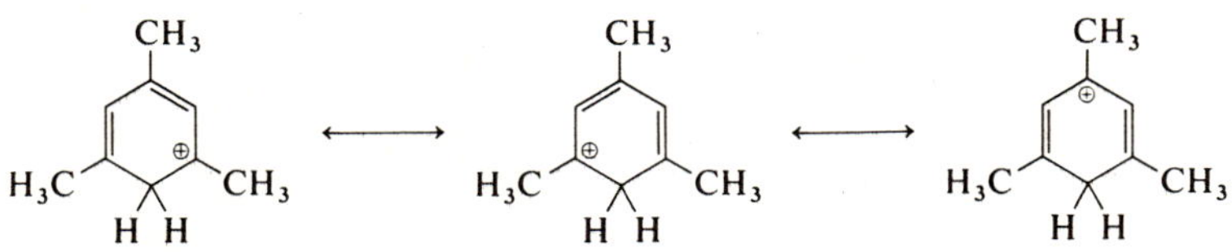

Note that the chemical shifts at $C_{2,4,6}$ and $C_{3,5}$ are made equivalent by a fast rearrangement transferring the proton from $C_1$ to $C_3$ to $C_5$.

1. Dihydropyrenes (1) can be reduced to the corresponding dianions (2):

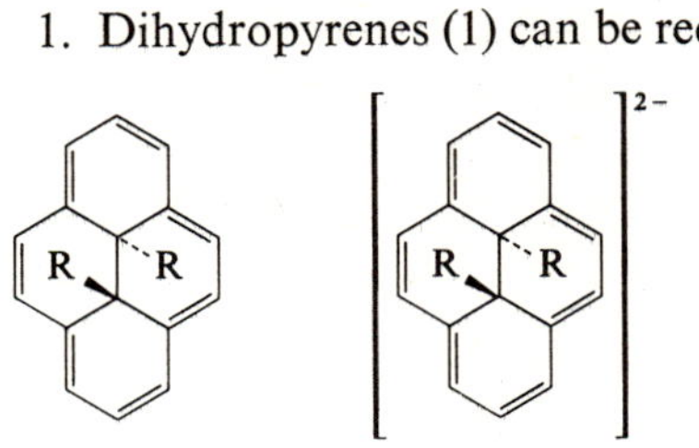

1 2

The table below summarizes the chemical shifts of protons on the periphery, and on the alkyl substituents R:

| | R | Interior protons α | (δ, ppm) β | γ | Exterior protons |
|---|---|---|---|---|---|
| 1 | $CH_3$ | −4.25 | — | — | 7.95–8.67 |
| 1 | $C_2H_5$ | −3.96 | −1.86 | — | 7.95–8.67 |
| 1 | $C_3H_7$ | −3.95 | −1.87 | −0.65 | 7.95–8.67 |
| 2 | $CH_3$ | 21.00 | — | — | −3.96 to −3.19 |
| 2 | $C_2H_5$ | 21.15 | 10.70 | — | −3.14 to −2.50 |
| 2 | $C_3H_7$ | 21.24 | 12.59 | 5.51 | −3.14 to −2.56 |

(Remember that resonances downfield from TMS have positive shifts and those upfield from TMS have negative values.) Explain these unusual chemical shifts.

2. R. C. Fort, Jr., G. W. H. Cheeseman, and E. C. Taylor, *J. Org. Chem.*, *29*, 2440 (1964) report a large deshielding of the indicated aromatic proton, by 1.03 ppm, when hydrogen is replaced by chlorine in the following molecule:

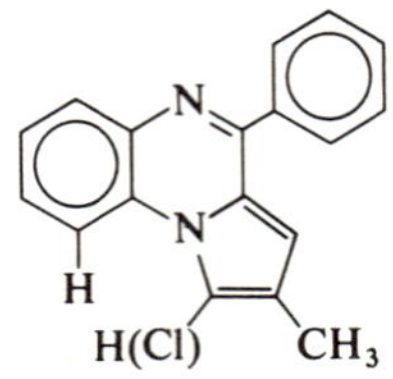


Calculate (a) the anisotropy, (b) the electric field, and (c) the steric compression contributions to the chemical shift. Use molecular models for the geometry, $\Delta\chi = 5 \times 10^{-30}$ cm$^3$/mole for the C—Cl bond [T. Schaefer and T. Yonemoto, *Can. J. Chem.*, *42*, 2318 (1964)] and zero for the C—H bond, simplify the Buckingham equation to $\Delta\sigma = 3 \times 10^{-12} E_z$, and use $Q = 11$ for the Yonemoto equation (*Can. J. Chem. 44*, 223 (1966)]. Do you get the correct sign and order of magnitude?

3. In derivatives of *endo*-2-methylnorbornane and -norbornene

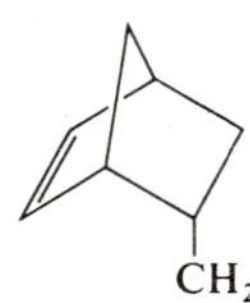


one of the protons is shifted to abnormally high field ($\delta = 0.6$ ppm). Which proton and why?

4. In carbon tetrachloride solution, chloroform protons are shifted downfield by 2–3 ppm upon formation of a complex with a halide anion $X^-$($X$ = F, Cl, Br, I). The iodoform protons have an opposite behavior, being shifted upfield by 0.3–0.4 ppm when bound in the pure complex as compared to the "free" value. The clue to the solution is that the two complexes differ in their structure; amplify this point, and account quantitatively for the observed shifts.

5. The following table summarizes chemical shifts for the methyl and methylene protons of ethyl derivatives (in hertz at 60 MHz from a $CH_4$ reference).

| X | $CH_2$ | $CH_3$ |
|---|---|---|
| Cl | 195 | 66 |
| Br | 189 | 86 |
| I | 176 | 99 |
| OH | 202 | 57 |
| CN | 127 | 65 |
| COOH | 128 | 56 |
| $C_6H_5$ | 144 | 59 |
| CHO | 134 | 54 |
| $NO_2$ | 249 | 81 |
| H | 38 | |

Can you establish a semiempirical correlation with a parameter representing the polarity of X? Dissect the chemical shifts into their various components and, obtain evidence for or against the existence of an inductive effect (through bonds) on the chemical shifts.

6. The olefinic protons of *cis*-stilbene resonate at $\delta = 6.55$ ppm, whereas the corresponding value for *trans*-stilbene is 7.10 ppm. Justify this difference. Is it possible to predict for which of the two isomers the aromatic protons will resonate at the highest field (look up the experimental answer in a spectral catalog)?
7. Assign, for each pair of molecules, the chemical shift appropriate to the underlined nuclei and give a reason for your answer:

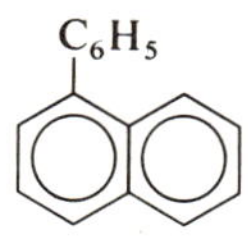


1

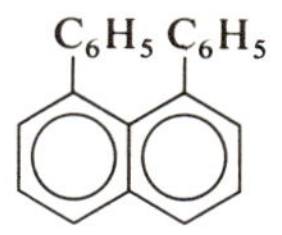


2

(a) $\delta = 6.85$
(b) $\delta = 7.35$

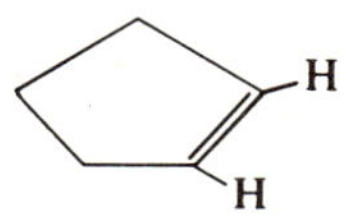


3

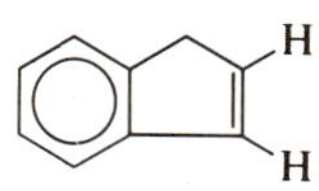


4

(a) $\delta = 5.60$
(b) $\delta = 6.50$
(c) $\delta = 6.80$

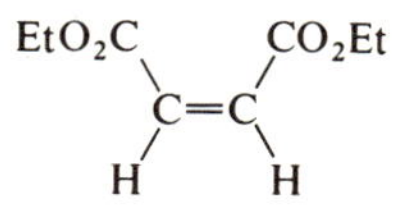


5

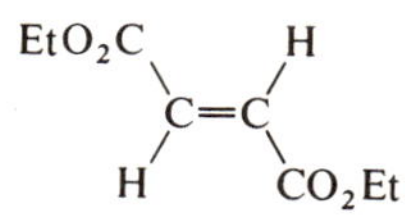


6

(a) $\delta = 6.28$
(b) $\delta = 6.83$

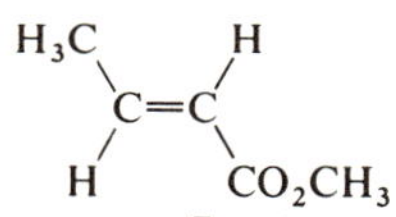


7

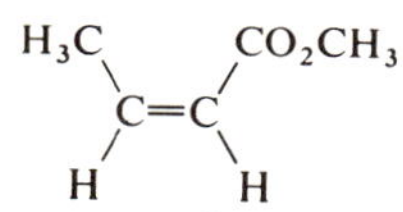


8

(a) $\delta = 1.88$
(b) $\delta = 2.14$

8. The nmr spectrum of cyclooctatetraene in strong acid is the following:
   $\tau$ 1.4 (5H)
   3.4 (2H)
   4.8 (1H)
   10.6 (1H)
   Comment.
9. In the nmr spectrum of the benzotriptycene (II) $H_a$ is at higher field than $H_b$ by 0.18 ppm, whereas for (III) $H_a$ is at lower field than $H_b$ by 0.38 ppm. Comment.

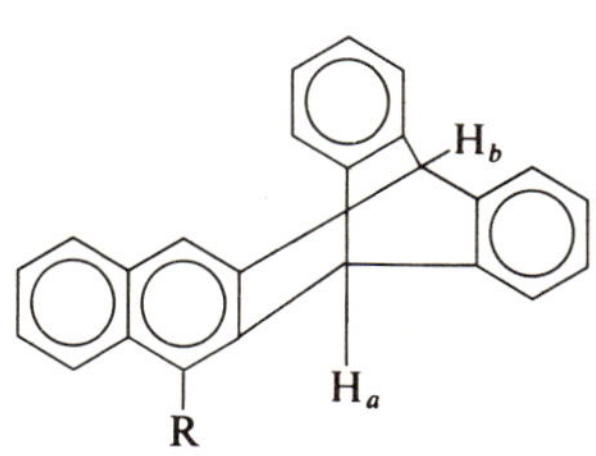


II R = Ph
III R = Me

10. Calculate the charge on the carbon atoms of the *t*-butyl cation from the $^{13}C$ chemical shift of the central carbon (273 ppm downfield from *t*-butyl chloride, when the methyl resonance of acetaldehyde is 149 ppm downfield from that of *t*-BuCl) and the proton chemical shift of 4.35 ppm (to be compared to a standard shift of ca. 1.5 ppm for a methyl on an uncharged trigonal carbon atom).

## 4.6

*APPENDIX I*
*Chemical shifts of protons in various environments (ppm downfield from TMS = 0.00)*

The indicated chemical shifts refer to dilute (5 percent or less) solutions in carbon tetrachloride or cyclohexane unless otherwise indicated. Use of other solvents, especially aromatic solvents, may lead to serious discrepancies.

These tables give spectral ranges as they are commonly encountered, and should be accepted only as being indicative of trends within a series of compounds. It is our belief and experience that interesting new molecules do not necessarily conform to such generalizations and empirical rules.

*TABLE A*

*Commonly Observed Absorption Range of Major Functional Groups and Structural Types*[a]

| Structural type | Chemical shift range (ppm, TMS = 0.00) |
|---|---|
| Cyclopropane $-CH_2-$ | 0.2–1.0 |
| $>C-CH_3$ (tetravalent C: $-\overset{\mid}{\underset{\mid}{C}}-CH_3$) | 0.6–1.5 |
| $X-\overset{\mid}{\underset{\mid}{C}}-\overset{\mid}{\underset{\mid}{C}}-CH_3$ (X = halogens, OH, OR, N) | 0.8–1.4 |
| $X-\overset{\mid}{\underset{\mid}{C}}-CH_3$ (X = halogens, OH, OR, N) | 1.0–2.0 |
| $-CH_2-$ (saturated) | 1.2–1.5 |
| $-\overset{\mid}{\underset{\mid}{C}}-H$ (saturated) | 1.4–1.8 |
| $>C=C<$ with $CH_3$ on C | 1.5–2.0 |
| $-\overset{O}{\overset{\|}{C}}-CH_3$ | 1.9–2.5 |
| $ArCH_3$ | 2.1–2.5 |
| $-S-CH_3$ | 2.2–2.8 |
| $>N-CH_3$ | 2.2–3.0 |
| $-C\equiv C-H$ (unconjugated) | 2.4–2.7 |
| $-C\equiv C-H$ (conjugated) | 2.7–3.1 |
| $-O-CH_3$ | 3.4–4.0 |
| $>C=CH_2$ (unconjugated) | 4.5–5.0 |
| $>C=C<$ with H on C (unconjugated) | 5.1–5.8 |
| $>C=CH_2$ (conjugated) | 5.3–5.8 |
| $>C=C<$ with H on C (conjugated) | 5.8–6.6 |
| ArH | 6.5–8.0 |
| $R-\overset{O}{\overset{\|}{C}}-H$ | 9.4–9.7 |
| $Ar-\overset{O}{\overset{\|}{C}}-H$ | 9.7–10.0 |

[a] For protons attached to carbon.

*TABLE B*

*Proton Chemical Shifts of Methyl Groups in Various Structural Environments*[a]

| Structural environment | Chemical shift range (ppm, TMS = 0.00) |
|---|---|
| $CH_3{-}CH_2{-}C{-}$ (C with bonds above and below) | 0.90–1.10 |
| $CH_3{-}C{-}C{=}C$, $CH_3{-}C{-}Ar$ | 1.05–1.20 |
| $CH_3{-}C{-}C({=}O){-}$, $CH_3{-}C{-}N{=}$ | 1.05–1.25 |
| $CH_3{-}C{-}N$ | 1.05–1.30 |
| $CH_3{-}C{-}O$ | 1.20–1.45 |
| $CH_3{-}C{=}C$, $CH_3{-}C{-}S{-}$, $CH_3{-}C{-}Cl$, $CH_3{-}C{-}Br$ | 1.55–1.70 |
| $CH_3C({=}O){-}R$ | 2.10–2.35 |
| $CH_3{-}C({=}O){-}NR_2$, $CH_3{-}NR_2$ | 2.20–2.35 |
| $CH_3{-}Ar$ | 2.10–2.45 |
| $CH_3{-}S{-}$ | 2.15–2.80 |
| $CH_3{-}C({=}O){-}Ar$ | 2.45–2.60 |
| $CH_3{-}OR$ | 3.20–3.50 |
| $CH_3{-}OAr$ | 3.70–3.95 |

[a] Methyl chemical shifts of the type $CH_3{-}M$ can be expressed as $\delta(\text{ppm}) = -2.3 + 1.3E_M + 0.4m$ where $E_M$ = Pauling electronegativity, $m$ = number of lone pairs on element $M$.

SOURCE: P. Bucci, *J. Amer. Chem. Soc.*, *90*, 252 (1968).

The chemical shifts of substituted alkanes, $Y{-}CH_2{-}X$, can be calculated according to Shoolery's[a] rules.

$$\delta\,(\text{ppm}) = 0.23 + \sigma_X + \sigma_Y$$

*TABLE C*

*Approximate Chemical Shifts of Methylene and Methine Protons in Substituted Alkanes*

| X,Y | $\sigma$ | X,Y | $\sigma$ |
|---|---|---|---|
| Br | 2.33 | $CH_3{-}$ | 0.47 |
| Cl | 2.53 | $C{=}C{-}$ | 1.32 |
| I | 1.82 | $C{\equiv}C{-}$ | 1.44 |
| $-NR_2$ | 1.57 | $C_6H_5{-}$ | 1.85 |
| $-OH$ | 2.56 | $CF_3{-}$ | 1.14 |
| $-OR$ | 2.36 | $-C{\equiv}N$ | 1.70 |
| $-OC_6H_5$ | 3.23 | $R{-}C({=}O){-}$ | 1.70 |
| $-OC({=}O)R$ | 3.13 | $ROC({=}O){-}$ | 1.55 |
| $-SR$ | 1.64 | $R_2N{-}C({=}O){-}$ | 1.59 |

[a] J. N. Shoolery, *Technical Information Bulletin*, *2*, No. 3, Varian Associates, Palo Alto, Calif.

SOURCE: B. P. Dailey and J. N. Shoolery, *J. Amer. Chem. Soc.*, *77*, 3977 (1955).

The chemical shifts of olefinic protons ($\delta$ C=C—H) in parts per million from internal TMS (see Table D) are given by

$$\delta_{C=C-H} = 5.25 + Z_{gem} + Z_{cis} + Z_{trans}$$

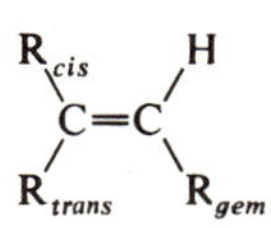

*TABLE D*

*Chemical Shifts of Olefinic Protons*

| Substituent R | $Z_i$ for R (ppm) $Z_{gem}$ | $Z_{cis}$ | $Z_{trans}$ |
|---|---|---|---|
| 1. −H | 0 | 0 | 0 |
| 2. −Alkyl | 0.45 | −0.22 | −0.28 |
| 3. −Alkyl-ring | 0.69 | −0.25 | −0.28 |
| 4. $-CH_2O$ | 0.64 | −0.01 | −0.02 |
| 5. $-CH_2S$ | 0.71 | −0.13 | −0.22 |
| 6. $-CH_2X$, X: F, Cl, Br | 0.70 | 0.11 | −0.04 |
| 7. $-CH_2N$ | 0.58 | −0.10 | −0.08 |
| 8. −C=C isolated | 1.00 | −0.09 | −0.23 |
| 9. −C=C conjugated | 1.24 | 0.02 | −0.05 |
| 10. −C≡N | 0.27 | 0.75 | 0.55 |
| 11. −C≡C | 0.47 | 0.38 | 0.12 |
| 12. −C=O isolated | 1.10 | 1.12 | 0.87 |
| 13. −C=O conjugated | 1.06 | 0.91 | 0.74 |
| 14. −COOH isolated | 0.97 | 1.41 | 0.71 |
| 15. −COOH conjugated | 0.80 | 0.98 | 0.32 |
| 16. −COOR isolated | 0.80 | 1.18 | 0.55 |
| 17. −COOR conjugated | 0.78 | 1.01 | 0.46 |
| 18. −C(H)=O | 1.02 | 0.95 | 1.17 |
| 19. −C(N)=O | 1.37 | 0.98 | 0.46 |
| 20. −C(Cl)=O | 1.11 | 1.46 | 1.01 |
| 21. −OR, R aliphatic | 1.22 | −1.07 | −1.21 |
| 22. −OR, R conjugated | 1.21 | −0.60 | −1.00 |
| 23. −OCOR | 2.11 | −0.35 | −0.64 |
| 24. $-CH_2-C=O$; $-CH_2-C\equiv N$ | 0.69 | −0.08 | −0.06 |
| 25. $-CH_2$-aromatic ring | 1.05 | −0.29 | −0.32 |
| 26. −Cl | 1.08 | 0.18 | 0.13 |
| 27. −Br | 1.07 | 0.45 | 0.55 |
| 28. −I | 1.14 | 0.81 | 0.88 |
| 29. −N−R, R aliphatic | 0.80 | −1.26 | −1.21 |
| 30. −N−R, R conjugated | 1.17 | −0.53 | −0.99 |
| 31. −N−C=O | 2.08 | −0.57 | −0.72 |
| 32. −Aromatic | 1.38 | 0.36 | −0.07 |
| 33. −Aromatic fixed | 1.60 | — | −0.05 |
| 34. −Aromatic *o*-substituted | 1.65 | 0.19 | 0.09 |
| 35. −SR | 1.11 | −0.29 | −0.13 |
| 36. $-SO_2$ | 1.55 | 1.16 | 0.93 |

SOURCE: A. E. Matter *et al.*, *Tetrahedron*, *25*, 691 (1969), pp. 693-694.

The chemical shift of aromatic protons are given by

$\delta$(ppm) = 7.27 ± effect of substituent

*TABLE E*

*Chemical Shifts of Aromatic Protons*

| Substituent | Ortho | Meta | Para | Substituent | Ortho | Meta | Para |
|---|---|---|---|---|---|---|---|
| $NO_2$ | −0.95 | −0.17 | −0.33 | $C(=O)CH_3$ | −0.64 | −0.09 | −0.3 |
| CN | −0.27 | −0.11 | −0.3 | C(=O)−H | −0.58 | −0.21 | −0.27 |
| $C_6H_5$ | −0.18 | 0.00 | +0.08 | *n*-Alkyl | −0.15 | +0.06 | +0.18 |
| F | +0.30 | +0.02 | +0.22 | $OCH_3$ | +0.43 | +0.09 | +0.37 |
| Cl | −0.02 | +0.06 | +0.04 | OH | +0.50 | +0.14 | +0.4 |
| Br | −0.22 | +0.13 | +0.03 | $OC(=O)CH_3$ | +0.21 | +0.02 | — |
| I | −0.40 | +0.26 | +0.03 | OTs | +0.26 | +0.05 | — |

SOURCE: L. M. Jackman and S. Sternhell, *Applications of Nuclear Magnetic Resonance Spectroscopy in Organic Chemistry*, Pergamon Press, Oxford, 1969, 2nd ed., p. 202.

*TABLE F*

*Chemical Shift of Protons Attached to Heteroatoms*

| Functional group | Chemical Shift (ppm, TMS = 0.00) |
|---|---|
| ROH (non-H-bonded, v. dil. sol.) | ~0.5 |
| ROH (in dimethylsulfoxide)[a] | 4.0–6.0 |
| ArOH (non-H-bonded, v. dil. sol.) | ~4.5 |
| ArOH (in dimethylsulfoxide)[b] | 9.0–12.0 |
| $RCO_2H$ (H-bonded) | 9.6–13.3 |
| $R_2NH$ (dil. inert solvent) | 0.5–1.5 |
| $RNH_2$ (dil. inert solvent) | 1.0–1.5 |
| $ArNH_2$, ArNHR, $Ar_2NH$ | 2.5–4.0 |
| $R_3NH^+$, $R_2NH_2^+$, $RNH_3^+$ (in $CF_3COOH$) | 7.0–8.0 |
| $Ar_3NH^+$, $Ar_2NH_2^+$, $ArNH_3^+$ (in $CF_3COOH$) | 8.5–9.5 |
| RSH | 1.0–1.6 |
| ArSH | 3.0–4.0 |

[a] O. L. Chapman and R. W. King, *J. Amer. Chem. Soc.*, *86*, 1256 (1964).
[b] M. T. Tribble and J. G. Traynham, *J. Amer. Chem. Soc.*, *91*, 379 (1969).

## 4.7

### *APPENDIX II* *Important nuclear magnetic resonance solvents and their properties*

A wide variety of solvents are used in nmr spectroscopy and it is important to know their properties. In particular, the "transparency" of the solvent and, for variable temperature work, the temperature range where the solvent is a liquid, are important. Another important consideration is the "inertness" of the solvent in terms of nonspecific van der Waals interactions with the solute, as measured, for instance, by the gas-to-liquid downfield shifts of the proton resonances of the solute. These gas-to-liquid shifts seem to increase as a function of the refractive index of the solvent and the most inert solvents appear to be hydrocarbons or fluorocarbons (Freons). The properties of a number of solvents are listed below.

| Solvent | Melting point °C | Boiling point °C | $n_D^{20°}$ | Chemical shift (ppm, TMS = 0) |
|---|---|---|---|---|
| Acetic acid[a] | 16.6 | 118.5 | 1.3721 | 2.06, 11.97 |
| Acetyl chloride | −112 | 51.5 | 1.3897 | 2.67 |
| Acetone[a] | −94 | 56.5 | 1.3592 | 2.07 |
| Acetonitrile[a] | −45.7 | 80.1 | 1.3442 | 1.96 |
| Aniline | −6.2 | 184.3 | 1.5863 | 3.6, aromatic region |
| Benzaldehyde | −56 | 178.1 | 1.5446 | 10.0, aromatic region |
| Benzene[a] | 5.5 | 80.1 | 1.5011 | 7.24 |
| Benzonitrile | −13 | 189 | 1.5289 | aromatic region |
| Carbon disulfide | −112 | 46 | 1.6255 | — |
| Carbon tetrachloride | 23 | 76.7 | 1.4630 | — |
| Chlorobenzene | −45 | 132 | 1.5284 | aromatic region |
| Chlorodifluoromethane | −160 | −40.8 | | 7.2 |
| Chloroform[a] | −63.5 | 61 | 1.4455 | 7.24 |
| Cyclohexane[a] | 6.5 | 81 | 1.4262 | 1.42 |
| Cyclopentane | −93.9 | 49.3 | 1.4064 | 1.50 |
| *p*-Cymene | −67.9 | 177.1 | 1.4908 | 1.22, 2.30, 2.87, 7.08 |
| Dichlorodifluoromethane | −155 | −29 | — | — |
| 1,1-Dichloro-2,2-difluoroethylene | −127.1 | 20.4 | 1.388 | — |
| 1,2-Dichloroethane | −35 | 84 | 1.4448 | 3.69 |
| Dimethylether | −136 | −24.9 | | ~3.3 |
| Dimethyl formamide[a] | −61 | 153 | 1.4319 | 2.79, 2.94, 7.90 |
| Dimethyl sulfoxide[a] | 18.4 | 100 (dec.) | 1.4787 | 2.50 |
| 1,4-Dioxane | 11 | 101 | 1.4232 | 3.56 |
| Ethanol[a] | −117 | 79 | 1.3611 | 1.22, 2.58, 3.70 |

*(Continued)*

*(Continued)*

| Solvent | Melting point °C | Boiling point °C | $n_D^{20°}$ | Chemical shift (ppm, TMS = 0) |
|---|---|---|---|---|
| Ethylether | −116.2 | 34.6 | 1.3526 | 1.16, 3.36 |
| Fluorotrichloromethane | −111 | 23.8 | — | — |
| Hexachlorobutadiene | | 215 | — | — |
| Hexamethylphosphorotriamide | 4.7 | 235 | 1.4584 | 2.60 |
| *n*-Hexane | −95.3 | 68.7 | 1.3749 | 0.90, 1.25 |
| Methanol[a] | −97.8 | 65 | 1.3288 | 3.34, 4.11 |
| Methylcyclohexane | −126.3 | 100 | 1.4230 | 0.90, 1.55 |
| Methylene chloride | −97 | 40 | 1.4237 | 5.28 |
| Nitrobenzene | 5.7 | 221 | 1.5562 | aromatic region |
| Nitromethane | −28.5 | 100.8 | 1.3819 | 4.29 |
| Pyridine | −42 | 115.5 | 1.5095 | aromatic region |
| Sulfolane | 28.4 | 290 | 1.4835 | 2.16, 2.92 |
| 1,1,2,2-Tetrachloroethane | −36 | 145 | 1.4821 | 5.9 |
| Tetrachloroethylene | −22 | 121 | 1.5018 | — |
| Tetrahydrofuran | −65 | 64 | 1.4076 | 1.79, 3.60 |
| Toluene[a] | −95 | 110.6 | 1.4893 | 2.31, 7.10 |
| Trifluoroacetic acid | −15.3 | 72.4 | | 11.34 |
| Vinyl chloride | −160 | −13.9 | | olefinic region |

[a] Available and commonly used in the perdeuterio form.

## 4.8 SUGGESTED READINGS

P. Laszlo, A. Speert, R. Ottinger, and J. Reisse, "Reconsideration of the Internal Tetramethylsilane Reference for Proton Magnetic Resonance Studies," *J. Chem. Phys., 48*, 1732 (1968).

T. Schaefer and W. G. Schneider, "Proton Magnetic Resonance Shifts and the Electron Density Distribution in Aromatic Systems," *Can. J. Chem., 41*, 966 (1963).

R. F. Zürcher, "The Cause and Calculation of Proton Chemical Shifts in Non-Conjugated Organic Compounds," in *Progress in NMR Spectroscopy*, J. W. Emsley, J. Feeney, and L. H. Sutcliffe, eds., 1967, vol. 2, pp. 205–257.

T. Yonemoto, "A Perturbation Calculation of the Effect of Electric Interaction on Proton Magnetic Shielding," *Can. J. Chem., 44*, 223 (1966).

A. D. Buckingham, "Chemical Shifts in the NMR Spectra of Molecules Containing Polar Groups," *Can. J. Chem. 38*, 300 (1960).

J. I. Musher, "Linear Variation of Proton Magnetic Shielding with Electric Field," *J. Chem. Phys., 37*, 34 (1962).

A. A. Bothner-By and J. A. Pople, "Diamagnetic Anisotropy of Electron Groups," *Ann. Rev. Phys. Chem., 16*, 43 (1965).

J. I. Musher, "On the Magnetic Susceptibility of Aromatic Hydrocarbons and 'Ring Currents'," *J. Chem. Phys., 43*, 4081 (1965).

J. M. Gaidis and R. West, "Evidence for Ring Currents in Hydrocarbons," *J. Chem. Phys., 46*, 1218 (1967).

J. I. Musher, "On the Non Existence of 'Ring Currents' in Aromatic Hydrocarbons," *J. Chem. Phys., 46*, 1219 (1967).

H. G. Ff. Roberts, "Uncoupled Hartree-Fock Calculations of the Ring Current Contribution to the Chemical Shifts in Conjugated Hydrocarbons," *Theoret. Chim. Acta* (Berl.), *15*, 63 (1969).

# CHAPTER 5 RELAXATION

## 5.1 *Microscopic Correlation Times*

In a gas or in a liquid, collisions of one molecule with another lead to energy transfers between the various degrees of vibrational motions as well as translations and rotations. These energy transfers offer a mechanism for relaxation which is exploited in the various spectroscopic processes, such as magnetic resonance. It will be remembered that when a dielectric is submitted to an electrostatic field $E_0$ between the plates of a condenser, the static dielectric constant $\varepsilon_0$ can be measured. If the electric field $E$ is made to oscillate at the frequency $\omega$ the molecular dipoles will orient themselves along the field $E$ in a finite time depending upon the frictional resistance of the medium. A complex dielectric constant $\varepsilon^*(\omega)$ is defined:

$$\varepsilon^*(\omega) = \varepsilon' - i\varepsilon'' = \varepsilon_\infty + \frac{\varepsilon_0 - \varepsilon_\infty}{1 + i\omega T} \qquad i = \sqrt{-1} \tag{1}$$

where $\varepsilon_\infty$ is identical to $n^2$, the square of the refractive index of the substance. The relaxation time $T$ is the characteristic time for the exponential decay of the system between two states, that is, between two successive values of the field. It reflects the time lag of the molecules in the dielectric as they strive to reorient themselves along the applied electric field $E(\omega)$. It can be shown that the dielectric relaxation time $T$ (macroscopic) is related to the true microscopic correlation time $\tau_c$ (during which the molecule turns around in any direction by one radian) by

$$\frac{T}{\tau_c} = \frac{\varepsilon_0 + 2}{\varepsilon_\infty + 2} \tag{2}$$

That is, this ratio is very close to unity in most liquids. Debye has developed a theory,[1] based upon Stokes' law, which expresses the correlation time as a function of the viscous drag imposed upon a *spherical* molecule of radius $a$ by a medium of macroscopic viscosity $\eta$, at the temperature $T$ (do not confuse temperature $T$ and relaxation times!):

$$\tau_c = \frac{V\eta}{KT} = \frac{4\pi a^3}{3}\frac{\eta}{KT} \tag{3}$$

This simple relationship is borne out by the data in Table 5.1. In general, the proportionality of $\tau_c$ to $\eta/T$ describes the results for elongated dipolar molecules such as alkyl alcohols and halides quite well. But there are major discrepancies. For instance *t*-butylchloride has its $\tau_c$ increase from $2 \times 10^{-12}$ in *n*-heptane solution ($\eta = 0.42$) to only $3 \times 10^{-12}$ in nujol—a high molecular weight paraffin—($\eta = 108$) at 20°C. The explanation is that small, spherical, relatively nonpolar molecules, such as *t*-butylchloride or camphor, can rotate on their lattice sites without having to push adjacent

*TABLE 5.1*

*Microscopic Correlation Time $\tau_c$ as a Function of Viscosity*

| Compound (pure liquid) | $\eta$ (centipoise) | $\tau_c$ ($10^{-12}$ sec) |
|---|---|---|
| $C_2H_5Br$ | 0.38 | 9 |
| *n*-$C_4H_9Br$ | 0.60 | 22 |
| *n*-$C_6H_{13}Br$ | 0.94 | 81 |
| *n*-$C_{10}H_{21}Br$ | 2.29 | 325 |
| *n*-$C_{16}H_{33}Br$ | 6.73 | 760 |

[1] P. Debye, *Polar Molecules*, The Chemical Catalog Company, New York, 1929.

molecules out of the way. An equivalent statement is that the Debye–Stokes formulation would be appropriate for a translation, but not for a rotation since collisions between hard spheres are frictionless just as for billiard balls and their rotation ought to be virtually independent of viscosity. Intuitively the relaxation times for non-spherical molecules in *polar* solvents should be longer than in *nonpolar* solvents, since their motion implies a geared or cooperative rotation of a number of molecules. It is indeed observed, in general, that solute-solvent complex formation increases the *microscopic* viscosity (which can be very different from the macroscopic viscosity $\eta$). What is the relationship of the effective microscopic viscosity $\eta_{eff}$ to the macroscopic viscosity $\eta$? One can define a microviscosity coefficient $0 < f_r < 1$ so that

$$\eta_{eff} = f_r \eta \tag{4}$$

It can be shown that the value of $f_r$ is dependent upon the matching of the molecular radii of the solute and the solvent. If the solvent molecules are very small compared to the solute molecules, the solvent indeed appears as a continuous medium and $f_r \rightarrow 1$. If however, solvent and solute molecules have approximately the same radius, then $f_r \sim \frac{1}{6}$. This value of $f_r$ has been observed experimentally for a number of pure liquids.

## 5.2 *Nuclear Magnetic Resonance Relaxation Processes*

Consider once more the magnetic resonance experiment: two energy states in the simplest case are separated by a quantum $h\nu$; the Heisenberg uncertainty principle states that the linewidth $\Delta\nu$ of the resonance line is of the order of $(2\pi\Delta t)^{-1}$, where $\Delta t$ stands for the lifetime of the nucleus (or electron) in the excited energy state. This lifetime $\Delta t$ is affected by a number of relaxation mechanisms. If the initial population of the upper level is $N_0$, it will vary with time according to differential equations of the type

$$T\frac{dN}{dt} + N = 0 \tag{5}$$

integrated as

$$N = N_0 \exp\left(-\frac{t}{T}\right) \tag{6}$$

In the absence of any additional energy absorption, if for instance the RF field $H_1$ used to induce absorption is abruptly switched off, the functional dependence of $N$ is thus an exponential decay with a characteristic time $T$ until the Boltzmann distribution is restored between the populations of the lower and upper states. The magnetic resonance experiment measures absorption of energy by the sample. This absorption is proportional to the difference between the population of the excited state and of the ground state. In a macroscopic assembly of nuclear (or electronic) spins, the magnetic moments $\boldsymbol{\mu}$ tend to line up with the direction of the directing magnetic field $\mathbf{H}_0$ in order to minimize the energy of interaction $E = -\boldsymbol{\mu}\mathbf{H}_0$. This amounts to a net magnetization of the sample. Thermal agitation tends to push the system off this equilibrium configuration. A spin can transfer its magnetic energy to the molecules around it in the medium as thermal energy. This spin-lattice relaxation occurs with a characteristic time $T_1$, called the longitudinal relaxation time since it describes the decay of magnetization along the direction of the applied field $\mathbf{H}_0$. This spin-lattice relaxation is a means for restoring a Boltzmann distribution of the $\alpha$ and $\beta$ energy levels after the RF field $H_1$ ,which tended to equalize their populations, has been removed. The spin-lattice relaxation mechanisms are thus important since they dictate the practical feasibility of the experiment. They tend to maintain an excess of spins in the lower energy level and therefore they check saturation (the equalization of the populations of the two states). In order to be efficacious, spin-lattice relaxation mechanisms must correspond to thermal fluctuations having large components at frequencies close to that of the magnetic resonance itself.

## 5.3 *Types of Spin-Lattice Relaxation Processes*

### *5.3.1 Relaxation by Dynamic Nuclear Coupling*

For nuclei with the spin number $I = \frac{1}{2}$, the only available relaxation mechanism is from dipolar coupling to other nuclei in the medium. It was indicated earlier that these couplings are time averaged to zero in an isotropic medium. Nevertheless, they do exist at every instant and fluctuate at a frequency corresponding to the molecular motion. If $\tau_c$ is the correlation time for random molecular reorientation, the correlation function

$$F(\nu) \propto \frac{2\tau_c}{1 + \nu^2\tau_c^2} \tag{7}$$

describes their distribution of intensity $F(\nu)$ about the value of $\tau_c$. As seen above, microscopic correlation times $\tau_c$ are of the order of $10^{-10}$–$10^{-11}$ sec for the usual organic molecules. This corresponds to a fluctuating magnetic field, due to the dipolar couplings, with a distribution of frequencies centered at $\sim 10^{11}$ Hz. In order to affect the magnetic resonance phenomenon, the fluctuating magnetic fields must have a component at the resonance frequency $\nu_0 = \gamma\hbar H_0 \sim 10^8$ Hz for most nuclei. Comparison of these two figures, $10^{11}$ and $10^8$, shows the inefficiency of dipolar coupling as a relaxation mechanism in a liquid.

Furthermore, dynamic nuclear coupling offers a relatively efficient relaxation mechanism only when two nuclei are close enough in space to interact significantly at any instant. $T_1$ is given by the expression

$$\frac{1}{T_1} = \hbar^2\gamma_1^2\gamma_2^2 \frac{1}{r_{12}^6} \tau_c \tag{8}$$

where all the quantities have been previously defined: $\hbar = h/2\pi$, $\gamma_1$ and $\gamma_2$ are the gyromagnetic ratios for the two nuclei, $r_{12}$ is the distance between them, and $\tau_c$ is the correlation time. This relaxation mechanism will make $T_1$ short (i.e., the relaxation will be made more efficient) for large values of $\tau_c$ at a given $r_{12}$ distance, when the viscosity of the medium is high. The effect decreases very rapidly with the sixth power of the distance between the two magnetic moments.

### *5.3.2 Coupling to Electronic Spins*

Whenever paramagnetic impurities, such as free radicals or dissolved oxygen, are present in a sample they offer extremely efficient relaxation mechanisms to the nuclei present. These mechanisms involve coupling of the **I** and **S** spin vectors. The spin-lattice relaxation time is then given by an expression of the type

$$\frac{1}{T_1} = \text{constant} \times N\mu_{\text{eff}}^2 \frac{\eta}{kT} \tag{9}$$

where $N$ is the number of paramagnetic ions of effective moment $\mu_{\text{eff}}$ in a unit volume of a solution with a bulk viscosity $\eta$ at the temperature $T$. The efficiency of this mode of relaxation comes from the proportionality of the mean magnetic field generated at the nuclei by the rapidly tumbling paramagnetic systems to the square of the electron magnetic moment. The magnetic moment of the electron is approximately three orders of magnitude greater than those of common nuclei. Hence, this relaxation mechanism is, in crude terms, about a million times more efficient than relaxation due to dipolar nuclear coupling. Very small values of $N$, the concentration of paramagnetic species, are sufficient to shorten considerably the spin-lattice relaxation time $T_1$. The great sensitivity of relaxation times to the presence of paramagnetic ions has rendered possible a number of useful studies, such as the binding of metal ions to biopolymers.

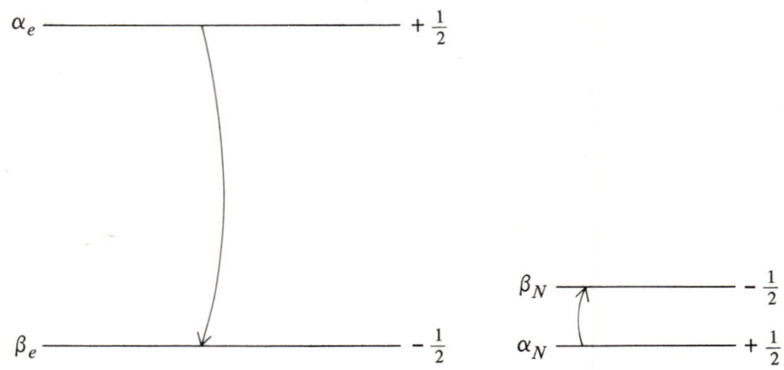


*FIGURE 5.1*
*"Flip-flop" relaxation.*

One such relaxation process involves an electronic spin going from its upper to its lower energy state, while at the same time a nuclear spin is excited from its lower to its upper energy state (Figure 5.1). It is termed a "flip-flop" relaxation mechanism, and corresponds to an allowed transition: the appropriate selection rule states that the $\Delta m_S = \pm 1$, $\Delta m_I = \mp 1$ transition is allowed, whereas the "flip-flip" process characterized by $\Delta m_S = \pm 1$, $\Delta m_I = \pm 1$ is forbidden.

### *5.3.3 Quadrupolar Relaxation*

For nuclei with spin numbers equal to or greater than unity, the distribution of charges within the nucleus no longer has spherical symmetry and there is a quadrupole moment $Q$. It can couple with the electrical field gradient $q = \delta E/\delta z$ at the quadrupolar nucleus. As for dipolar couplings, this interaction is time averaged to zero by the fast tumbling in a liquid, but is modulated instantaneously at the frequency of molecular motion. The relaxation thus afforded to such a quadrupolar nucleus *or to a nucleus coupled to it*, is rather efficient. It is expressed as

$$\frac{1}{T_1} = \frac{3}{8}\frac{(e^2qQ)^2}{\hbar^2}\tau_c \qquad (10)$$

where $\tau_c$ is again the microscopic correlation time. With ordinary values of the quadrupolar coupling $e^2qQ$, $T_1$ is typically of the order of a millisecond.

### *5.3.4 Inter- and Intramolecular Components of Relaxation*

It is convenient to break down a composite relaxation time into its separate components. A useful distinction is that between inter- and intramolecular contributions. In the simplest case, when neither paramagnetic impurities nor quadrupolar nuclei are present, the relaxation time $T_1$ is due solely to dynamic dipolar coupling of the nuclei. There is an intramolecular component arising from the magnetic dipole moments of the other nuclei in the same molecule; as indicated by (8) it fluctuates at a frequency determined by the rotational motion of the molecule. The intermolecular contribution, due to nuclei in other molecules, depends on the translational motion and is usually much smaller due to the $r^{-3}$ dependence of the dipolar couplings. However, in the general case one would write

$$\frac{1}{T_1} = \left(\frac{1}{T_1}\right)_{\text{intra}} + \left(\frac{1}{T_1}\right)_{\text{inter}}$$

$$= \left(\frac{1}{T_1}\right)_{\text{rot}} + \left(\frac{1}{T_1}\right)_{\text{transl}} \propto \frac{A}{\tau_c} + \frac{B}{D} \qquad (11)$$

in which $\tau_c$ is given by the Debye–Stokes functional dependence

$$\tau_c = \frac{4\pi}{3} a^3 \frac{\eta}{KT} \tag{12}$$

and the reciprocal of the diffusion coefficient $D$ is also given by a Stokes-like term:

$$\frac{1}{D} = \frac{6\pi\eta a}{KT} \tag{13}$$

Both terms are linear with respect to $\eta/KT$.

## 5.4 *Spin-Spin Relaxation*

Not only is there a longitudinal relaxation time $T_1$ describing the relaxation of the nuclear magnetization along the direction $z$ of the applied field $H_0$, but there is also a transverse relaxation time $T_2$ which describes the relaxation of the magnetization in the $(x, y)$ plane perpendicular to $H_0$. To a first approximation, the magnitude of $T_1$ determines the rate at which a given resonance line will be saturated (i.e., the rate of equalization of the $\alpha$ and $\beta$ populations) whereas the magnitude of $T_2$ determines the line width. The line width is defined by the quantity $\nu_{1/2}$, the width of the spectral line at half-height, and is equal to $1/2\pi T_2$. $T_2$ originates in transfers of magnetic energy among the nuclei, so it is referred to as the spin-spin relaxation time. In normal solvents at ambient temperatures, $T_1$ and $T_2$ are determined by the same factors. The factors which have just been discussed in relationship to $T_1$ affect $T_2$ to the same extent. However, $T_1$ and $T_2$ diverge at low temperatures when the viscosity of the solvent becomes very high. $1/T_1$ is proportional to

$$F(\nu) = \frac{\tau_c}{1 + \nu_0^2\tau_c^2} \tag{14}$$

where $\nu_0$ is the resonance frequency, whereas $T_2$ is proportional to the reciprocal of the correlation time $\tau_c$

$$T_2 \propto \frac{1}{\tau_c} \tag{15}$$

Hence at low viscosities $\tau_c$ is small and $\nu_0^2\tau_c^2$ is smaller than 1, so that both $T_1$ and $T_2$ are directly proportional to the correlation time $\tau_c$ (and in fact equal). At high viscosities, $\nu_0^2\tau_c^2$ becomes greater than unity and $1/T_1$ starts being proportional to $1/\tau_c$ while $1/T_2$ is still proportional to $\tau_c$.

Another source of differentiation between $T_1$ and $T_2$, even in liquids of low viscosity, is due to the inhomogeneity of the magnet: different parts of the same macroscopic sample "see" slightly different values of the applied magnetic field $H_0$. This leads to a line broadening superimposed upon the "natural" line width due *only* to the "internal" structure of the sample, that is, caused by the fluctuating local fields from the neighboring nuclear magnets. Usually, unless there is a pronounced viscosity broadening, the observed line width is due to this instrumental imperfection rather than to the true $T_2$.

## 5.5 *Dynamic Nuclear Polarization and the Overhauser Effect*

As was stated earlier, relaxation by electronic spins of nuclear states by "flip-flop" processes is superior to nuclear relaxation. In general, due to the greater magnetic moment of the electron, relaxation of electronic states is considerably faster (more efficient) than that of nuclear states. Based upon these natural properties, a set of experimental techniques called "dynamic nuclear polarization" has been devised.

Aside from their elegance, these methods are capable of extremely useful organic applications. We shall discuss them using the classical example of the system constituted of one electronic and one nuclear spin which are mutually coupled. The energy levels can be represented as follows:

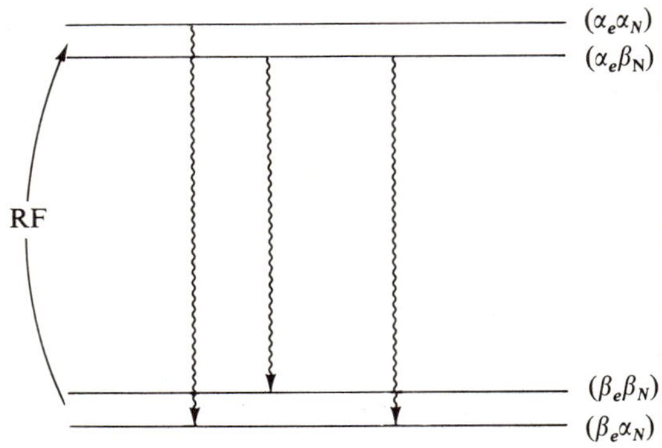


To a first approximation the magnetic resonance splitting $g_N\beta_N H_0$ can be neglected compared to the esr splitting $g_e\beta_e H_0$ and the nuclear-electronic coupling term, $aSI$, since it is much smaller than either of these. The spacings of the energy levels are then given by

$$(\alpha_e\,\alpha_N) - (\beta_e\,\beta_N) = (\alpha_e\,\beta_N) - (\beta_e\,\alpha_N) = g_e\,\beta_e\,H_0 \tag{16}$$

$$(\beta_e\,\beta_N) - (\beta_e\alpha_N) = (\alpha_e\alpha_N) - (\alpha_e\,\beta_N) = \frac{a}{2} \tag{17}$$

Under normal conditions, the populations of the $\alpha_N$ and $\beta_N$ *nuclear* states are in the Boltzmann ratio:

$$\frac{N_0(\alpha_N)}{N_0(\beta_N)} = 1 + \frac{g_N\,\beta_N\,H_0}{KT} \tag{18}$$

Let us now saturate the $(\beta_e\alpha_N) \rightarrow (\alpha_e\alpha_N)$ *electronic* transition with a strong RF field, thus equalizing the populations of these two states. Due to the absence of the reverse "flip-flop" relaxation process, the system will use the available electronic and "flip-flop" relaxation pathways in order to reach a steady state characterized by the following populations:

$$N(\beta_e\,\alpha_N) = N(\alpha_e\,\alpha_N) \tag{19}$$

$$\frac{N(\alpha_e\,\beta_N)}{N(\beta_e\,\alpha_N)} = 1 - \frac{g_e\,\beta_e\,H_0}{KT} \tag{20}$$

$$\frac{N(\beta_e\,\beta_N)}{N(\alpha_e\,\beta_N)} = 1 + \frac{g_e\,\beta_e\,H_0}{KT} \tag{21}$$

The *new* ratio of the populations of the $\alpha_N$ and $\beta_N$ nuclear spin states, after imposition of the double resonance RF field, will thus be

$$\frac{N(\alpha_N)}{N(\beta_N)} = \frac{N(\beta_e\,\alpha_N) + N(\alpha_e\,\alpha_N)}{N(\beta_e\,\beta_N) + N(\alpha_e\,\beta_N)} = \frac{2\,N(\alpha_e\,\beta_N)}{[1 - (g_e\,\beta_e\,H_0/KT)]N(\alpha_e\,\beta_N)[2 + (g_e\,\beta_e\,H_0/KT)]} \tag{22}$$

$$= \frac{2}{2 - 2(g_e\,\beta_e\,H_0/KT) + (g_e\,\beta_e\,H_0/KT)} \cong 1 + \frac{g_e\,\beta_e\,H_0}{2KT} \tag{23}$$

These approximations are permissible because $g_e\beta_e H_0 \ll KT$. The ratio of the nuclear polarizations, after and before saturation of the electronic transition, is then given by

$$\frac{[N(\alpha_N)/N(\beta_N)] - 1}{[N_0(\alpha_N)/N_0(\beta_N)] - 1} = \frac{1}{2}\frac{g_e \beta_e}{g_N \beta_N} = \frac{1}{2}\frac{\gamma_e}{\gamma_N} = 350 \quad (24)$$

In other words, the intensity of the nmr signal, which is proportional to the polarization $[N(\alpha_N) - N(\beta_N)]/N(\beta_N)$ is now boosted by a factor 350, and the only requirement for such dynamic nuclear polarization to be feasible is that there exists coupling of the nuclear and electronic spins.

The effect is qualitatively similar in the case of two nuclei interacting through dipolar coupling, that is, two nuclei close enough in space so that the term $(3 \cos^2 \theta - 1)/r^3$ is large at any instant. Then the Overhauser enhancement is given by $(1/2)(\gamma_1/\gamma_2)$ if $\gamma_1$ and $\gamma_2$ are the gyromagnetic ratios. For a pair of protons, $\gamma_1 = \gamma_2$ and the Overhauser enhancement has a theoretical limit of 50 percent. Measurement is performed by monitoring the intensity of the $H_1$ resonance while irradiating proton $H_2$: the Overhauser enhancement is observed if and only if proton $H_2$ is close enough in space to proton $H_1$ to provide it with efficient dipolar relaxation (the effect goes down with the sixth power of internuclear distance). This is in the absence of indirect coupling: its presence leads to negative Overhauser effects, and thus runs contrary to that of direct coupling. This technique is extremely useful for ascertaining, step by step which nuclei are in close spatial relationships in molecules. Obviously, this can be applied to the determination of configurations, cases of *cis-trans* isomerism, and so on. For instance, Anet and Bourn[2] have studied these nuclear Overhauser effects–often abbreviated as NOEs–in $\beta,\beta$-dimethylacrylic acid:

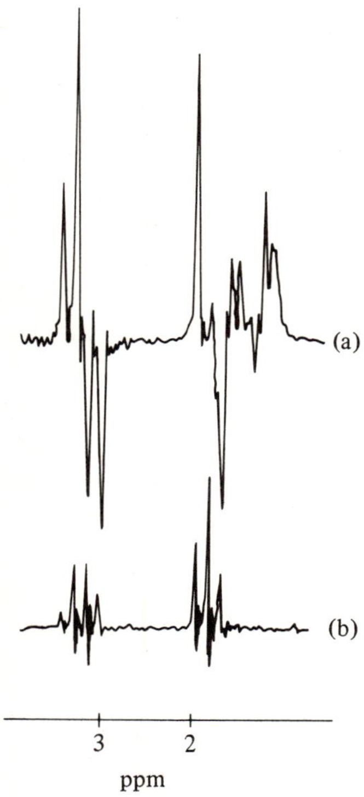


*FIGURE 5.2*
***Reaction of ethyllithium with ethyl iodide in benzene. (a) Spectrum taken during the reaction, showing the methylene (δ 3.2) and methyl protons (δ 1.85) of ethyl iodide and butane (1.0–1.6) (3.5–2.5-ppm region scanned with a spectrum amplitude twice that of the remainder of the spectrum). (b) Reference spectrum of ethyl iodide.***

NOE = + 12 percent

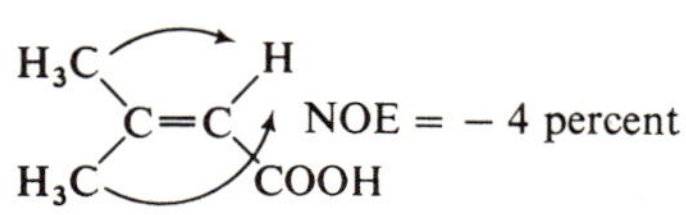


Not only can dynamic nuclear polarization occur using a strong RF field as a "polarizer," but it is also possible to polarize nuclear spin states by chemical means. An unambiguous way of detecting free radicals as they are produced in solution is to observe the peculiar phenomena typical of this chemically induced dynamic nuclear polarization (CIDNP). These are characterized by enhanced intensities for certain lines, stimulated emission leading to the inversion of other lines, also with an intensity increase, and apparently great selectivity in determining which lines are enhanced in intensity. For instance, the reaction

$$C_2H_5Li + C_2H_5I \longrightarrow C_4H_{10} + LiI$$

can be conducted inside an nmr tube in the probe of the spectrometer. The butane product gives a spectrum (Figure 5.2) with inverted lines and enhanced intensities, which is symptomatic of a population inversion such as described above, and which must occur through the intermediacy of a species with an unpaired spin.

## 5.6 *PROBLEMS*

1. Why is it important to conduct NOE experiments on carefully degassed solutions and to use magnetically inert solvents?
2. What are the theoretical NOE enhancements between hydrogen and deuterium; hydrogen and carbon-13? How would you conduct such experiments?

---

[2] F. A. L. Anet and A. J. R. Bourn, *J. Amer. Chem. Soc.*, *87*, 5250 (1965).

3. Upon irradiation of the methyl groups in 1,2,3,4-tetramethylphenanthrene, certain aromatic protons suffer substantial NOE's. Which are they?
4. The line-shape of a proton spin-spin coupled to a deuteron is determined by $T_1$ for the deuteron and is a triplet for long $T_1$ and a singlet for short $T_1$. When a deuteron is introduced into the benzylic positions of fluorene (1) and phenanthrindene (2), and the signal for the geminal proton is analyzed, the indicated correlation times are deduced:

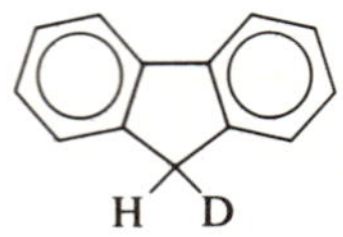


1

$\tau_c = 10.6$ psec (1 psec = $10^{-12}$ sec)

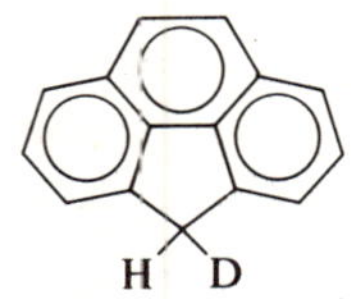


2

$\tau_c = 13.0$ psec (at 35°C)

Account for the difference in $\tau_c$ for (1) and (2).
5. In the nuclear Overhauser experiment performed on $\beta,\beta$-dimethylacrylic acid and described in the text, no effect can be seen if the intensity of the *cis* methyl is observed while saturating the resonance of the olefinic proton. Explain this apparent paradox.
6. Two isomers of 3-ethylidenequinuclidine (below) are known, only one of which shows a 31 percent NOE of the bridgehead proton upon irradiation of the methyl resonance. Which is which?

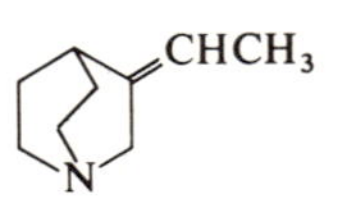


## 5.7 *SUGGESTED READINGS*

C. P. Smyth, "Dielectric Polarization and Relaxation," *Ann. Rev. Phys. Chem.*, *17*, 433 (1966).

D. E. Woessner, B. S. Snowden, Jr., R. H. McKay, and E. T. Strom, "Proton and Deuteron Spin-Lattice Relaxation in *n*-Dodecane," *J. Magnetic Resonance*, *1*, 105 (1969).

F. A. L. Anet and A. J. R. Bourn, "Nuclear Magnetic Resonance Spectral Assignments from Nuclear Overhauser Effects," *J. Amer. Chem. Soc.*, *87*, 5250 (1965).

P. D. Kennewell, "Applications of the Nuclear Overhauser Effect in Organic Chemistry," *J. Chem. Ed.*, *47*, 278 (1970).

H. Fisher and J. Bargon, "Chemically Induced Dynamic Nuclear Polarization During Thermal Decomposition of Peroxides and Azo Compounds," *Accounts Chem. Res.*, *2*, 110 (1969).

H. R. Ward and R. G. Lawler, "Nuclear Magnetic Resonance Emission and Enhanced Absorption in Rapid Organometallic Reactions," *J. Amer. Chem. Soc.*, *89*, 5518 (1967).

R. G. Lawler, "Chemically Induced Dynamic Nuclear Polarization," *J. Amer. Chem. Soc.*, *89*, 5519 (1967).

H. R. Ward, R. G. Lawler, and R. A. Cooper, "Chemically Induced Dynamic Nuclear Polarization Evidence for One-Electron Transfer During Some Halogen-Metal Exchange Reactions," *J. Amer. Chem. Soc.*, *91*, 746 (1969).

A. R. Lepley and R. L. Landau, "Free Radical Iodide-Lithium Interchange," *J. Amer. Chem. Soc.*, *91*, 748 (1969).

A. R. Lepley, "Reaction Dynamics in Nuclear Polarization," *J. Amer. Chem. Soc.*, *91*, 749 (1969).

G. L. Closs *et al.*, "Theory of Chemically Induced Nuclear Spin Polarization," *J. Amer. Chem. Soc.*, *92*, 2183, 2185, 2186 (1970).

J. Jacobus, "Chemically Induced Dynamic Nuclear Spin Polarization: A Caveat," *Chem. Comm.*, *11*, 709 (1970).

# II

# DERIVATION OF MOLECULAR PARAMETERS FROM SPECTRA

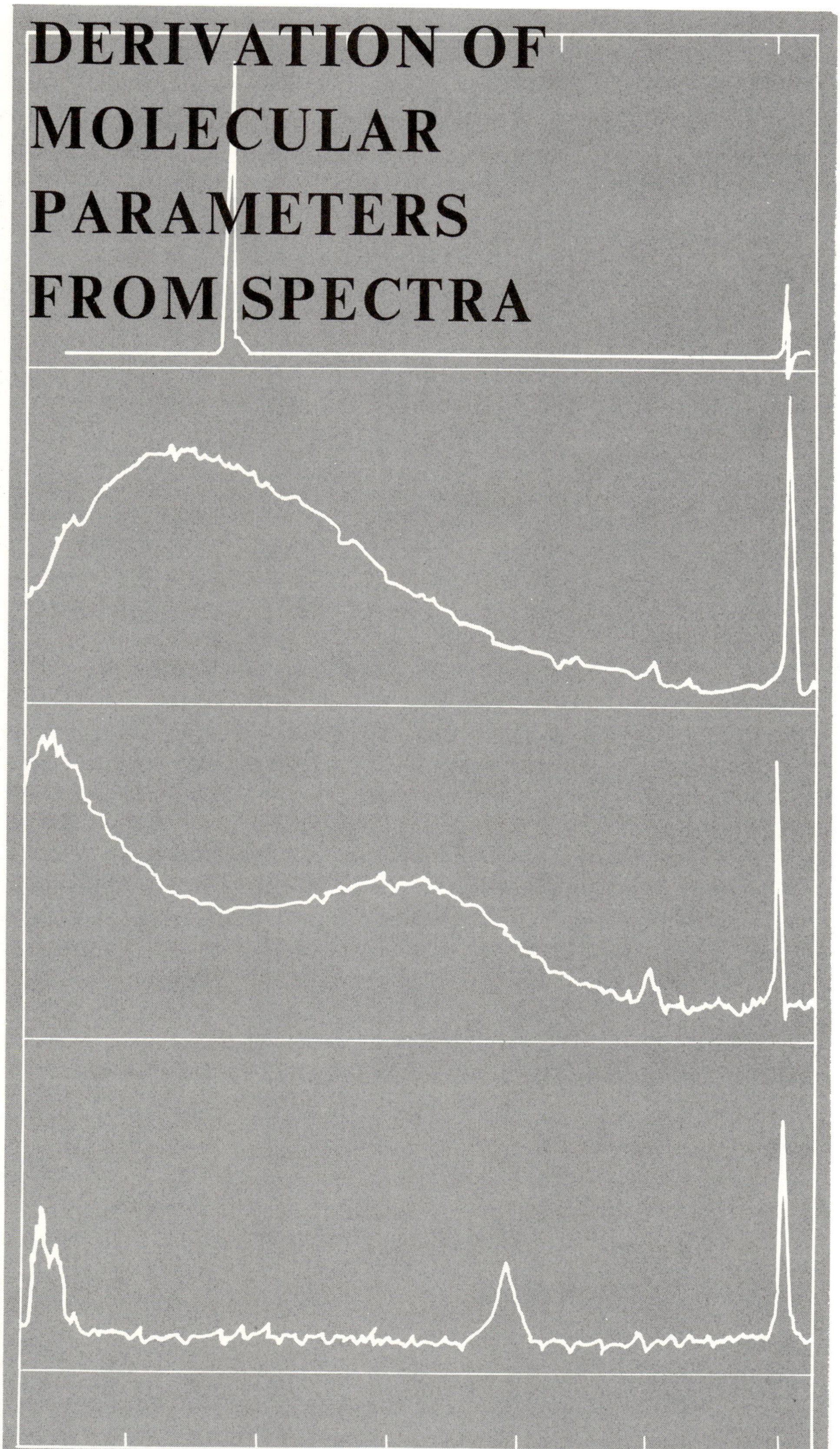

# CHAPTER 6 SIMPLE MOLECULAR ORBITAL THEORY AND ITS APPLICATION TO SPECTROSCOPY

Modern organic chemistry is becoming increasingly more quantitative and therefore more theoretical. Today's organic chemistry student is exposed early in the curriculum to simple quantum mechanics and its application to organic chemistry, molecular orbital theory. In its simplest form, the Hückel method, molecular orbital theory is far from a rigorous explanation of chemical reality. It provides, however, a useful and beautiful insight almost at an intuitive level, into the working of much more sophisticated theories. Often the theoretician will communicate his achievements by comparing them with the results of simple Hückel theory. It is then a constant source of amazement to the organic chemist that this rudimentary method achieves such a degree of success in semiquantitatively correlating empirical data.

The purpose of any molecular orbital theory is to give a description of the electrons from a description of the position of the nuclei in a molecular assembly, and in particular to provide (1) the total electronic energy and (2) the distribution of the electrons (electron density). From the former quantity one can derive the relative stability of various conformations or configurations of molecules and their absorption spectra. The electron density in turn, as we shall see, is related to dipole moments, chemical shifts, and nmr and esr coupling constants.

Let us then recall that electrons in molecules can be arbitrarily separated into $\sigma$ and $\pi$ electrons. The Hückel method considers only the latter and makes the further assumption that interelectronic repulsions can be lumped into the general electrostatic potential to which a given electron is being submitted. The approach of molecular orbital theory is to build molecular orbitals (MO) as linear combinations of atomic orbitals (LCAO) and to obtain the values of the coefficients of the resultant molecular orbitals:

$$\psi_i = \sum_r c_{ir} \phi_r$$

where $\phi_r$ designates an atomic orbital at atom $r$, and $\psi_i$ is one of the molecular orbitals. By definition, the number of molecular orbitals is equal to the number of atomic orbitals, that is, of carbon centers in a conjugated unsaturated hydrocarbon.

## 6.1 *Ethylene*

The MO coefficients for ethylene are obtained by solving the Hückel secular determinant, in which $\alpha$ is the energy of an electron in a $p_z$ atomic orbital on carbon and $\beta$ is a measure of the overlap between the $p_z$ atomic orbitals on adjacent carbon atoms. Thus, for ethylene, numbering the carbon atoms as 1 and 2, the secular determinant is

$$\begin{vmatrix} \overset{11}{\alpha - E} & \overset{12}{\beta} \\ \underset{21}{\beta} & \underset{22}{\alpha - E} \end{vmatrix} = 0$$

for which the solutions are $E_1 = \alpha + \beta$, $E_2 = \alpha - \beta$ and the MO's are given by $\psi_1 = (1/\sqrt{2})\ (\phi_1 + \phi_2)$, $\psi_2 = (1/\sqrt{2})\ (\phi_1 - \phi_2)$. (The $1/\sqrt{2}$ coefficients arise from the normalization condition that $\sum_i c_i^2 = 1$; $\frac{1}{2} + \frac{1}{2} = 1$.) This calculation shows that ethylene has two $\pi$-electronic energy levels, and that in its ground state the two $\pi$ electrons will be accommodated in the lower energy level corresponding to bonding of carbons 1 and 2. That is, the electrons occupy the molecular orbital $\psi_1$ which puts electron density in between the two centers (*bonding*), rather than $\psi_2$ which removes electron density from this region (*antibonding*)

$$\psi_1^2 = \tfrac{1}{2}\phi_1^2 + \phi_1\phi_2 + \tfrac{1}{2}\phi_2^2$$

$$\psi_2^2 = \tfrac{1}{2}\phi_1^2 - \phi_1\phi_2 + \tfrac{1}{2}\phi_2^2$$

These electron densities may be compared with the electronic density resulting from simple superposition of the atomic orbitals: $\psi_{\text{superposition}} = \phi_1^2 + \phi_2^2$. The corresponding energy diagram is shown in Figure 6.1a. One can superimpose on this simple $\pi$ system the $\sigma$ framework (shown in Figure 6.1b) which can be obtained by the extended Hückel method or various self-consistent-field MO techniques. The relative ordering of the $\sigma$ and $\pi$, $\sigma^*$ and $\pi^*$ MOs indicated here is still a matter of discussion.

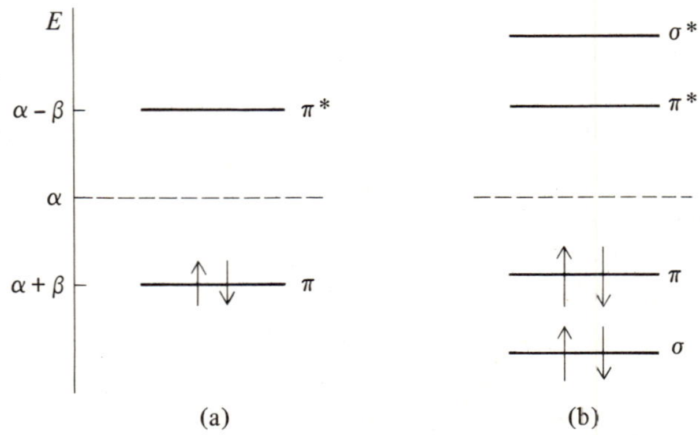


*FIGURE 6.1*
*Energy of ethylene molecular orbitals; (a) $\pi$ framework; (b) $\pi$ and $\sigma$ orbitals.*

It is then readily seen that the longest wavelength (lowest energy) electronic transition corresponds to promotion of a bonding electron from the highest occupied MO to the lowest unoccupied MO. The $\pi \rightarrow \pi^*$ transition for ethylene is experimentally observed at 162.5 nm.

## 6.2 *Butadiene*

Let us see what happens as unsaturation is increased and conjugation extended, for example, in butadiene. For this system the secular determinant is

$$\begin{vmatrix} \alpha - E & \beta & 0 & 0 \\ \beta & \alpha - E & \beta & 0 \\ 0 & \beta & \alpha - E & \beta \\ 0 & 0 & \beta & \alpha - E \end{vmatrix} = 0$$

$$(\alpha - E)\begin{vmatrix} \alpha - E & \beta & 0 \\ \beta & \alpha - E & \beta \\ 0 & \beta & \alpha - E \end{vmatrix} - \beta \begin{vmatrix} \beta & \beta & 0 \\ 0 & \alpha - E & \beta \\ 0 & \beta & \alpha - E \end{vmatrix}$$

$$= (\alpha - E)^2 \begin{vmatrix} \alpha - E & \beta \\ \beta & \alpha - E \end{vmatrix} - \beta \begin{vmatrix} \beta & \beta \\ 0 & \alpha - E \end{vmatrix} - \beta^2 \begin{vmatrix} \alpha - E & \beta \\ \beta & \alpha - E \end{vmatrix}$$

$$+ \beta^2 \begin{vmatrix} 0 & \beta \\ 0 & \alpha - E \end{vmatrix} = 0$$

or

$$E_1 = \alpha + 1.618\beta$$
$$E_2 = \alpha + 0.618\beta$$
$$E_3 = \alpha - 0.618\beta$$
$$E_4 = \alpha - 1.618\beta$$

The corresponding energy diagram is shown in Figure 6.2.

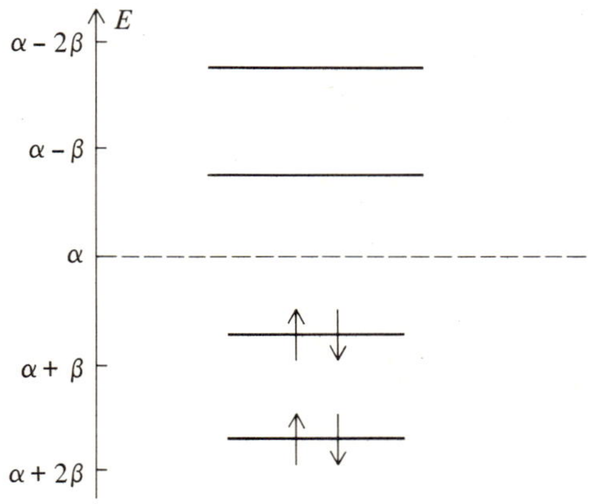


*FIGURE 6.2*
*Molecular orbital energy diagram for butadiene.*

The additional double bond has resulted in a splitting of the original ethylene bonding and antibonding levels into two levels each with a reduction in the energy difference between the highest occupied and lowest unoccupied MO. The promotion of an electron requires less energy so the absorption of light is shifted to higher wavelengths. Indeed the $\pi \rightarrow \pi^*$ transition of butadiene is found at 217 nm.

Similar effects are observed as conjugation is extended even further, as illustrated diagrammatically in Figure 6.3. Indeed a good correlation is obtained between the first observed $\pi \rightarrow \pi^*$ transition and the difference in energy between the highest occupied and lowest unoccupied MO, as given by simple Hückel MO theory and illustrated in Figure 6.4.

As seen from these examples the electronic energy obtained from simple Hückel MO theory gives excellent qualitative predictions of absorption spectra. By means of more sophisticated methods like extended Hückel or self-consistent field methods the calculation and prediction of electronic transitions can be made quantitative. These techniques, however, are beyond the scope of this book.

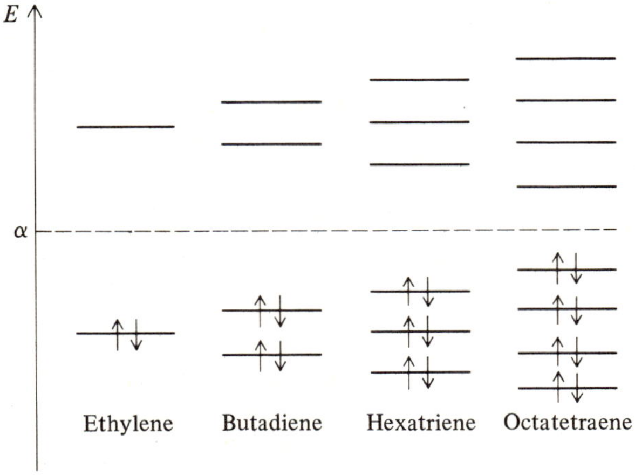


*FIGURE 6.3*
*Molecular orbitals of some polyenes.*

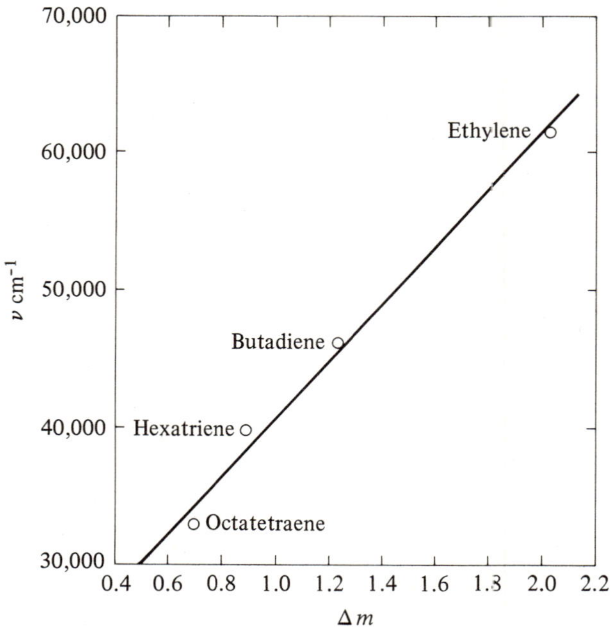


*FIGURE 6.4*
*Frequency of the first $\pi \rightarrow \pi^*$ transition of polyenes versus Hückel MO energy difference.*

## 6.3 *Bond Orders and Charge Densities*

Simple MO theory is capable of yielding further information by consideration of the electronic density either on an individual atom or between the nuclei. The electronic density can be obtained from the coefficients of the molecular orbitals. To calculate the coefficients for the appropriate molecular orbital we use the system of equations corresponding to the Hückel determinant, replacing the energy by its value:

$$\begin{aligned} c_1(\alpha - E_1) + c_2\beta &= 0 \\ c_1\beta + c_2(\alpha - E_1) + c_3\beta &= 0 \\ c_2\beta + c_3(\alpha - E_1) + c_4\beta &= 0 \\ c_3\beta + c_4(\alpha - E_1) &= 0 \end{aligned}$$

with

$$E_1 = \alpha + 1.618\beta$$

We obtain

$$\begin{aligned} c_2 &= 1.618c_1 \\ c_1 + c_3 &= 1.618c_2 \\ c_2 + c_4 &= 1.618c_3 \\ c_3 &= 1.618c_4 \end{aligned}$$

Hence

$$c_1 = c_4 \qquad \text{and} \qquad c_2 = c_3$$

and the normalization condition:

$$\begin{aligned} c_1^2 + c_2^2 + c_3^2 + c_4^2 &= 1 \\ [2c_1^2 + (1.618c_1)^2 \times 2] &= 1 \\ c_1 = c_4 &= 0.3717 \\ c_2 = c_3 &= 0.6015 \end{aligned}$$

*TABLE 6.1*

*Coefficients of the Molecular Orbitals of Butadiene*

| | $\phi_1$ | $\phi_2$ | $\phi_3$ | $\phi_4$ |
|---|---|---|---|---|
| $\psi_1$ | 0.3717 | 0.6015 | 0.6015 | 0.3717 |
| $\psi_2$ | 0.6015 | 0.3717 | −0.3717 | −0.6015 |
| $\psi_3$ | 0.6015 | −0.3717 | −0.3717 | 0.6015 |
| $\psi_4$ | 0.3717 | −0.6015 | 0.6015 | −0.3717 |

As an exercise, the reader should have no difficulty in calculating the remaining coefficients, listed in Table 6.1. From these coefficients one can obtain useful quantities, such as the charge densities and bond orders. Remember that the wave function $\psi$ has no direct physical meaning, but that the square of its amplitude $(\psi)^2$ is a measure of the electron density at a given point in space. Likewise, the coefficients $c_{ir}$ have no physical connotation, but products of the type $c_{ir}c_{ir}$ and $c_{ir}c_{is}$ do have an important significance. If $r$ and $s$ are bonded atoms, the product $p_{rs} = c_{ir}c_{is}$ measures the extent of $\pi$ bonding between carbon atoms $r$ and $s$. It is called the $\pi$ bond order for that bond. Likewise the product $q_r = c_{ir}c_{ir}$ measures the extent of the "self-bonding" of an atom to itself, that is, the charge density on that nucleus $r$. Let us now see how these calculated quantities relate to the experimental observations, starting with butadiene, a four-$\pi$-electron system. The filled orbitals are $\psi_1$ and $\psi_2$. Let us calculate the bond orders between carbon atoms 1 and 2, and 2 and 3:

$$p_{12} = 2\sum_i^{\text{occ}} c_{i1}c_{i2} = 0.3717 \times 0.6015 + 0.6015 \times 0.3717 = p_{21} = p_{34} = p_{43}$$

$$= 2 \times 0.3717 \times 0.6015 = 0.894$$

$$p_{23} = 2\sum_i^{\text{occ}} c_{i2}\,c_{i3} = 2(0.6015)^2 - 2(-0.3717)^2 = 0.447$$

This is already sufficient knowledge to provide some insight into nmr coupling constants in unsaturated molecules. For instance, the *cis* $^3J_{HH}$ coupling constant between the olefinic protons of butadiene is experimentally measured[1] as 10.2 Hz. Using the bond order-coupling constant relationship indicated earlier for such protons on adjacent trigonal carbons,

$$J_{HH} = 11.6p - 0.2$$

we calculate $J_{HH} = 11.6(0.894) - 0.2 = 10.2$ Hz, in very close agreement with the experimental value. One can go a step further, and consider a related molecule, 1,3-cyclohexadiene,

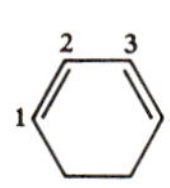


for which the butadiene moiety is constrained into a cisoid rather than a transoid arrangement.

The Hückel method ignores stereochemistry, and will make identical prediction for the cisoid or for the transoid conformation. Let us use it as a first approximation and calculate the values for the vicinal coupling constants between the olefinic protons:

$$J_{12} = J_{34} = 11.6(0.894) - 0.2 = 10.2 \text{ Hz}$$
$$J_{23} = 11.6(0.447) - 0.2 = 5.0 \text{ Hz}$$

[1] R. T. Hobgood, Jr., and J. H. Goldstein, *J. Mol. Spectrosc.*, *12*, 76 (1964).

The experimental values are 9.7 Hz and 5.1 Hz,[2] in reasonable agreement. It is seen from this simple example that the results of MO theory can be used to rationalize nmr coupling constants between the proton nuclei on different $\pi$ centers. Likewise MO theory serves to calculate esr coupling constants from the electronic densities at the different carbons in a conjugated organic free radical. It will be remembered that the unpaired electronic spin is coupled to each of the protons in the molecule and the magnitude depends upon the amount of *unpaired* charge density $\rho_C$ at the carbon atom bearing the proton, as given by the McConnell relationship:

$$a_H = Q_C \rho_C \qquad \text{where } Q_C \sim 22.5 \text{ Oe}$$

Let us consider for instance the butadiene radical anion $C_4H_4^{\overline{\cdot}}$, which has five $\pi$ electrons. The unpaired electron is placed into the $\psi_3$ MO. Therefore, the values of $\rho_C$ are simply the squares of the coefficients $c_{3r}$ at the carbon atoms $r = 1, 2, 3, 4$:

$$\rho_C(1) = \rho_C(4) = (0.6015)^2 = 0.362$$
$$\rho_C(2) = \rho_C(3) = (0.3717)^2 = 0.138$$

Using $Q_C = 22.5$ Oe, we can therefore calculate:

$$a_H(1) = a_H(4) = 8.15 \text{ Oe}$$
$$a_H(2) = a_H(3) = 3.1 \text{ Oe}$$

in reasonable agreement with the experimental values[3] of 7.62 and 2.79 Oe, while for 1,3-cyclohexadiene the experimental values are $a_H = 8.21$ and $a_H = 2.0$ Oe.

The agreement can be improved in the case of butadiene by using the value of $Q_C$ *appropriate to butadiene.* It turns out that one can derive it simply from the data since

$$2\rho_C(1) + 2\rho_C(2) = 1 \qquad \text{(the unpaired electron density)}$$

$$\rho_C(1) = \frac{7.62}{Q}$$

$$\rho_C(2) = \frac{2.79}{Q}$$

$$\rightarrow \frac{15.24}{Q} + \frac{5.58}{Q} = 1 \qquad Q = 20.82 \text{ Oe}$$

corresponding to

$$\rho_C(1) = \rho_C(4) = 0.366 \qquad \text{(experimental)}$$
$$\rho_C(2) = \rho_C(3) = 0.133 \qquad \text{(experimental)}$$

Again the agreement between experiment and the Hückel theory is rather satisfactory.

## 6.4 *Benzene*

The six MOs obtained from the application of Hückel's method to benzene are shown diagrammatically in Figure 6.5a, with the corresponding atomic orbital coefficients $c_{ir}$ at each carbon $r$. Their energies are indicated by the horizontal lines in Figure 6.5b. These MOs differ in their symmetry properties. If we denote by $C_1$ and

[2] S. L. Manatt, D. D. Elleman, *Abstracts, 48th Meeting of the Societa Italiana per il progresso delle scienze*; M. A. Cooper, *et al.*, *J. Chem. Phys.*, *53*, 2343 (1970).
[3] D. H. Levy and R. J. Myers, *J. Chem. Phys.*, *41*, 1062 (1964).

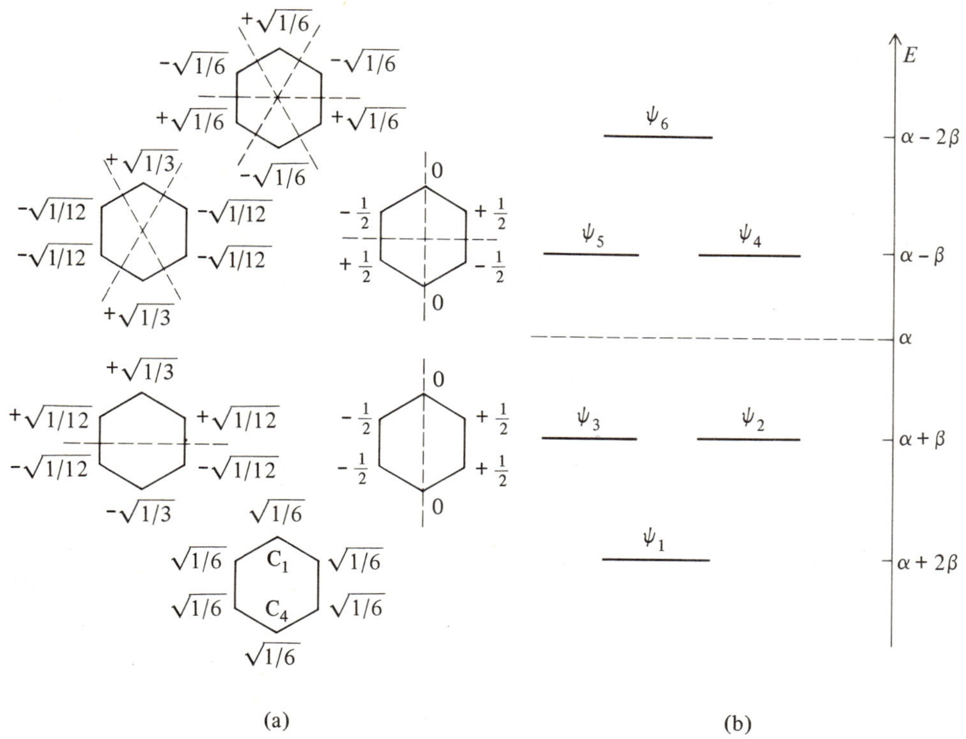


*FIGURE 6.5*
*Benzene molecular orbitals (a) and their relative energies (b).*

$C_4$ the carbon atoms lying along the vertical in Figure 6.5a, the axis $C_1$-$C_4$ is an axis of antisymmetry (plus sign changes to minus sign) for $\psi_2$ and $\psi_4$. The corresponding nodes are indicated by the dotted lines (the energy of the MOs increases with the number of nodes). Benzene itself, a closed-shell molecule, has its six $\pi$ electrons in the MOs of lowest energy, that is, $\psi_1$, $\psi_2$, and $\psi_3$. An interesting problem arises, however, when one more electron is available, as in the benzene radical anion $C_6H_6{}^{\overset{\cdot}{-}}$. Where should the extra electron go, into $\psi_4$ or into $\psi_5$? The simple answer is the intuitive one: the electron, having no grounds to discriminate between $\psi_4$ and $\psi_5$, will spend 50 percent of its time in $\psi_4$ and 50 percent in $\psi_5$. Hence, the electron density at each carbon, as given by $c_{ir}^2$, will be

$$0.5\left(\sqrt{\frac{1}{3}}\right)^2 + 0.5(0)^2 = \tfrac{1}{6} \text{ at position 1 (or 4)}$$

$$0.5\left(\sqrt{\frac{1}{12}}\right)^2 + 0.5(\tfrac{1}{2})^2 = \tfrac{1}{6} \text{ at position 2 (or 3, 5, 6)}$$

This accounts for the esr spectrum of the benzene radical anion consisting of seven equidistant lines with a splitting $a_H = 22.5 \times \frac{1}{6} = 3.75$ Oe.

In reality the explanation is somewhat more involved. The $\psi_4$ and $\psi_5$ MOs have identical energies and are said to be degenerate. The benzene molecule will tend to distort in order to remove this degeneracy, and thereby make the energies of the two orbitals different. This is the so-called Jahn–Teller distortion. The benzene molecule distorts according to its normal modes of vibration in order to put the unpaired electron into either—but not both—orbital. Such a geometrical distortion can be shown to correspond to a small energy stabilization. What the benzene radical anion does is to flip between these two distorted hexagons with a very high frequency, of the order of $10^{-7}$ sec. This effectively averages out the coupling constants to the different protons (hence the regular septet) and is only manifested through a slight line broadening.

A more obvious way of removing the degeneracy is through chemical substitution. Introduction at position 1 of an electropositive substituent will lower the energy of $\psi_4$ relative to $\psi_5$, so that the unpaired electron will preferentially elect the $\psi_4$ orbital which carries zero electron density at the substituted carbon. An electronegative substituent will do the converse. Hence the prediction is that the esr spectrum of the $X{-}C_6H_5^{\overline{\cdot}}$ radical anion should have five lines, with splittings $a_H = Q/4 \sim 6$ Oe for X more electropositive than hydrogen, and a triplet of quartets, with splittings $a_H \sim 8$ and 2 Oe, respectively, if X is more electronegative than hydrogen (since the electron densities on carbon are given by the squares of the appropriate atomic orbital coefficients).

This is indeed observed for *para*-xylene: five lines with $a_H = 5.34$ Oe. For toluene the situation is a trifle more complicated; $a_H$ (*para*) is not zero. This is because the MOs differ slightly from those of benzene. But the quintet splitting still predominates.

Even more remarkable is the case of the monodeuteriobenzene radical anion.[4] Deuterium turns out to be electropositive with respect to hydrogen. As a result of the minute isotopic change—a secondary deuterium isotope effect[5]—the esr spectrum is analyzed for the following coupling constants:

$$a_H(2,3,5,6) = 3.92 \pm 0.01 \text{ Oe}$$
$$a_H(4) = 3.41 \pm 0.01 \text{ Oe}$$
$$a_D = 0.55 \pm 0.01 \text{ Oe}$$

The latter corresponds—after correction by a factor of 6.55, for the difference between the gyromagnetic ratios of H and D—to an $a_H$ of 3.6 Oe. Hence, the degeneracy of orbitals is already partly lifted in deuteriobenzene. One can go the other way too, and stabilize the symmetrical MO rather than the antisymmetric orbital, as for instance in *meta*-xylene. The experimental spectrum gives $a_H(2) = 6.85, a_H(5) = 7.72, a_H(4,6) = 1.46$.

Hückel MO theory has also had remarkable success in explaining the products of the Birch reduction of aromatic hydrocarbons, which are thought to arise from radical anion intermediates.[6]

## 6.5 *Photoelectron Spectroscopy*

It appears that esr evidence supports the idea that the effect of substitution in benzene radicals is to remove the orbital degeneracy and to stabilize either a symmetrical or an antisymmetric MO. This is not only reasonable, it is also true. Before illustrating this statement with an example, let us indicate the principles of photoelectron spectroscopy, an elegant new method that allows the direct determination of the energies of individual molecular orbitals.

Consider a diatomic molecule A—B in the ground vibrational state with the Morse potential curve shown in Figure 6.6. Molecule A—B is submitted to the impact of a beam of photons, and by losing one electron ionizes into the $A{-}B^+$ species; a different molecule therefore endowed with a different Morse curve (Figure 6.6). Such an electronic transition obeys the Franck–Condon principle, that is, it occurs much

[4] R. G. Lawler, J. R. Bolton, G. K. Fraenkel, and T. H. Brown, *J. Amer. Chem. Soc.*, *86*, 520 (1964); R. G. Lawler, G. K. Fraenkel, *J. Chem. Phys.*, *49*, 1126 (1968).
[5] P. Laszlo and Z. Welvart, *Bull. Soc. Chim. Fr.*, 2412 (1966).
[6] A. J. Birch, *Quart. Rev.* (London), *12*, 17 (1958).

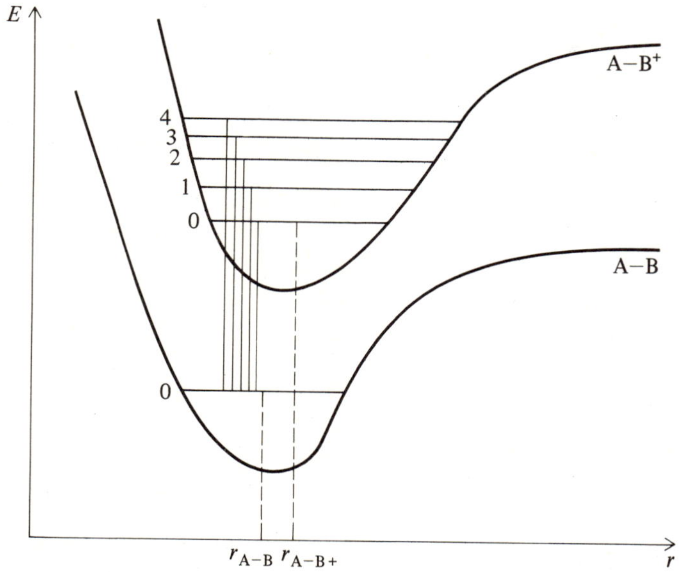


*FIGURE 6.6*
*Morse potential curve of* A—B *and A*—B+.

faster ($10^{-15}$ sec) than a molecular vibration (about $10^{-12}$ sec). Therefore, the ionization is of the so-called vertical type: the nuclei do not have time to reorient themselves into the new equilibrium configuration for the A—B+ ion. Three possibilities exist for the electron removed by the ionization.

1. It is a nonbonding electron (Figure 6.7a). The most probable transition is from the ground vibrational state of A—B into the ground vibrational state of A—B+. Hence a very sharp peak is observed for this particular electron when the ionization potential is varied.
2. It is a bonding electron (Figure 6.7b); upon its removal, the A—B bond stretches. The effect of vertical ionization is to populate a large number of A—B+ vibrational states, hence the ionization curve is expected to be broad.
3. The ionization curve should be broad for the converse case of an antibonding electron being expelled (Figure 6.7c).

In practice[7,8] one usually photoionizes molecules with the helium line at 584 Å, corresponding to a quantum $h\nu_0 = 21.21$ eV. This allows the determination of all the

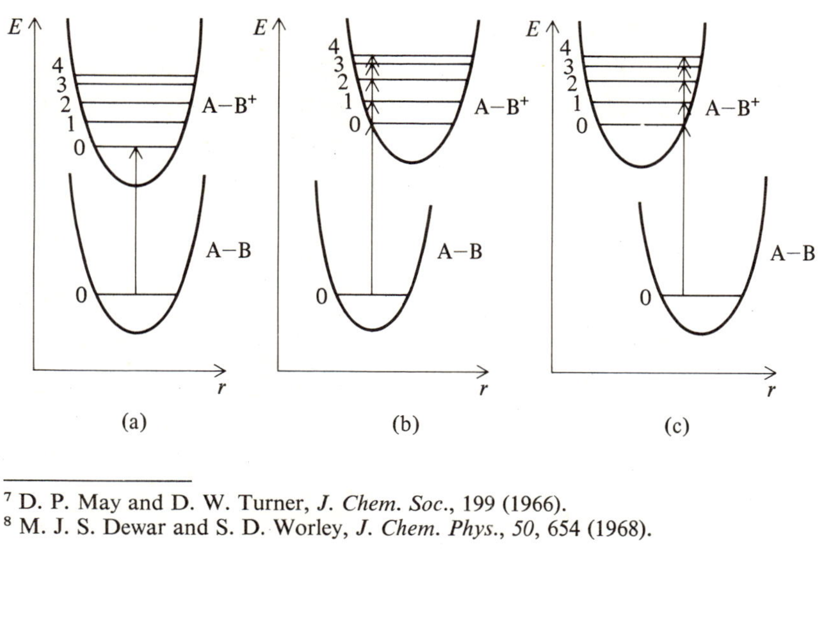


*'IGURE 6.7*
*'ypes of ionization of* A—B *into* —B+.

[7] D. P. May and D. W. Turner, *J. Chem. Soc.*, 199 (1966).
[8] M. J. S. Dewar and S. D. Worley, *J. Chem. Phys.*, *50*, 654 (1968).

ionization potentials smaller than this value of 21.21 eV, by measuring the kinetic energy $\frac{1}{2}mv^2$ of the electrons expelled. These obey the relation

$$h\nu_0 = \text{I.P.} + \tfrac{1}{2}mv^2$$

where I.P. is a given ionization potential.

A typical high-resolution photoionization spectrum, that of chlorobenzene, is shown in Figure 6.8. In addition to the peaks corresponding to removal of the $3p_x$ and $3p_y$ nonbonding electrons on chlorine, it is apparent that separate peaks are observed for the $\pi_2$ and $\pi_3$ molecular orbitals, thus demonstrating the lifting (by ~0.5eV) of the degeneracy between these hitherto equienergetic MOs.

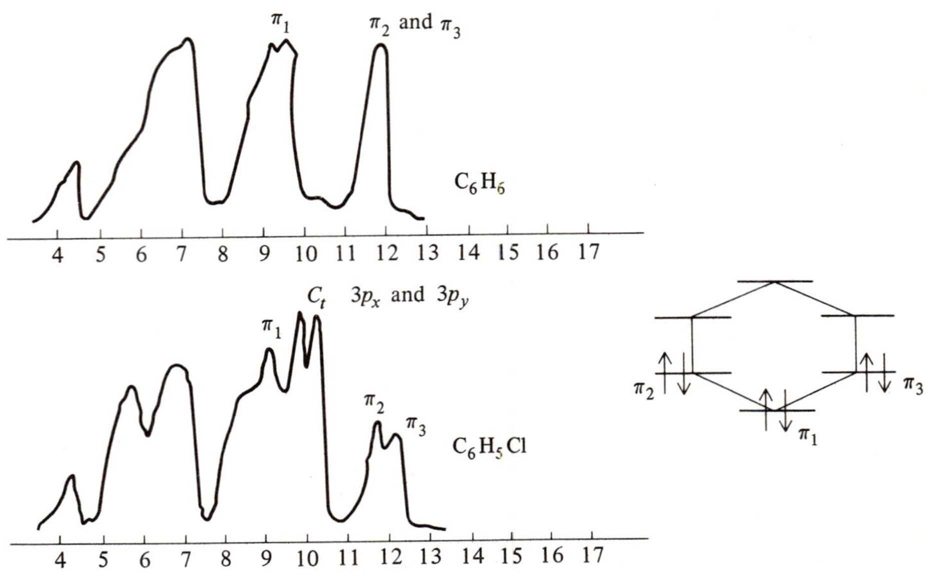


*FIGURE 6.8*
*Photoelectron spectra of benzene and chlorobenzene. The unlabeled bands correspond to removal of innershell electrons.*

## 6.6 *PROBLEMS*

1. The allyl radical has the following esr hyperfine constants:

   $a_H = 4.06$ Oe  to the proton on carbon 2

   $a_H = 13.93$ Oe  to one pair of terminal protons

   $a_H = 14.83$ Oe  to the other pair of terminal protons

   How close are these values to those predicted by Hückel theory?
2. The benzyl radical displays a complex esr spectrum from which the following hyperfine constants can be extracted:

   $a_H = 16.4$ Oe (2H)

   $a_H = 6.2$ Oe (1H)

   $a_H = 5.2$ Oe (2H)

   $a_H = 1.8$ Oe (2H)

   Assign them to the corresponding protons in the molecule.
3. The following compound absorbs in the uv at 185 nm. Does this seem reasonable? Why?

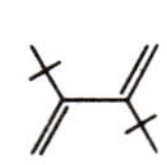 $\lambda_{max} = 185$ nm

4. The esr spectrum of methyl cyclooctatetraene radical anion shows the following hyperfine constants to the ring protons:

   $a_H = 4.8$ Oe (3H)

   $a_H = 1.6$ Oe (4H)

   Explain this observation, given the following molecular orbitals for the $C_8H_8^{\overline{\cdot}}$ ion:

   $$\psi_4 = \tfrac{1}{2}\phi_2 - \tfrac{1}{2}\phi_4 + \tfrac{1}{2}\phi_6 - \tfrac{1}{2}\phi_8$$
   $$\psi_5 = \tfrac{1}{2}\phi_1 - \tfrac{1}{2}\phi_3 + \tfrac{1}{2}\phi_5 - \tfrac{1}{2}\phi_7$$

   The unsubstituted cyclooctatetraene radical anion shows nine lines with $a_H = 3.2$ Oe.
5. It is well known that the resonance energy of benzene is 36 kcal/mole, as derived from comparison of the experimental heat of combustion with the calculated heat of formation. The $\pi \rightarrow \pi^*$ transition for ethylene occurs at 162.5 nm. Derive from these data the value of Hückel's $\beta$. Are you surprised by the result? How do you explain it?
6. In the uv spectra of the cations indicated, the absorption maximum of longest wavelength (with an absorption coefficient $\varepsilon$ between 1000 and 10,000) occurs at

   | | |
   |---|---|
   | allyl $C_3H_5^+$ | 314 nm |
   | pentadienyl $C_5H_7^+$ | 397 nm |
   | heptatrienyl $C_7H_9^+$ | 470 nm |
   | tropylium $C_7H_7^+$ | 275 nm |

   Interpret these variations in the framework of Hückel MO theory. What can you conclude concerning $\beta$?
7. The Birch reduction uses sodium in liquid ammonia, together with a source of protons such as ethanol, to effect the transformation of an aromatic hydrocarbon ArH into the dihydro derivative $ArH_3$. It is thought to proceed through the intermediacy of either or both the monoanion $ArH^-$ and the dianion $ArH^=$. For instance, benzene yields the 1,4-cyclohexadiene, even though the 1,3-cyclohexadiene would be the more stable product thermodynamically. Explain this result by means of MO theory.
8. Fulvene can be written as either resonance form (1) or (2), the latter including an aromatic electron sextet. Which of the two would you favor as the prevailing ground-state structure, given that $J_{12} = 5.2$ Hz and $J_{23} = 2.2$ Hz? (Assume that a bond order-coupling constant relationship can be empirically devised for five-membered rings in a manner similar to six-membered rings). Is this conclusion in keeping with predictions from Hückel theory?

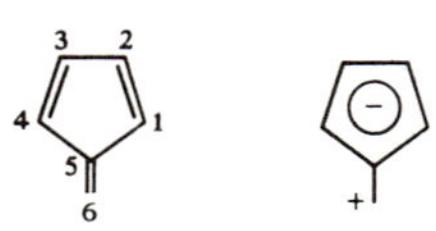


1 2

## 6.7 SUGGESTED READINGS


A. Streitwieser, Jr., *Molecular Orbital Theory for Organic Chemists*, John Wiley and Sons, Inc., New York, 1961, chap. 8.

J. D. Roberts, *Notes on Molecular Orbital Calculations*, W. A. Benjamin, Inc., New York, 1962.

M. J. S. Dewar, *The Molecular Orbital Theory of Organic Chemistry*, McGraw-Hill Book Co., New York, 1969.

A. J. Merer and R. S. Mulliken, *Chem. Rev.*, *69*, 639 (1969).

K. W. Bowers, "Orbital Degeneracy in Benzene and Substituent Effects," in *Radical Ions*, E. T. Kaiser and L. Kevan, eds., Wiley-Interscience Publishers, Inc., New York, 1968, chap. V.

A. Carrington, *Orbital Degeneracy and Spin Resonance in Free Radical Ions*, Royal Institute of Chemistry, 1969.

A. D. Baker, "Photoelectron Spectroscopy," *Accounts Chem. Res., 3*, 17 (1970).

D. M. Hercules, "Electron Spectroscopy," *Anal. Chem.*, *42*, 20A (1970).

D. Betteridge and A. D. Baker, "Analysis Potential of Photoelectron Spectroscopy," *Anal. Chem.*, *42*, 43A (1970).

W. C. Price and D. W. Turner, eds., "A Discussion on Photoelectron Spectroscopy," *Phil. Trans. Roy. Soc. London*, *286A*, 1–175 (1970).

# CHAPTER 7 ULTRAVIOLET, OPTICAL ROTATORY DISPERSION, CIRCULAR DICHROISM, AND THE CARBONYL CHROMOPHORE

Simple molecular orbital theory suggests that the $\pi \rightarrow \pi^*$ transition is of lowest energy for unsaturated molecules such as ethylene or butadiene; it corresponds to promotion of an electron from the highest occupied to the lowest unoccupied orbital. Experimentally the magnitude of the energy gap is given by the position of the longest wavelength absorption. As seen from the previous section, theory has been very successful in qualitative as well as quantitative analysis of the experimental results for olefins and polyenes.

It is tempting to try and expand this success to related compounds. An important class is that of carbonyl-containing derivatives. The simplest case is that of the formaldehyde molecule. It is derived from ethylene in a formal sense, the oxygen heteroatom replacing one of the $=CH_2$ moieties. The example of the carbonyl chromophore will be studied here with reference to MO theory. This theory provides a convenient conceptual frame. Not all the details are unequivocally established; one cannot say that the correspondence between theory and the underlying facts is secure. Theory merely provides a language which is useful as a guide to reality.

## 7.1 *Electronic Transitions*

The oxygen atom has six valence electrons in three different orbitals. There are two sets of nonbonding electrons (the *n* electrons). These lone pairs should be higher in energy than the bonding $\pi$ electrons, and they ought to have the same energy. In the MO picture of the formaldehyde molecule (Figure 7.1) this degeneracy is removed. While one electron in a $p_z$ atomic orbital of oxygen combines with its counterpart on carbon into a bonding molecular orbital, one oxygen lone pair is accommodated into a lower energy orbital and the other ends up higher than the carbonyl $\pi$-molecular

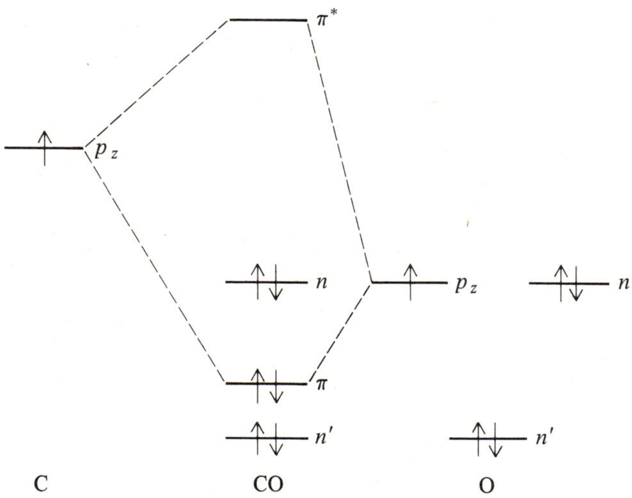


*FIGURE 7.1*
*Molecular orbitals of* $\rangle C=O$

orbital. The lower-energy orbital has more *s* character: if oxygen is given *sp* hybridization, the $n'$ orbital can be taken as one of the two *sp* hybrids. The *n* orbital is then the $p_y$ atomic orbital on oxygen, since the $p_z$ is used for the C=O $\pi$ bond. The remaining electron on oxygen must obviously be used for the C—O $\sigma$ bond. One could infer from this picture that the electronic transition of lowest energy (highest wavelength) corresponds to promotion of a nonbonding *n* electron into the antibonding $\pi^*$ orbital. It is observed at 290 nm ($\varepsilon = 18$). The $\pi \rightarrow \pi^*$ transition would give rise to the maximum at 165 nm ($\varepsilon = 10^{4.5}$), a position almost identical to that of the ethylene $\pi \rightarrow \pi^*$ transition at 162.5 nm. This identification is tempting and has often been made, but it has not been demonstrated satisfactorily. The relative ordering of the various energy levels in Figure 7.1 is not at all secure, and may be modified in time with additional knowledge. The difference in the intensities of the $n \rightarrow \pi^*$ and $\pi \rightarrow \pi^*$ absorptions also has a theoretical explanation based upon the MOs depicted in Figure 7.1, and would thus appear to support the assignment shown. The $n \rightarrow \pi^*$ is a transition of low probability (a "forbidden" transition) while the $\pi \rightarrow \pi^*$ has a high intensity, hence a high probability (allowed transition). The symmetry of the orbitals involved in the transitions explains the difference. The bonding $\pi$ orbital and the antibonding $\pi^*$ orbital are in the same plane, whereas the nonbonding orbitals are in a plane orthogonal to that of the $\pi$ orbitals. Hence promotion of an electron from the nonbonding orbital is not possible without some geometrical change. That absorption occurs nevertheless is a result of vibrational interactions which slightly distort the molecule, so that the $n \rightarrow \pi^*$ transition acquires a finite probability.

Additional insight is gained into the nature of the 290 nm transition through solvent effects. This transition is observed at 290 nm in a nonpolar, nonhydrogen-bonding solvent such as cyclohexane or isooctane; if the spectrum is run instead in a polar, hydrogen-bonding solvent like ethanol or methanol, a shift to shorter wavelength (a hypsochromic or blue shift) is observed. The reason is that ethanol hydrogen bonds to the oxygen lone pairs of formaldehyde, and it takes extra energy to break this hydrogen bond in order to promote a lone-pair electron to the excited state. Such solvent effects are general for carbonyl-containing compounds. The $\lambda_{max}$ of acetone ($CH_3COCH_3$), for instance, is at 279 nm in hexane, but at 272 nm in ethanol and 265 nm in water. Water forms a stronger hydrogen bond than ethanol, hence it takes more energy to break the hydrogen bond and greater overall energy to promote the lone-pair electron to the excited state and a shift to shorter wavelengths occurs. The amount of shift can be used as an approximate measure of the strength of the hydrogen bond.

Further use of solvent effects can be made to determine the presence of basic or acidic sites in the molecule. The $\lambda_{max}$ of aniline ($C_6H_5\ddot{N}H_2$) occurs at 280 nm ($\varepsilon =$ 1500) in aqueous ethanol. In the presence of a drop of acid, however, the $\lambda_{max}$ shifts to 255 nm ($\varepsilon = 200$). The lone pair on nitrogen is protonated, accounting for the large blue shift as well as the decrease in the intensity of the absorption. The situation is reversed in the case of phenol, $C_6H_5\ddot{O}H$, whose $\lambda_{max}$ in aqueous ethanol is at 270 nm ($\varepsilon = 1500$). In the presence of a small amount of base such as $^-\ddot{O}H$ the $\lambda_{max}$ of phenol undergoes a shift to higher wavelength (a bathochromic or red shift); $\lambda_{max} = 287$ nm ($\varepsilon = 2600$). The base removes the acidic proton, generating the phenoxide ion, $C_6H_5O^-$, and an extra set of lone-pair electrons on oxygen accounting for the shift to higher wavelength or lower energy and the increased intensity.

## 7.2 Optical Rotatory Dispersion and Circular Dichroism[1]

Let us now examine the symmetry properties of the MOs of the carbonyl group on formaldehyde (Figure 7.2) which are connected by the 290 nm $n \rightarrow \pi^*$ transition. One can therefore consider that these three nodal surfaces (only two of which are also

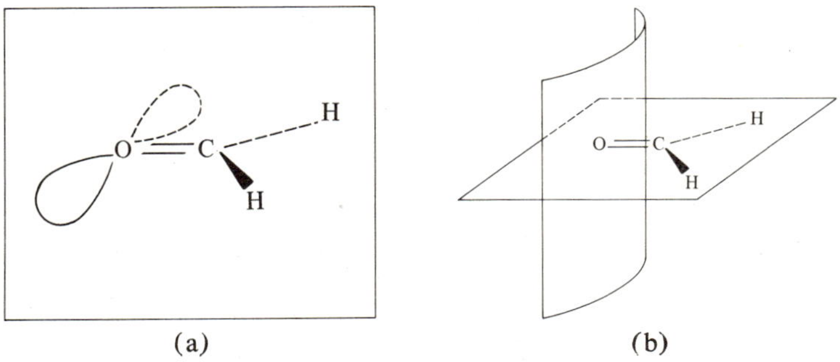


*FIGURE 7.2*
*Nodal surfaces of formaldehyde. (a) The nodal plane bisects the HCH angle and is perpendicular to the carbonyl plane. (b) The plane of the carbonyl group is a nodal plane and there is another nodal surface—not necessarily a plane—normal to the C—O axis and intersecting it between the carbon and oxygen.*

symmetry planes) for the carbonyl group combine together to divide the entire space into eight regions or "octants," as shown in Figure 7.3 for cyclohexanone. Consequences of considerable practical importance can be derived from examination of such an *octant diagram*:

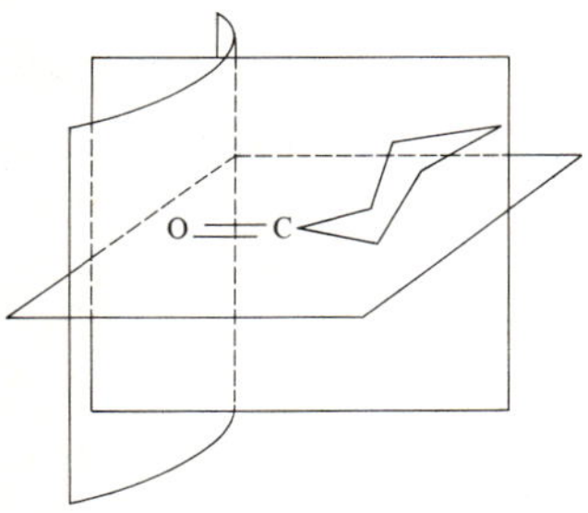


*FIGURE 7.3*
*The octants for cyclohexanone.*

1. In principle a substituent placed in one of these nodal planes or surface—for instance an axial or an equatorial bromine at position 4—should not perturb the $n \rightarrow \pi^*$ electronic transition.
2. By contrast, a substituent away from these nodal surfaces should affect the $n \rightarrow \pi^*$ electronic transition.

Let us specify further this rather abstract statement, by considering a cyclohexanone molecule having an axial methyl group at either the 2 or 6 position, adjacent to the carbonyl:

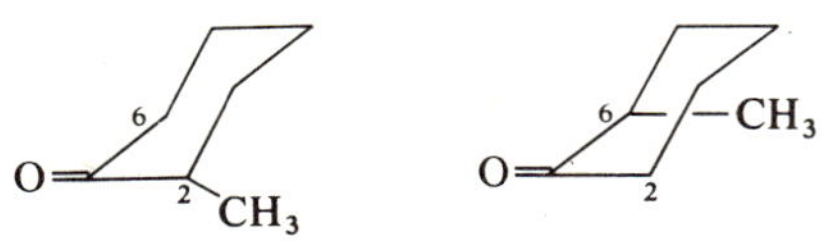


These two nonsuperimposable molecules are enantiomers, and such chiral (optically active) molecules have the property of rotating the plane of a linearly polarized monochromatic light beam of wavelength $\lambda$ by an angle $\alpha$:

$$\alpha_{\text{rad}} = \pi(n_{\text{L}} - n_{\text{R}}) \frac{l_{\text{cm}}}{\lambda_{\text{cm}}}$$

in the cgs system; or

$$\alpha_{\text{deg}} = \frac{1800}{\lambda_{\text{cm}}} (n_{\text{L}} - n_{\text{R}}) l_{\text{cm}}$$

in more common units, where $l$ is the sample path length, $n_{\text{L}}$ and $n_{\text{R}}$ are the indices of refraction for left- and right-circularly polarized light, respectively, and $\lambda$ is the wavelength of the incident light. This relationship holds because linear polarization of light is equivalent to the superposition of a *left* (L) and a *right* (R) circularly polarized beam, and because the speed of these two beams L and R, as measured by the corresponding refractive indices $n_{\text{L}}$ and $n_{\text{R}}$, must be different in an optically active substance. Similarly, the absorption coefficients $\varepsilon_{\text{L}}$ and $\varepsilon_{\text{R}}$ for the L and the R radiation must also be

[1] These have been termed *chiroptical* methods.

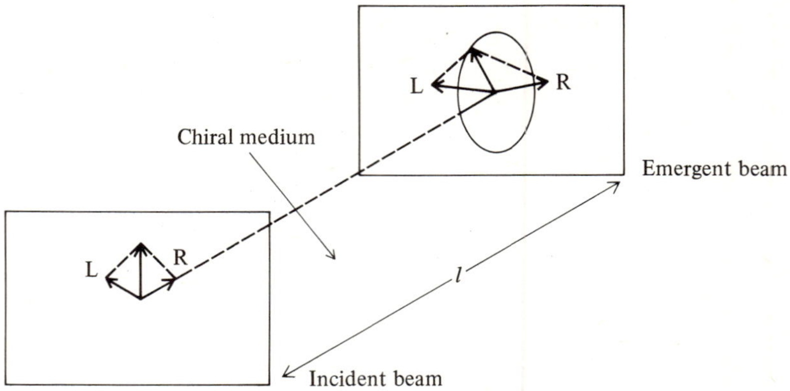


*FIGURE 7.4*
*Rotation of polarized light by a chiral medium.*

different for this chiral substance. Hence, it is to be expected a priori that a chiral molecule such as either of the above methylcyclohexanones in solution will transform the linearly polarized incident beam into an elliptically polarized vibration (Figure 7.4).

What about the magnitude of these effects? They should be the greatest in the vicinity of an $n \rightarrow \pi^*$ transition for which the absorption coefficient $\varepsilon$, as well as $\varepsilon_L - \varepsilon_R$, will be large. This is indeed the case, and such *Cotton effects* can be observed experimentally. In the region of an electronic absorption band (here the $n \rightarrow \pi^*$ transition), the difference $\Delta\varepsilon = \varepsilon_L - \varepsilon_R$ of the absorption coefficients for the left-handed and the right-handed radiations can be plotted as a function of the wavelength. It is found, and this can be theoretically justified, that the axial 2-methylcyclohexanone will give a positive curve corresponding to a positive Cotton effect (Figure 7.5a) whereas the enantiomeric 6-methylcyclohexanone will give the same curve but with the opposite sign (or negative Cotton effect, Figure 7.5b). This experiment (and the corresponding theoretical considerations) allow one to attribute a sign—that of the corresponding Cotton effect—for each of the eight octants defined previously (Figure 7.3).

Clearly, a molecule such as *cis*-diaxial-2,6-dimethyl cyclohexanone, which could be made to exist in such a conformation by an additional equatorial 4-*t*-butyl group, is meso (achiral). The contributions of each methyl group to the Cotton effect are equal but of opposite sign and thus cancel. More generally, the contribution of a substituent to the total Cotton effect will depend upon its octant for the sign and upon the distance to the chromophore (here a carbonyl) for its magnitude. Let us restate this conclusion in its generality:

*FIGURE 7.5*
*(a) Positive Cotton effect curve of $\Delta\epsilon$ versus $\lambda$ for 2-methylcyclohexanone. (b) Negative Cotton effect curvefor6-methylcyclohexanone.*

A dissymetrically substituted chromophore will display a Cotton effect whose magnitude and sign will be determined by the geometrical relationships of the various groups in the molecule to the particular chromophore.

Likewise, a fortiori a dissymetric chromophore will display a Cotton effect whose sign and magnitude will reflect the direction and extent of its chirality. However, one must of course have a *resolved* optically active compound in order to observe a Cotton effect. See Table 7.1 for some examples of chromophores exhibiting Cotton effects.

*TABLE 7.1*

*Examples of Chromophores Exhibiting Cotton Effects.*

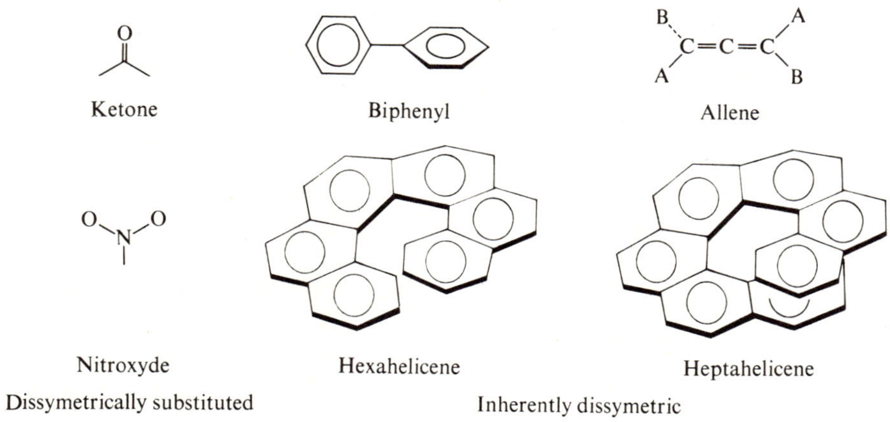


These principles are obviously capable of very useful applications to many problems of stereochemistry; we shall consider some illustrative examples. Before doing so, however, it is to be noted that there exists another experimental method, optical rotatory dispersion (ORD) for deducing the sign of the Cotton effect. While circular dichroism (CD) consists of the study of the ellipticity as a function of wavelength, $\Delta\varepsilon = f(\lambda)$, optical rotatory dispersion is a measure of the optical rotation as a function of wavelength, $\alpha = f(\lambda)$. Again, two types of curves can be observed, corresponding to either a positive or a negative Cotton effect (Figure 7.6). The inflection point in these ORD curves corresponds to the uv absorption maximum $\lambda_{max}$. In the absence of such a maximum, no Cotton effect curve would be observed. However, even far from *another* absorption maximum $\lambda_0$, the optical rotation is a function of the wavelength and varies smoothly with $\lambda$ according to the following equation, called the Drude relation:

$$\alpha = \frac{A}{\lambda^2 - \lambda_0^2} \qquad \text{where } A \text{ is a constant}$$

Therefore, Cotton effect curves are usually superimposed over the plain curves due to distant $\lambda_0$ absorptions.

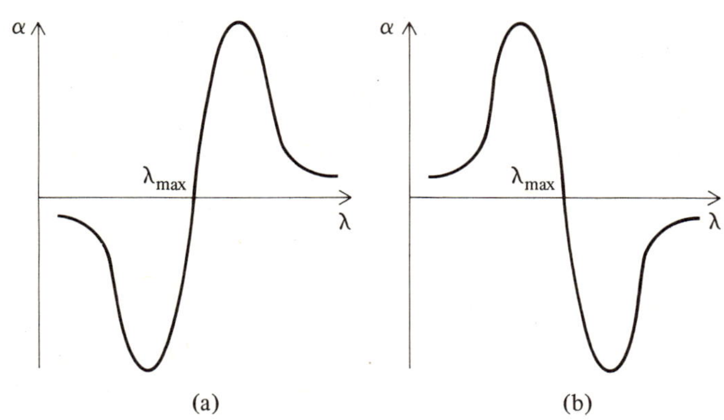


*FIGURE 7.6*

*Optical rotatory dispersion curves; (a) positive, and (b) negative Cotton effects.*

The ORD and CD measurements are performed experimentally as follows[2]: a beam of light is made monochromatic after dispersion by a prism; another prism with a perpendicular optical axis is used to polarize the incident beam, so that two perpendicularly plane-polarized beams making an angle with each other leave the prism. With such a design, prisms of synthetic quartz transparent below 2100 Å (in contrast to natural quartz), and a photomultiplier as the detector, commercial instruments achieve a sensitivity of $10^{-5}$ optical density unit $\Delta\varepsilon/\varepsilon$ per millimeter of recording chart. The sample container is made of silica with thickness below 0.1 mm. It is amusing that, although optical activity was discovered by Arago during Napoleon's reign, ORD and CD remained largely unknown, and their potential untapped, until 1953 (C. Djerassi). It has been claimed that this historical lapse is due to the invention of the Bunsen burner (1790s) which afforded an easy and adequate source of monochromatic light (the sodium D line) so that scientists did not bother to measure the optical rotation at other wavelengths; it is indeed amazing that so much stereochemical information could be obtained from such sparse data.

## 7.3 *Application*

Let us first consider the optically active cyclohexanones. Since most substituents are usually on the same side of the nodal surface separating the carbon from the oxygen of the carbonyl as the carbon atom, the octant diagram can be simplified by considering only the four rear octants. (In fact, very little is known about the front octants.) These are conveniently represented from the perspective of an observer on the carbonyl axis looking towards carbon 1 and the rest of the cyclohexanone ring (Figure 7.7).

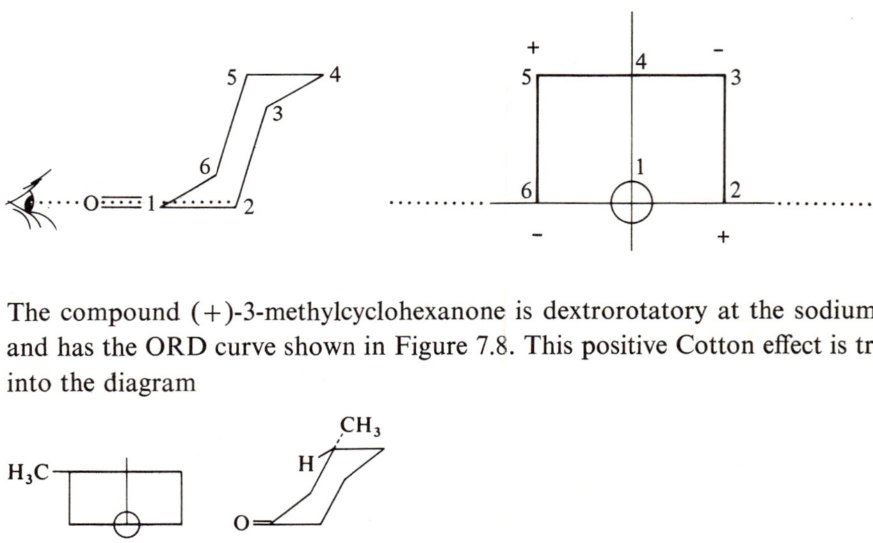


*FIGURE 7.7*
*The four rear octants of cyclohexanone.*

The compound (+)-3-methylcyclohexanone is dextrorotatory at the sodium D line, and has the ORD curve shown in Figure 7.8. This positive Cotton effect is translated into the diagram

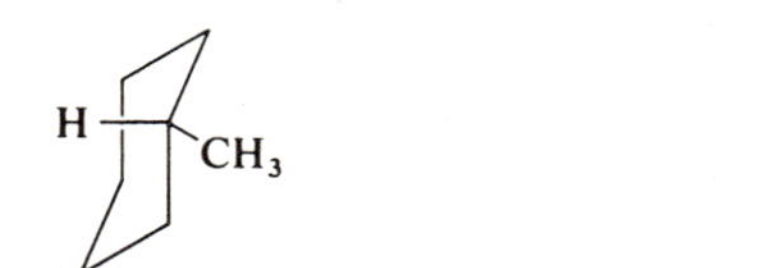


with the methyl group predominantly equatorial. Hence the absolute configuration of this molecule is deduced to be *S*, rather than the enantiomeric molecule with the *R* configuration.

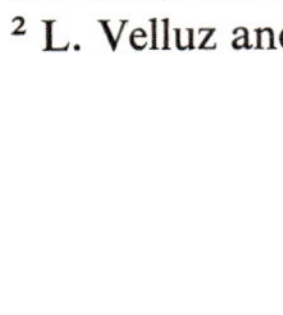


[2] L. Velluz and M. Legrand, *Angew. Chem. Int. Ed.*, *4*, 838 (1965).

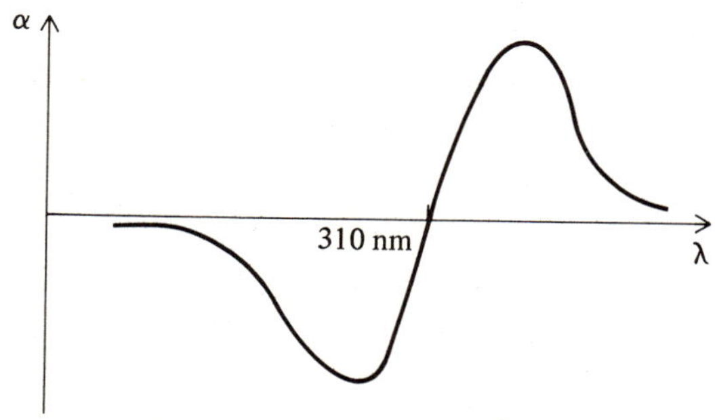


FIGURE 7.8
*Optical rotatory dispersion of (+)-3-methylcyclohexanone.*

Let us consider finally another implication of the octant rule in the effect of α substitution in cyclohexanone molecules. An important aspect of their geometry is that an equatorial substituent, α to the carbonyl, is very nearly in the plane of the carbonyl whereas an axial substituent is almost perpendicular to it. Therefore, it is to be expected that an α halogen atom in the equatorial position, that is, in the nodal plane of the carbonyl, will exert no (or very little) perturbation on the position of the absorption maximum $\lambda_{max}$, as measured either from the uv, ord, or cd spectrum, and will not alter the sign of the Cotton effect. Conversely, an axial halogen will markedly change the Cotton effect according to the octant in which it is placed. The following examples illustrate these predictions:

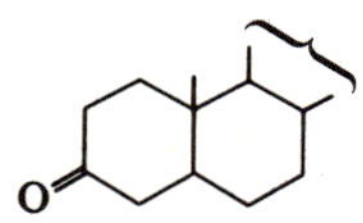


3-Cholestanone
$\lambda_{max}$ = 286 nm
$\alpha_D^{MeOH} = +2700°$

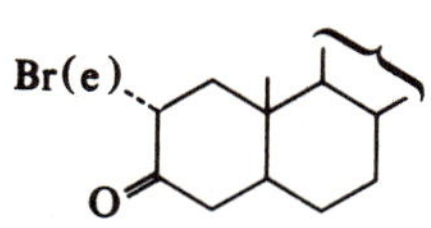


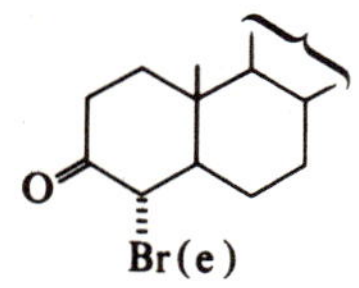


$\alpha_D^{MeOH} = +2500°$ $\qquad$ $\alpha_D^{MeOH} = +800°$

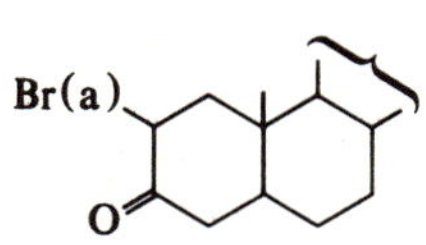


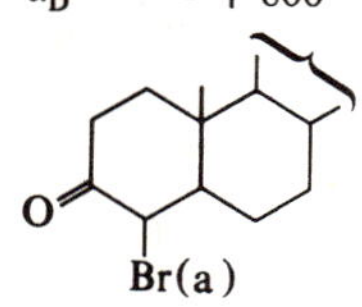


$\alpha_D^{MeOH} = +8000°$ $\qquad$ $\alpha_D^{MeOH} = -7500°$

The fit is not perfect; these bromoketones do not have identical conformations because of the constraints within the steroidal skeleton.

It must be noted, however, that in the same way as one speaks routinely of $n \rightarrow \pi^*$ and $\pi \rightarrow \pi^*$ transitions for the carbonyl group even though the molecular orbital picture is not that secure, one should beware that "octants," which have become an integral part of the chemist's language and thinking, are figments of our imagination which happen to fit most of the available data. One may speak rigorously only of quadrants for the simple formaldehyde molecule, as determined by its high symmetry. The occurrence of regional rules for optical activity, and their formulation in terms of octants, are bold extrapolations from that starting point. In fact, an interesting anomaly to the octant rule has appeared recently.[3] The octant rule predicts a positive Cotton effect for axial substituents in the 3 position of cyclohexanone (1).

[3] G. Snatzke and G. Eckhardt, *Tetrahedron*, *24*, 4543 (1968); ibid., *26*, 1143 (1970).

The same prediction holds true for the conformationally rigid adamantanones. Yet an axial Cl, Br, or I produces a net negative Cotton effect in adamantanone (2), *contrary to the predictions of the octant rule*. In contrast to the behavior of the axial substituents, equatorial Cl, Br, I, and $CH_3$ all behave normally and obey the octant rule in producing a positive Cotton effect in 2. Furthermore, an axial F produces a positive Cotton effect in 2, whereas an equatorial F produces a negative effect. Although the peculiarities just described may have special explanations, they nevertheless demonstrate the lack of complete general validity for the octant rule.

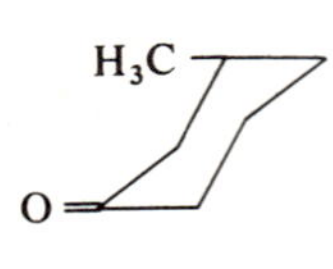


1

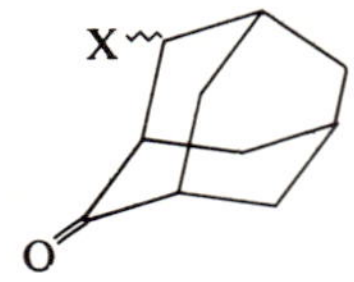


2

X = Cl, Br, I, F

## 7.4 *PROBLEMS*

SAMPLE PROBLEM

It is found experimentally that the Cotton effects for 2- and 3-cholestanone have the same positive sign, but that the former molecule displays a considerably larger effect than the latter. Explain.

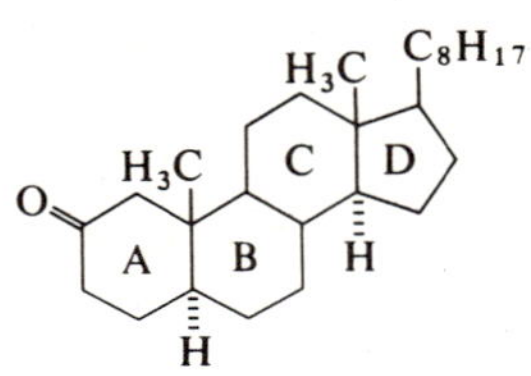


2-Cholestanone

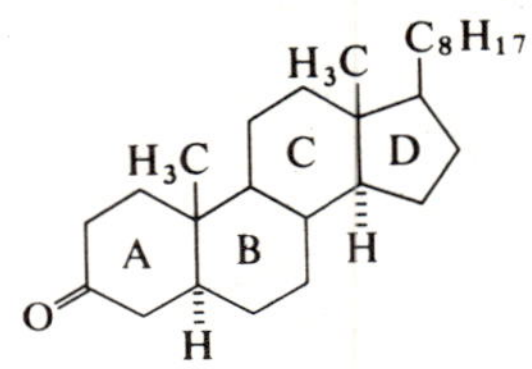


3-Cholestanone

ANSWER

Let us construct the octant diagrams for the two isomers:

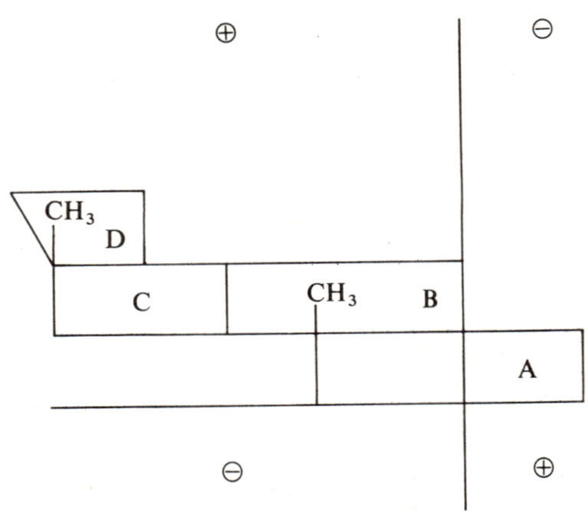


Octant diagram for 2-cholestanone; the positive contributions are overwhelming.

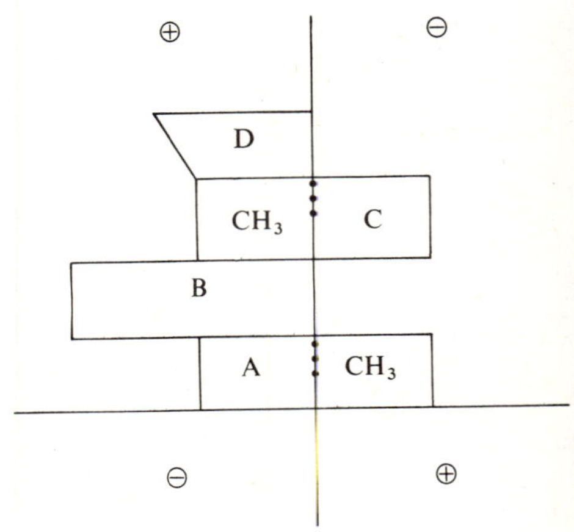


Octant diagram for 3-cholestanone; the positive contributions are less dominant.

1. Mesityl oxide, $(CH_3)_2C{=}CHCOCH_3$, has two absorption maxima in the uv region, both of which are strongly solvent dependent. Explain this solvent effect on the spectrum of mesityl oxide.

| Solvent | $\lambda_{max}$ (nm) | $\epsilon$ | $\lambda_{max}$ (nm) | $\epsilon$ |
|---|---|---|---|---|
| Hexane | 229.5 | 12,600 | 327 | 97.5 |
| Ether | 230 | 12,600 | 326 | 96 |
| Ethanol | 237 | 12,600 | 315 | 78 |
| Methanol | 238 | 10,700 | 312 | 74 |
| Water | 244.5 | 10,000 | 305 | 60 |

2. The tetracyclic diterpene (1) has a negative Cotton effect, but its isomer (2) has a positive Cotton effect. Explain this observation.

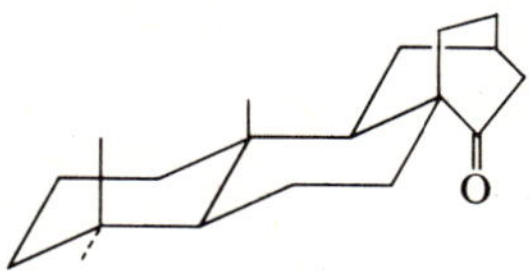


1

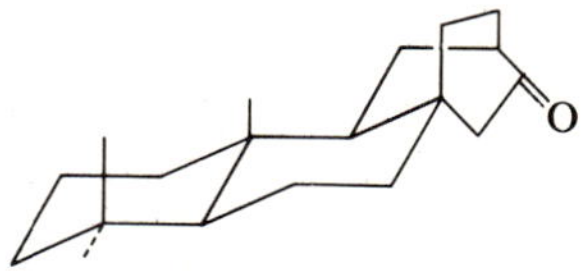


2

3. Cholestan-3-one (1) displays a positive Cotton effect, whereas its 5$\beta$ epimer coprostan-3-one (2) has a negative Cotton effect. Explain.

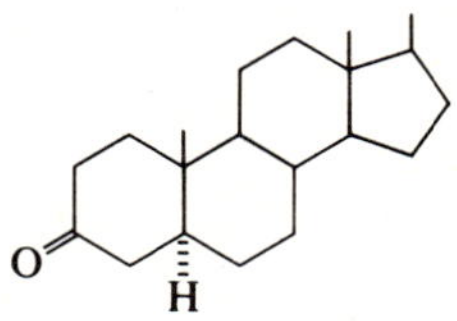


1

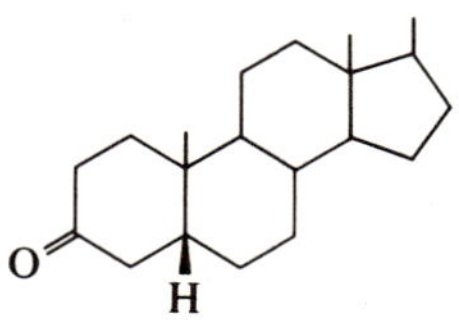


2

4. Senegenic acid (1) can be transformed into a methyl keto ester with the structure shown (2):

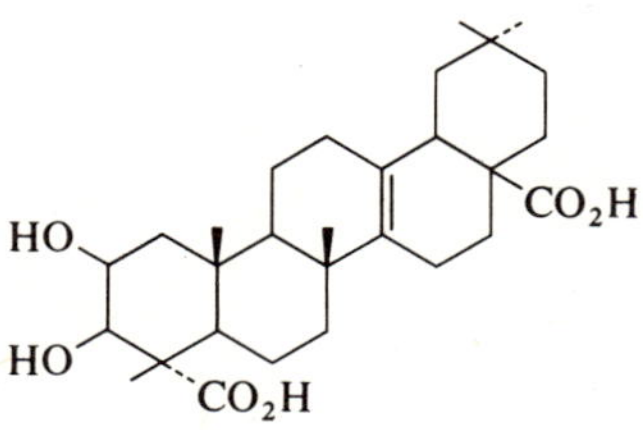


1

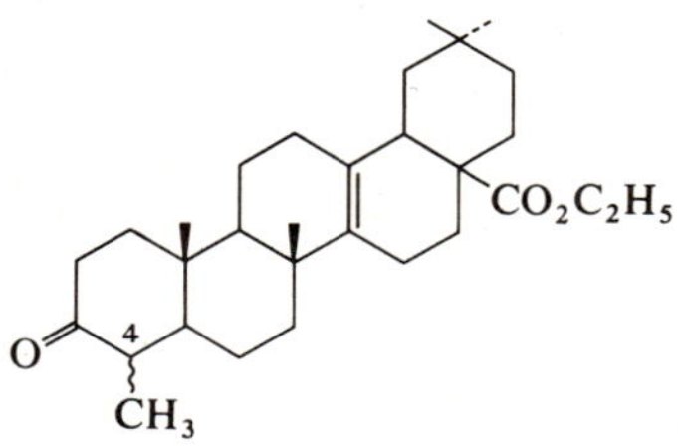


2

It displays a positive Cotton effect. What is the stereochemistry at $C_4$?

5. The four-membered ring of cyclobutanones is puckered. Draw the octant rule for cyclobutanones. As an example of an application, what is the configuration at $C_3$ of (−)-4,4-dimethyl-3-ethyl-cyclobutanone, which has a negative CD curve, knowing that (+)-2,2,3-trimethylcyclobutanone pertains to the 3-*S* series and displays a positive Cotton effect?
6. What is the sign of the Cotton effect for 8$\beta$,9$\beta$-dimethyl-*trans*-2-decalone?

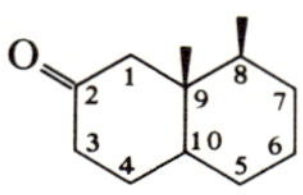


7. A very intense and negative Cotton effect is shown by *trans*-tetrahydroeremophilone.

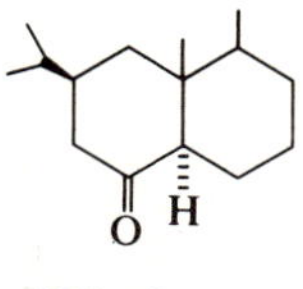


Why?

## 7.5 *SUGGESTED READINGS*

G. Snatzke, "Circular Dichroism and Optical Rotatory Dispersion—Principles and Application to the Investigation of the Stereochemistry of Natural Products," *Angew. Chem., Int. Ed., 7*, 14 (1968).

K. Mislow, *Introduction to Stereochemistry*, W. A. Benjamin, Inc., New York, 1965.

H. H. Jaffé, and M. Orchin, *Symmetry in Chemistry*, John Wiley and Sons, Inc., New York, 1965.

M. Orchin and H. H. Jaffé, *The Importance of Antibonding Orbitals*, Houghton Mifflin Company, Boston, 1967.

C. W. Deutsche, D. A. Lichtner, R. W. Woody, and A. Moscowitz, "Optical Activity," *Ann. Rev. Phys. Chem., 20*, 407 (1969).

R. R. Gould and R. Hoffmann, "The Direct Calculation of Optical Rotatory Strengths," *J. Amer. Chem. Soc., 92*, 1813 (1970).

W. D. Chandler and L. Goodman, "Allowed and Forbidden Character in $\pi^* \leftarrow n$ Spectra of Cycloalkanones," *J. Mol. Spectry., 35*, 232 (1970).

A. I. Scott and A. D. Wrixon, "A Symmetry Rule for Chiral Olefins," *Tetrahedron, 26*, 3695 (1970).

# CHAPTER 8 ANALYSIS OF MAGNETIC RESONANCE SPECTRA

Analysis of magnetic resonance spectra is a phrase so misused that its accepted meaning now runs contrary to its original meaning. When confronted with a complex magnetic resonance spectrum, the first task of the organic chemist is one of recognition: to establish to which type this spectrum, or part of a spectrum, belongs. Once this is done, he will know whether it is possible to break down the spectrum into simpler subspectra. This is spectral analysis, a phenomenological activity whose starting point is the experimental spectrum. On the contrary, what we shall call spectral synthesis is the theoretical calculation, based upon quantum mechanics, of the interaction of a number of various magnetic moments and the prediction of the corresponding line frequencies and intensities. The result of spectral synthesis is then matched with that of spectral analysis and the parameters varied until the mutual fit is satisfactory. Usually spectral analysis as just defined has not been well treated whereas spectral synthesis has been well covered, perhaps because it constitutes one of the most elegant applications of quantum mechanics to chemistry. In ordcr to compensate for this inequality in the coverage of spectral analysis and spectral synthesis we provide here a thorough discussion of spectral analysis *sensu stricto*. Our presentation of spectral synthesis will be limited to its principles since a considerable body of generally excellent literature is available on this topic.[1]

The parameters to be extracted from magnetic resonance spectra, the chemical shifts and coupling constants, are extremely important because of the information they shed on spin or charge densities and on molecular geometries. The procedures for extracting this useful information from the experimental spectra will now be outlined.

These procedures vary substantially according to the type of problem and the desired accuracy of the parameters. Most often, an organic chemist confronted with a complicated magnetic resonance spectrum does not know the structure of the corresponding molecule in terms of constitution, configurations, or conformations. Actually, he may want to analyze this spectrum in order to gain information about the structure! Therefore, it would be absurd at this stage to feed such crude data into a large digital computer and to hope that it would disgorge all the accurate chemical shifts and coupling constants without some additional input. The additional input is the structure of the problem. More precisely, it turns out that the calculations are greatly simplified and often become feasible by hand when information about the composition of the systems of magnetic moments is inferred from the spectrum.

The first mental step in any spectral analysis is obviously a hard look at the spectrum. In the case of magnetic resonance spectra, it is a rapidly acquired training in recognizing and distinguishing individual groups of lines present in a spectrum. Integral curves come as a powerful help in this recognition; magnetic resonance absorptions are directly proportional to the number of resonating nuclei, and it must be checked whether a well-defined group of lines indeed contains an integer number of nuclei. A concrete example will come in handy at this stage. Consider the pmr spectrum of pyridine, reproduced in Figure 8.1 with its integral. We find three groups corresponding to 2, 2, and 1 protons, one of which is well separated from the other

[1] R. A. Hoffman, *Pure Appl. Chem.*, *11*, 543 (1965).

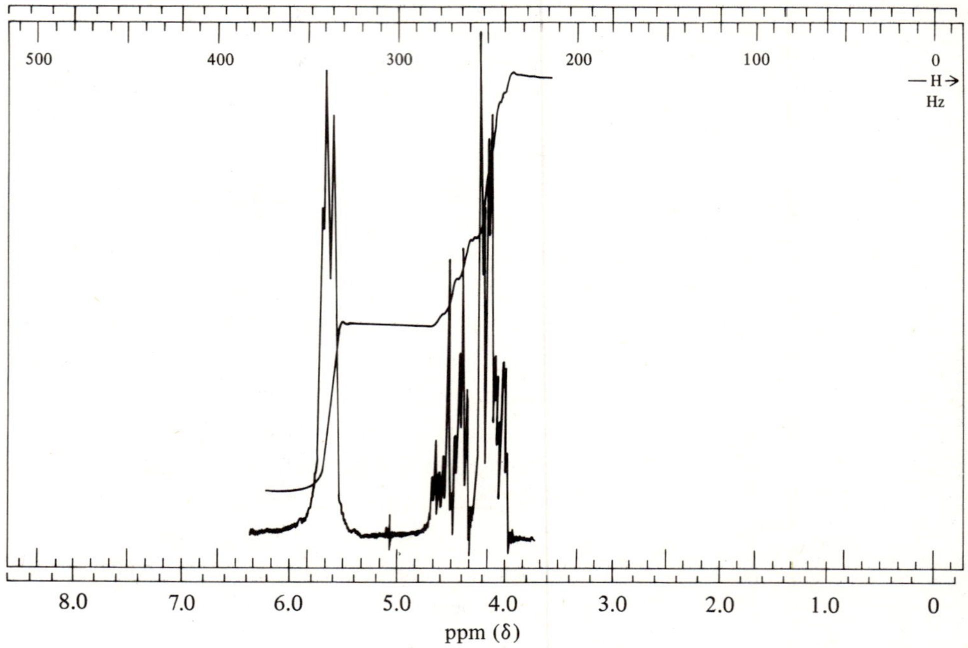


*FIGURE 8.1*
*Spectrum of pyridine.*

two. Therefore, *even if we did not know that the substance is pyridine* we would recognize that the structure of the problem is of the formal type (AB)(CD)(E). Let us now indicate how the appearance of the spectrum depends upon the interactions of the various spins belonging to one class or to another.

## 8.1 *Classification of Magnetic Resonance Spectra*

Even for relatively simple problems, such as the interaction of four different nuclear moments, the number of resulting lines can be very large and the appearance of the spectrum rather complex. The problem is simplified by recognition of the type of spectrum, and the symmetry elements present.

One starts by distinguishing between "strongly coupled" and "weakly coupled" cases. Let us consider, by way of an example, a three-spin system, which we shall temporarily label with Greek letters:

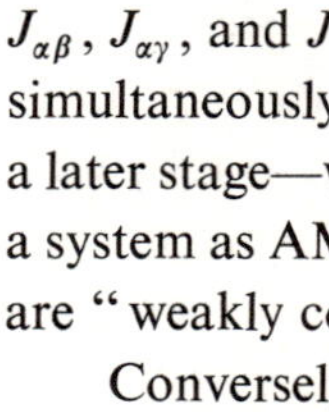


The six parameters that enter the expressions for the magnetic interactions among $\alpha$, $\beta$, and $\gamma$ are the three chemical shifts $\nu_\alpha$, $\nu_\beta$, and $\nu_\gamma$, and the three coupling constants $J_{\alpha\beta}$, $J_{\alpha\gamma}$, and $J_{\beta\gamma}$. If the quantities $(\nu_\alpha - \nu_\beta)/J_{\alpha\beta}$, $(\nu_\alpha - \nu_\gamma)/J_{\alpha\gamma}$, and $(\nu_\beta - \nu_\gamma)/J_{\beta\gamma}$ are *all* simultaneously greater than a factor 10, then some simplification—to be examined at a later stage—will occur in their spectrum. By way of definition we shall designate such a system as AMX, using letters far apart in the alphabet to remind us that these nuclei are "weakly coupled" (i.e., that the above ratios are large—much larger than unity).

Conversely, if all of the above ratios were of the order of unity we would speak of a "strongly coupled" system, and we would call it ABC. These are nothing but definitions, ABC and AMX being just shorthand for the above ratios being either small or large.

Such definitions might appear to prejudge the result of the analysis: how is it possible to label an experimental spectrum for a three-spin case as AMX *before* the chemical shifts and coupling constants have been calculated? This is precisely why recognizing spectral patterns is important, since weakly coupled spectra are synonymous with well-individualized groups of lines corresponding to a given nucleus (or nuclei) with little or no "mixing."

A tentative scheme for classifying the spectrum of pyridine would thus be (AB) (XYZ), standing for a five-spin system, all of the same species (protons), two of which appear to be weakly coupled to the other three, which are themselves more strongly coupled.

Note that the ratios $(\nu_i - \nu_j)/J_{ij}$ can be manipulated somewhat. By changing the solvent, or by changing the spectrometer frequency, one can increase this ratio by perhaps up to a factor of two.

By this scheme the seven-spin system of the benzene radical anion is written as ABCDEF-X, since we possess the advance knowledge that the electron spin X necessarily resonates at a frequency very different from that of the protons A through F. Likewise, methyl fluoride will correspond to an ABC-X system, and in general different nuclear species having different nuclear moments will thus be separated by hyphens in this nomenclature scheme.

*8.1.1 Nuclei Equivalent by Symmetry*

In the nomenclature developed so far we have not made use of our advance knowledge that the three protons in the methyl group of methyl fluoride *must* have the same chemical shift since they are equivalent by symmetry. Likewise with the six protons of the benzene radical anion. In these two cases the nuclei that are equivalent will have not only the same chemical shift and therefore be termed *isochronous*, but will *also* have identical relationships towards the lone X spin. Therefore, in addition to being isochronous, they are also *isogamous* since they display the same coupling constants towards the other sets of isochronous nuclei in the molecule, as graphically displayed by the equal edges in the pyramids below:

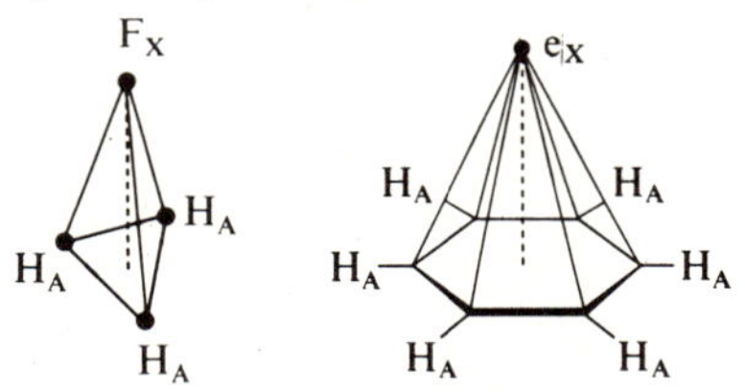


(Note: These formal representations of the magnetic interactions have nothing to do with the actual molecular geometry.)

Hence, it will be simpler to refer to the methyl fluoride four-spin system as $A_3$–X, since we know of the existence of only one coupling constant $J_{AX}$. Similarly, the spectrum of the benzene radical anion is of the $A_6$–X type, since there is only one constant, called $a_H$, due to the spin interaction of the electron.

Let us return to the case of pyridine. Obviously, we cannot label it as an $A_2X_2Y$ system, since A spins are a priori known to couple with two distinct sets of nuclei, X and Y, with different coupling constants. Furthermore, even within the X set, the two nuclei have different relationships with either of the A nuclei, for instance of the *ortho* and *para* type:

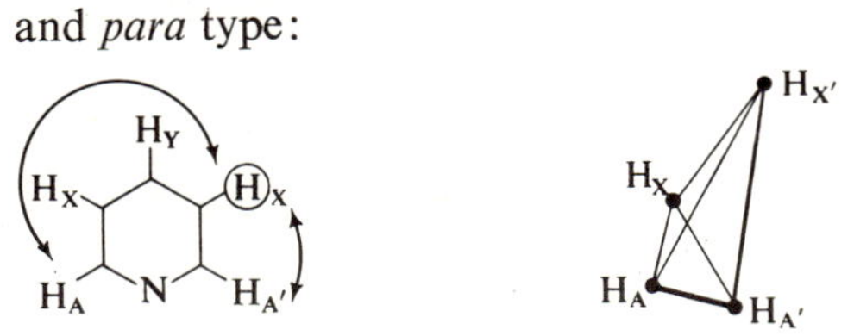

*TABLE 8.1*

*Table of Equivalencies*

| | 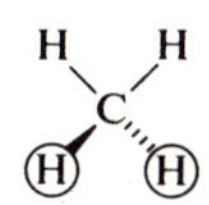 | 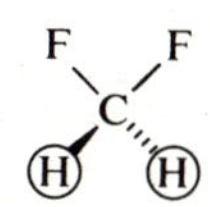 | 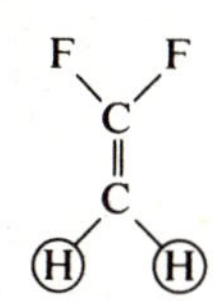 | 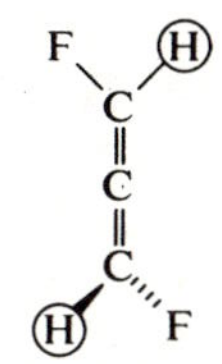 | 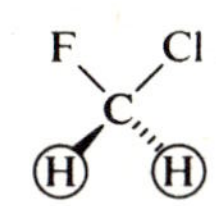 | 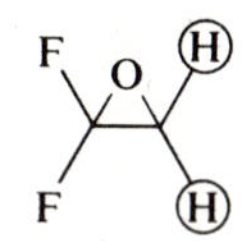 | 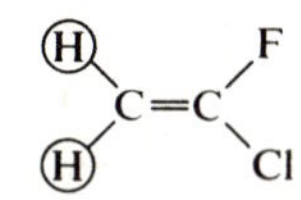 |
|---|---|---|---|---|---|---|---|
| Symmetry element for the nuclei considered | $C_2$ | $C_2$ | $C_2$ | $C_2$ | $\sigma$ | $\sigma$ | none |
| Protons | homotopic | homotopic | homotopic | homotopic | enantiotopic | enantiotopic | diastereotopic |
| *Chemical shifts* | | | | | | | |
| Achiral solvent | isochronous | isochronous | isochronous | isochronous | isochronous | isochronous | anisochronous |
| Chiral solvent | isochronous | isochronous | isochronous | isochronous | anisochronous | anisochronous | anisochronous |
| *Coupling constants* | | | | | | | |
| Achiral solvent | isogamous | isogamous | anisogamous | anisogamous | isogamous | anisogamous | anisogamous |
| Chiral solvent | isogamous | isogamous | anisogamous | anisogamous | anisogamous | anisogamous | anisogamous |
| *nmr classification* | | | | | | | |
| Achiral solvent | $A_4$ | $A_2$–$X_2$ | AA′–XX′ | AA′–XX′ | $A_2$–X | AA′–XX′ | AB–X |
| Chiral solvent | $A_4$ | $A_2$–$X_2$ | AA′–XX′ | AA′–XX′ | AB–X | AB–XY | AB–X |
| Maximum number of lines in the proton nmr | 1 | 3 | 10 | 10 | 2 (ach.)<br>8 (ch.) | 10 (ach.)<br>16 (ch.) | 8 (ach.)<br>8 (ch.) |

Nomenclature recognizes such an important difference by labeling pyridine as an AA′XX′Y case; that is, the A's and the X's are only isochronous *without* being isogamous, since $J_{AX}$ can be different from $J_{A'X}$. We are using the same letters for nuclei having the same chemical shift (isochronous), and superscripts whenever they display more than one kind of coupling constant towards the other nuclei in the molecule.

### *8.1.2 Distinction of Homotopic, Enantiotopic, and Diastereotopic Nuclei*

Whenever a molecule contains a pair of isochronous nuclei the equality of the chemical shifts *must* result from the existence of a symmetry operation interconverting the nuclei. This symmetry operation can a priori be performed via a symmetry axis of order $n$, $C_n$, or via a rotation-reflection axis of order $n$,$S_n$—equivalent to symmetry with respect to an axis $C_n$ followed by reflection through a $\sigma$ plane perpendicular to $C_n$. In the former case, the two nuclei are termed *homotopic*, their relationship corresponds to the identity sign, and the two chemical shifts are equal in a chiral as well as in a nonchiral environment. In the latter case, the two nuclei which are termed *enantiotopic* are interchanged by reflection in a plane of symmetry. Their chemical shifts will be equal only in an achiral medium, but will be different in a chiral solvent. It is thus clear that asymmetric molecules, which do not have any such symmetry elements, cannot contain either homotopic or enantiotopic isochronous nuclei, but only diastereotopic and hence anisochronous nuclei. Chiral molecules, since they cannot have a plane of symmetry, may have homotopic nuclei but not enantiotopic nuclei. Finally it is possible to recognize whether a pair of nuclei, for instance protons, belongs to either of these three types by performing a test of isotopic substitution. Replacement of one of the protons by a deuteron would generate diastereomers if the protons are diastereotopic, enantiomers if they are enantiotopic, and the identical molecule if they are homotopic.

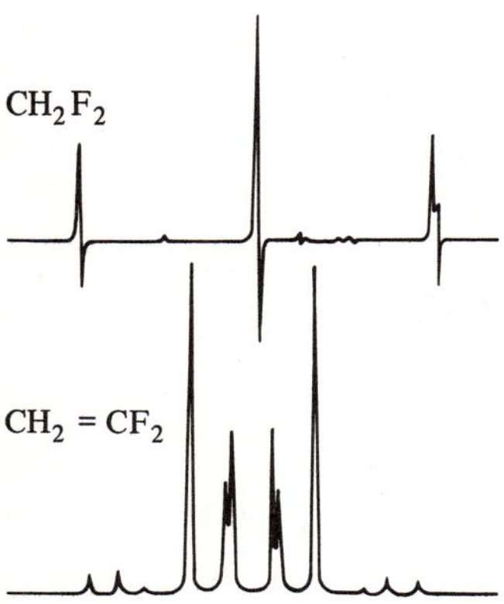


*FIGURE 8.2*
*Proton resonance spectra of $CH_2F_2$ and $CH_2{=}CF_2$ at 60 MHz. The $^{19}F$ resonance would be identical.*

Table 8.1 lists a number of examples (together with some illustrative spectra shown in Figure 8.2) and emphasizes that enantiotopic or homotopic nuclei can be subdivided into two more classes according to whether they are isogamous or anisogamous. *Note that anisochronous nuclei are necessarily anisogamous.* Further examples are presented in the problems at the end of this chapter.

## 8.2 *Spin-Spin Decoupling*

Another approach to spectral analysis is afforded by spin-spin decoupling. Application of a second intense RF field at the resonance frequency of a given nucleus will collapse its spin-spin interactions with other nuclei in the same molecule. The simplest example is that of a two-spin AX system. In the normal magnetic resonance experiment, the molecule is submitted to a permanent magnetic field $H_0$ which orients both nuclear spins along its $Z$ axis, and to a RF field $H_1$ which serves to induce transitions between the quantized energy states. If, in addition, another RF field $H_2$ is applied at the resonance frequency for nucleus X,

$$\nu_2 \equiv \nu_x = \frac{\gamma_x H_2}{2\pi}$$

with a strong amplitude such that $\gamma\, H_2/2\pi \gg J_{AX}$, then nucleus X is forced to be quantized along an $X$ axis perpendicular to the direction of the magnetization of the A nucleus (which also happens to be the direction for the original magnetization of X). Decoupling is achieved and the A nuclei will be insensitive to the orientation of the X nuclei, whether parallel or antiparallel to the field, since the A and the X nuclear moments are now quantized along orthogonal axes. If one observes the appearance of the A resonance while increasing gradually the magnitude of $H_2$, it is progressively

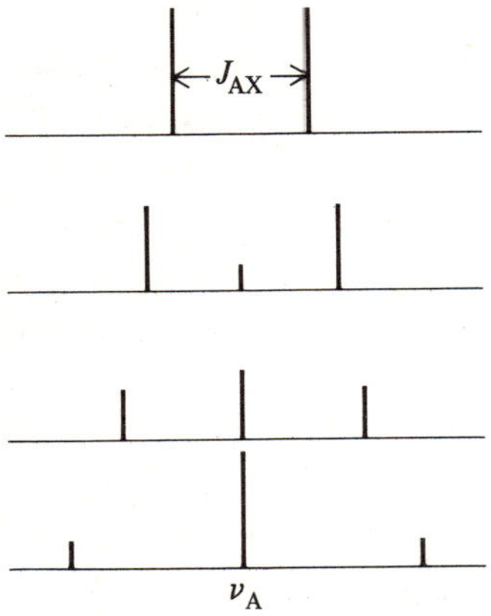


*FIGURE 8.3*
*Spectra of an AX system as a function of irradiating field $H_2$ increasing from top to bottom.*

"decoupled" from nucleus X (Figure 8.3). In practice, this decoupling can be accomplished if and only if $|\nu_A - \nu_X| > 5|J_{AX}|$, that is, in cases of weak coupling. The experiment is performed either by *field sweep*, in which $\nu_1$ and $\nu_2$ remain constant and $H_0$ is varied, or by *frequency sweep*, in which $\nu_2$ and $H_0$ are kept constant and $\nu_1$ is varied. Although this latter mode requires an excellent stability of the field-frequency ratio $\gamma H_0/2\pi\nu_2$, it has the advantage that the perturbation of the spin system remains constant in time, since the irradiating field $H_2$ is maintained at a fixed spectral position. It has the further advantage that *all* the nuclei coupled to the irradiated nucleus X are simultaneously decoupled from it. Such double-resonance techniques are used to measure chemical shifts. For example, the chemical shift of proton X, whose resonance is hidden beneath a complex multiplet, can be determined by observing in a less-crowded region of the spectrum the signals for proton A, which is coupled to X, while varying the frequency until the hidden protons are decoupled from A. The relationship

$$(\nu_1 - \nu_2) = (\nu_A - \nu_X) - \frac{\gamma H_2^2}{2(\nu_A - \nu_X)}$$

is then used to determine $\nu_X$, knowing $\nu_A$ and the experimental parameters $\nu_1$ and $\nu_2$. Usually the second term $\gamma H_2^2/2(\nu_A - \nu_X)$ is small and can be omitted. Likewise heteronuclear decoupling, referring to irradiation at the frequency of a different nuclear species, can be used to measure chemical shifts. Historically, this procedure was applied first to the determination of the chemical shifts for carbon-13 nuclei coupled to protons, since at the time it was more accurate than the direct measurement of the $^{13}C$ chemical shift. Another very important application is to the determination of the relative signs of coupling constants. For example, the unsaturated lactone shown below has an $AKX_2$ proton system whose appearance is depicted in Figure 8.4.

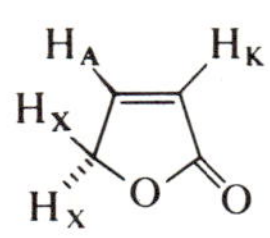


This spectrum is the superposition of two $ax_2$ subspectra corresponding to either Figure 8.5a or 8.5b. If $J_{AX}$ and $J_{KX}$ have the same sign, both low-field parts will correspond to the $\alpha$ spin state of K, and both high-field parts to the $\beta$ spin state of K and conversely (Figure 8.5b) if the product $J_{AX}J_{KX}$ is negative.

Let us assume that one can irradiate selectively at the central line of the low-field triplet without affecting the high-field triplet. Then the result will be that the corresponding doublet of the *same* sub-spectrum will collapse to a singlet. Experimentally,

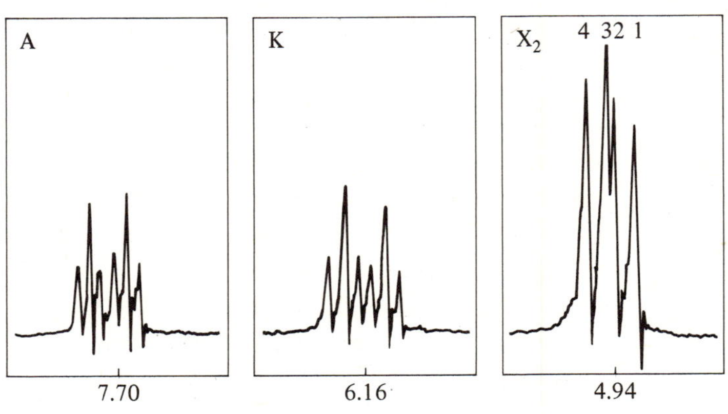


*FIGURE 8.4*
*$AKX_2$ spectrum of the lactone under discussion.*

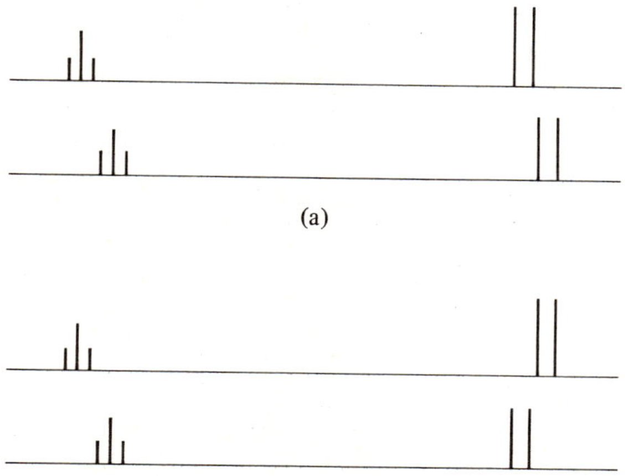

(a)

(b)

*FIGURE 8.5*
*(a) $ax_2$ subspectra if $J_{AX}$ and $J_{KX}$ are of the same sign.*
*(b) $ax_2$ subspectra if $J_{AX}$ and $J_{KX}$ have opposite signs.*

irradiation of the central line of the low-field triplet collapses the high-field doublet, and vice versa. This demonstrates unambiguously that the arrangement of the two subspectra is as shown in Figure 8.5b and $J_{AX}$ and $J_{KX}$ have opposite signs. In a similar way, $J_{AX}$ is found to have the same sign as $J_{AK}$ so that the final values are[2]

$J_{AK} = +5.8$ Hz

$J_{AX} = +1.7$ Hz

$J_{KX} = -2.1$ Hz

## 8.3 *PROBLEMS*

1. Label the spectra of the following compounds: $CH_2F_2$, $H_2C{=}CF_2$, $H_2C{=}C{=}CF_2$, $HCF{=}CHF$ (*cis* or *trans*), $HFC{=}C{=}CHF$.
2. Consider a single-resonance AX-type spectrum where A and X are coupled to each other ($J_{AX} \neq 0$). Is it possible to determine the sign of the coupling from the spectrum alone?
3. In pyrazolidine (1) the ring protons are observed to give an AA′XX′ pattern, and only one N-methyl and one C-methyl resonance are observed. What can one conclude from this information regarding the stereochemistry of this compound?

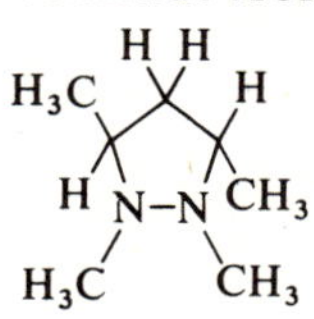


1

4. Is it a fair statement to say that an isopropyl group $(CH_3)_2CH$ will show two methyl doublets in the proton magnetic resonance spectrum *only* if there is restricted rotation around the bond connecting it to the rest of the molecule?
5. Given a molecule of the general type $X^*CH_2X^*$ in which $X^*$ is a chiral group having no plane or center of symmetry, what is the nature (homotopic, enantiotopic, diastereotopic) of the methylene protons?
6. The proton magnetic resonance spectrum of the diphenylmethanide carbanion is temperature dependent: whereas the two *para* protons appear as one triplet at all temperatures, the four *ortho* protons appear as one signal at −20°C, but as two doublets at −60°C. Explain this observation.

[2] R. Freeman, *Mol. Phys.*, 5, 499 (1962).

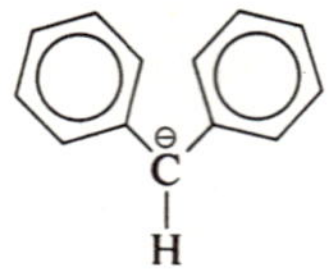

7. The dimethylcyclopropylcarbinyl carbonium ion displays two methyl singlet resonances separated by 0.54 ppm at ca. 2.7 and 3.2 ppm. Explain this observation.

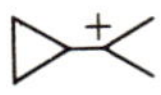

## 8.4 *SUGGESTED READINGS*

K. Mislow and M. Raban, "Stereoisomeric Relationships of Groups in Molecules," in *Topics in Stereochemistry*, N. L. Allinger and E. L. Eliel, eds., Wiley-Interscience Publishers, Inc., New York, 1967, vol. 1.

O. Sciacovelli, W. von Philipsborn, C. Amith, and D. Ginsburg, "An Application of Homonuclear INDOR Spectroscopy: The Structure of a Dimer of 11, 13-Dioxo-12-Methyl-12-Aza[4.4.3] Propellane," *Tetrahedron*, *26*, 4589 (1970).

# CHAPTER 9 SYNTHESIS OF MAGNETIC RESONANCE SPECTRA

As indicated in Chapter 8, synthesis of magnetic resonance spectra consists of the calculation of theoretical spectra from quantum mechanical principles. The simplest example is afforded by the interaction of two nuclear spins. As discussed in the previous section, this interaction can be of the AX, the AB, or the $A_2$ type. Let us now examine these various types of spectra and their interrelationships.

## 9.1 *The AX, AB, and $A_2$ Spin Systems*

The first step in any such calculation is to write the spin-product functions appropriate to the problem at hand. For the AX case, they are

$\alpha\alpha \quad \alpha\beta \quad \beta\alpha \quad \beta\beta$

Hence there are four energy states which are ordered as in Figure 9.1. Let us perturb this energy diagram by introducing the mutual coupling of these A and X nuclear spins: Each energy level is either lowered or raised by $\frac{1}{4}J$ according to the sign of the product $m_I(\mathrm{A})m_I(\mathrm{X})$. Assuming that the coupling constant $J$ is positive, the effect of coupling can be represented as in Figure 9.2. According to the selection rules $\Delta m_{\mathrm{A}} = \pm 1$, $\Delta m_{\mathrm{X}} = 0$ and $\Delta m_{\mathrm{A}} = 0$, $\Delta m_{\mathrm{X}} = \pm 1$ there will be two A transitions and two X transitions; each pair is separated by $J$ and centered about the frequency $\nu_{\mathrm{A}}$ or $\nu_{\mathrm{X}}$. The resulting spectrum is shown in Figure 9.3. Let us now go to the other extreme case, the $A_2$ system in which the two nuclei are made equivalent by symmetry. The symmetry operation interchanging these two nuclei leaves the product functions $\alpha\alpha$ and $\beta\beta$ unaffected. However, the two functions $\alpha\beta$ and $\beta\alpha$ are no longer appropriate for this system since the two A particles are now indistinguishable by symmetry: one cannot know whether it is particle 1 or particle 2 which possesses the $\alpha$ spin. The spin-product functions must reflect this indistinguishability. The difficulty is removed by generating new product functions to replace $\alpha\beta$ and $\beta\alpha$:

$$\frac{1}{\sqrt{2}}(\alpha\beta + \beta\alpha) \qquad \text{(s)}$$

$$\frac{1}{\sqrt{2}}(\alpha\beta - \beta\alpha) \qquad \text{(a)}$$

where $1/\sqrt{2}$ is a normalization coefficient. The first is symmetric (s) and the second is antisymmetric (a) with respect to the symmetry operation exchanging $\alpha$ and $\beta$. Hence

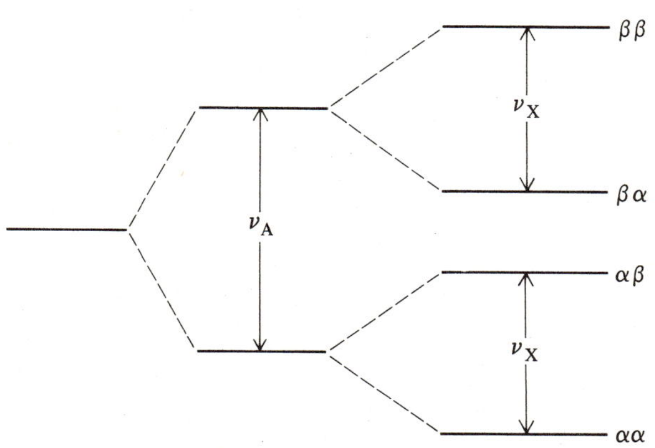


*FIGURE 9.1*
*Energy states of an AX system.*

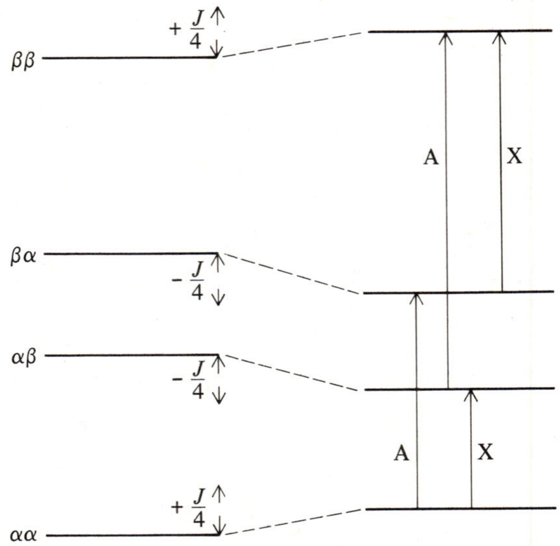


*FIGURE 9.2*
*Energy states of an AX system perturbed by mutual coupling between the A and X nuclei.*

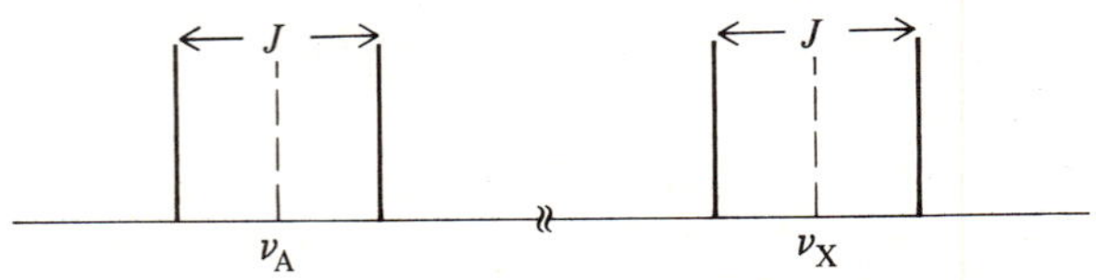


*FIGURE 9.3*
*An AX spectrum where A and X are mutually coupled.*

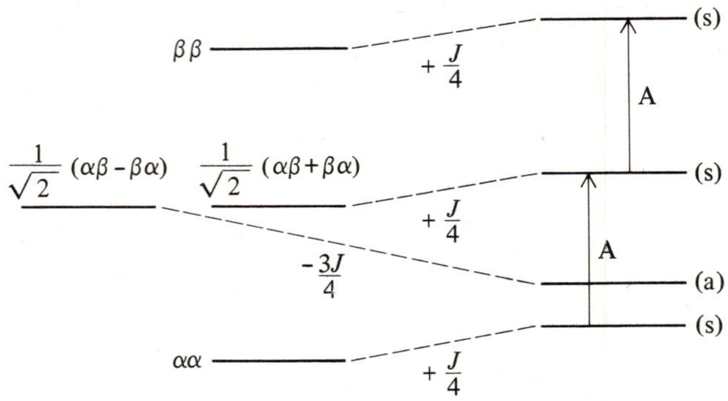


*FIGURE 9.4*
*Energy diagram of an A system.*

the energy diagram is now constructed as in Figure 9.4, where s and a denote the symmetric or antisymmetric character of these spin functions. Again coupling raises or lowers each energy level. However, the $m_I = I_z$ projections are "good" spin numbers only in weakly coupled cases such as the AX system. In strongly coupled systems, such as the AB or $A_2$ cases, it is mandatory to use the spin operator $I$ itself and therefore to consider also the $I_x$ and $I_y$, not only the $I_z$, components. Each energy level is raised or lowered according to the sign of the product $JI_1I_2$. It can be shown[1] that the incremental changes are

$$\left.\begin{aligned} \alpha\alpha &: +\frac{J}{4} \\ \frac{1}{\sqrt{2}}(\alpha\beta+\beta\alpha) &: +\frac{J}{4} \\ \beta\beta &: +\frac{J}{4} \end{aligned}\right\} \quad \text{for the symmetric functions}$$

[1] J. D. Roberts, *An Introduction to the Analysis of Spin-Spin Splitting in High-Resolution Nuclear Magnetic Resonance Spectra*, W. A. Benjamin, Inc., New York, 1961, Chap. 2.

and correspondingly $-(3/4)J$ for the antisymmetric $(1/\sqrt{2})$ $(\alpha\beta - \beta\alpha)$ function. The selection rules for such a system state that allowed transitions will have $\Delta m = \pm 1$ *and that they will connect states having the same symmetry*. Hence there will be two allowed transitions:

$$\alpha\alpha \rightarrow \frac{1}{\sqrt{2}}(\alpha\beta + \beta\alpha)$$

$$\frac{1}{\sqrt{2}}(\alpha\beta + \beta\alpha) \rightarrow \beta\beta$$

Since both have the same energy the spectrum will consist of a single line. *This finding is general: Any isolated spin system of type $A_p$ will have a single resonance line.*

*9.1.1 The Circumvention of Magnetic Equivalence*

Whenever a set of nuclei are both isochronous and isogamous, they are said to be magnetically equivalent and the spectrum is not influenced by their mutual coupling constant(s). For instance, in the spectrum of the benzene radical anion, an $A_6$–X system, the coupling constants within the aromatic A protons (*ortho*, *meta*, and *para*) do not manifest themselves in the spectrum. This result is dictated by symmetry considerations, and is merely a consequence of those same coupling constants within the A nuclei not influencing the $A_6$ spectrum of benzene itself, for which the pmr spectrum consists of a single line. Likewise, in methyl fluoride it is impossible to deduce from the spectrum alone the value of $^2J_{HH}$, or in ethyl chloride to find the values of the coupling constants between any two protons of the methyl group.

All that has been described in this chapter so far applies to indirect spin-spin coupling constants, to the exclusion of direct coupling constants. If, for instance, the benzene molecule is in an ordered solution such as a nematic solvent then the aromatic protons are directly coupled to one another by $D_{12}(ortho) \neq D_{13}(meta) \neq D_{14}(para)$, which effectively breaks their magnetic equivalence and also allows the $J_{ij}$ indirect coupling constants to influence the spectrum, which therefore becomes very complicated.

It is sometimes possible to obtain the desired indirect coupling constant by examining the spectrum of an isotopically substituted species: the spectrum of $H_2CF_2$ does not yield the interproton coupling constant, but the spectrum of $HDCF_2$ is an A–K–$X_2$ system to an excellent approximation. By isotopic substitution one has transformed a symmetrical into an unsymmetrical tetrahedron:

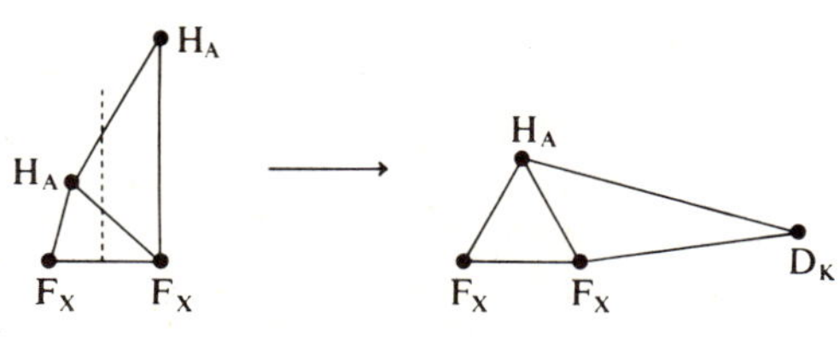


It is an easy matter to extract from the A–K–$X_2$ spectrum the hydrogen-deuterium coupling constant $J_{HD}$, which is simply related to the desired $J_{HH}$ by

$$\frac{J_{HH}}{J_{HD}} = \frac{\mu_H}{\mu_D} \sim 6.52 \qquad \text{(ratio of the magnetic moments)}$$

*9.1.2 The AB Spectrum*

The appearance of the spectrum for the AB case is intermediate between that for the AX and the $A_2$ case as shown in Figure 9.5. The spectrum continues to be symmetrical about its center of gravity at $\frac{1}{2}(\nu_A + \nu_B)$. There, however, is a concentration

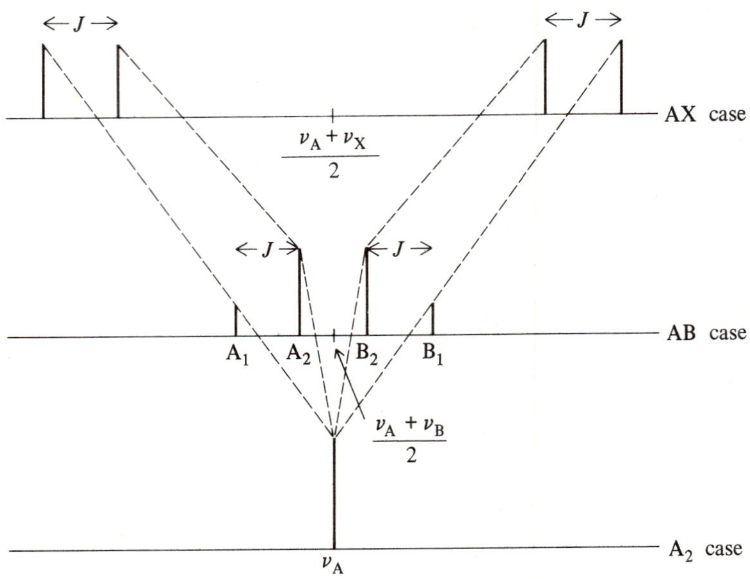


*FIGURE 9.5*
*AX, AB, and $A_2$ spectra.*

of spectral intensity in the center of the spectrum; furthermore, the inner lines are more intense than the outer lines but their separation still remains $J$.

Before going further, let us indicate the results which can be worked out with some effort for the strongly coupled AB case (note the contrast with the AX case, with only weak coupling and solved immediately). Four lines are observed, as seen in Figure 9.5, symmetrical with respect to $(\nu_A + \nu_B)/2$. The coupling constant $J_{AB}$ is given by the interval recurring twice, of lines $A_1$ and $A_2$, or $B_1$ and $B_2$. The chemical shift $(\nu_A - \nu_B)$ is the square root of the product of the large interval $L$ between the small outer lines ($A_1$ and $B_1$) by the small interval $s$ for the strong inner lines ($A_2$ and $B_2$): $(\nu_A - \nu_B) = [(A_1 - B_1)(A_2 - B_2)]^{1/2} = \sqrt{Ls}$. As a further check on the analysis, the ratio $(A_1 - B_1)/(A_2 - B_2)$ should be the same as that of the intensity of the inner lines to the intensity of the outer lines:

$$\frac{(A_1 - B_1)}{(A_2 - B_2)} \equiv \frac{I_{A_2}}{I_{A_1}} \equiv \frac{I_{B_2}}{I_{B_1}}$$

## 9.2 *Rules for the Appearance of First-Order Spectra*

Whenever one deals with a system of coupled magnetic moments of the type $A_mX_r$, that is, molecules or parts of molecules containing only sets of isochronous *and* isogamous nuclei which are weakly coupled, the following rules describe the appearance of the spectrum (a so-called first-order spectrum):

1. The A part contains $(2rI_X + 1)$ lines, the X part contains $(2mI_A + 1)$ lines, in which $I_A$ and $I_X$ are the spin quantum numbers for the A and the X species. These lines are equally spaced, and their splitting is identical to the coupling constant $J_{AX}$.
2. The intensities of the lines in one group correspond to all the possible arrangements of the spin numbers $m_z$ in the other set.

These rules are merely generalizations of what was discussed above. For instance, the proton spectrum of ethyl chloride is not affected by coupling to the chlorine nucleus (because of its very fast relaxation), and is an $A_2X_3$ system, in which $I_A = I_X = \frac{1}{2}$. Hence, the A group consists of four equally spaced lines and the X group consists of three equally spaced lines. For each orientation of the A nuclear spins, the three X nuclear spins can be either parallel ($\alpha$) or antiparallel ($\beta$) with respect to the applied field $H_0$, corresponding to the arrangements $\beta\beta\beta$ with probability 1, $\alpha\beta\beta$, and $\beta\alpha\beta$, and $\beta\beta\alpha$ with a probability of 3 (since the three nuclei are indistinguishable particles), likewise $\alpha\alpha\beta$, $\alpha\beta\alpha$, and $\alpha\alpha\beta$ (3), and finally $\alpha\alpha\alpha$ with probability 1. Such are the observed intensities. A similar argument explains the 1:2:1 intensities within the X group of lines. The resulting spectrum is shown in diagrammatic form and compared to the experimental spectrum in Figure 9.6.

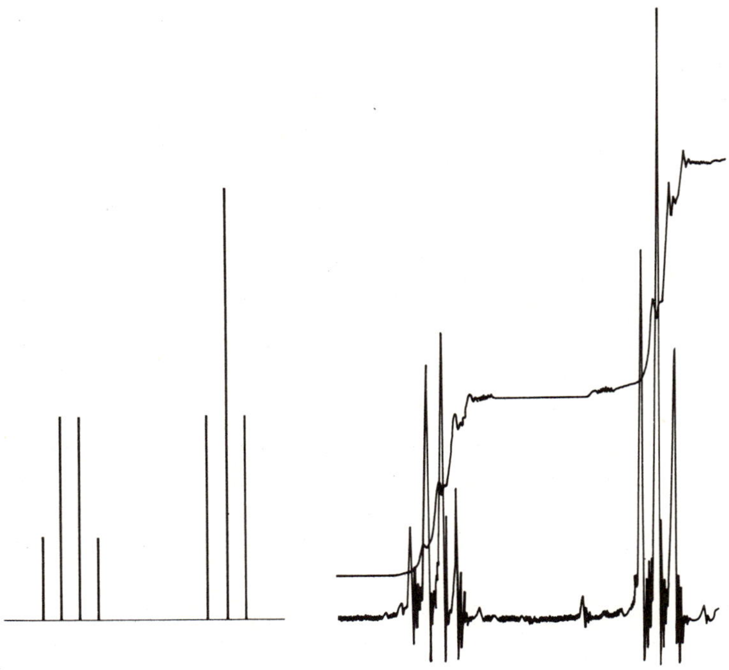

(a) (b)

*FIGURE 9.6*
***Diagrammatic (a) and experimental (b) spectra of ethyl chloride.***

When only particles with spin of $\frac{1}{2}$ are involved, the intensities of the lines in a first-order spectrum are given by the binomial expansion of $(1 + 1)^n$, the coefficients of which are easily found by construction of the Pascal triangle:

1
1 1
1 2 1
1 3 3 1
1 4 6 4 1
1 5 10 10 5 1
1 6 15 20 15 6 1
1 7 21 35 35 21 7 1

Pascal triangle

Let us now consider a different molecule, ethyl chloride$-\alpha,\alpha'$-$d_2$; it is an $A_2$–$X_3$ case, in which the As are deuterons and the Xs are protons. The A part consists of $2(3)(\frac{1}{2}) + 1 = 4$ equally spaced lines, with relative intensities 1:3:3:1 for the same reason as above; the X part consists of $2(2)(1) + 1 = 5$ equally spaced lines. To find its distribution of intensities, one must write the different possible arrangements of two deuterium spins with a quantum number $I = 1$; each can have the possible values (projections $m_N$) $+1$, 0, and $-1$. The possibilities of orientation are

| $m_N$ | $m_N$ | Total spin | Probability |
|---|---|---|---|
| +1 | +1 | +2 | 1 |
| +1 | 0 | +1 | 2 |
| 0 | +1 | +1 | |
| +1 | −1 | 0 | |
| −1 | +1 | 0 | 3 |
| 0 | 0 | 0 | |
| −1 | 0 | −1 | 2 |
| 0 | −1 | −1 | |
| −1 | −1 | −2 | 1 |

The spectrum is shown in schematic form in Figure 9.7.

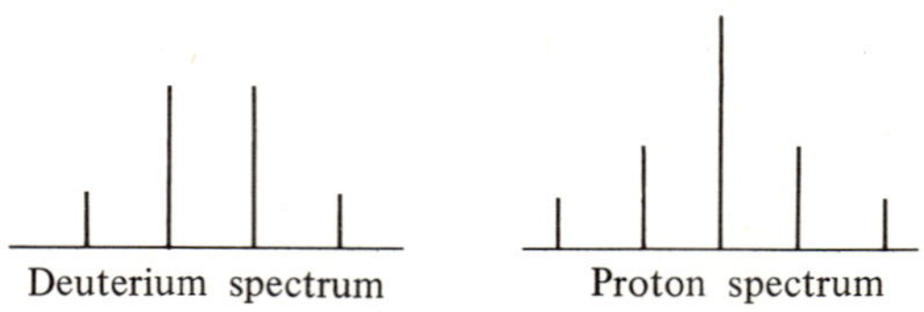


*FIGURE 9.7*
*Schematic representation of the nmr spectrum of* $CH_3CD_2Cl$.

An equivalent rationalization of the splitting of the deuterium resonance is to consider it as being split in succession by each proton with the same coupling constant $J_{HD}$, as shown in Figure 9.8. It follows that the chemical shift of the nuclei appearing as such a first-order multiplet corresponds simply to its center of gravity: either the central most intense line if there is an odd number of lines or the average of the two most intense lines if the total number of lines is even. Note also that the distance

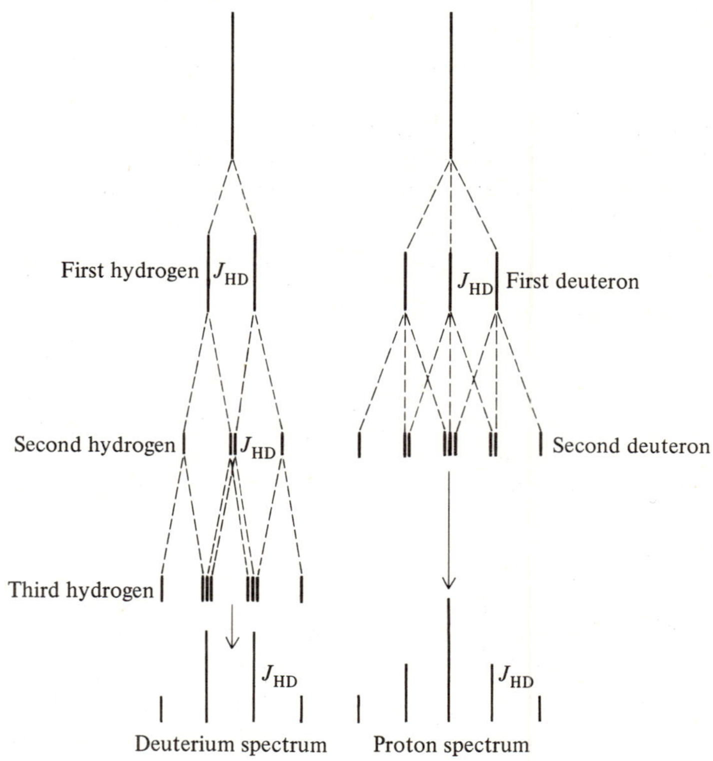


*FIGURE 9.8*
*Coupling in the nmr spectrum of* $CH_3CD_2Cl$.

between the outer lines is always equal to the sum of the coupling constants to the coupled nuclei. This property is trivial when all the coupling constants are equal, but it can be extended to more complex cases as will be indicated later.

## 9.3 *Subspectral Analysis and Extensions of the First-Order Rules*

The rules discussed above for systems of the $A_mX_r$ type derive their simplicity from two distinct sources:

1. Weak coupling, due to $(\nu_A - \nu_X) \gg J_{AX}$.
2. Magnetic equivalence, since within each set all the nuclei are both isochronous and isogamous.

We shall now examine cases of intermediate complexity resulting from fulfillment of only one of these two possible conditions.

### 9.3.1 *Further Cases of Weak Coupling*

The question now raised is that of the appearance of a system of nuclear (and eventually electronic) spins which is weakly coupled throughout. The simplest case is the three-spin AMX system. Due to the large differences in chemical shifts, it is again possible to "generate" the spectrum by splitting each resonance in turn with the coupling of the corresponding nucleus to each of the two other nuclei, as shown in Figure 9.10. An example of an AMX spectrum is given in Figure 9.11. Therefore, it is again possible to "read" the coupling constants and chemical shifts directly from the spectrum, the only difference being that, in general, unequal splittings are observed. Note again that for each multiplet the distance of the outer lines is equal to the sum of (all) the coupling constants involving that nucleus. For nucleus A, for instance, $J_{AM} + \frac{1}{2}J_{AX} + \frac{1}{2}J_{AX} = J_{AM} + J_{AX}$. It is for this reason that the width at half-height $(\nu_{1/2})$ of a peak is often taken as a good approximation to the sum of the coupling constants affecting the corresponding nucleus. If a peak is a doublét the distance of the two components is smaller than the total width at half-height, $\nu_{1/2}$, by a quantity which is the linewidth of the individual components. For a complex unresolved multiplet the error made by this assimilation will be of the order of the linewidth of an uncoupled line in the spectrum (see Figure 9.9).

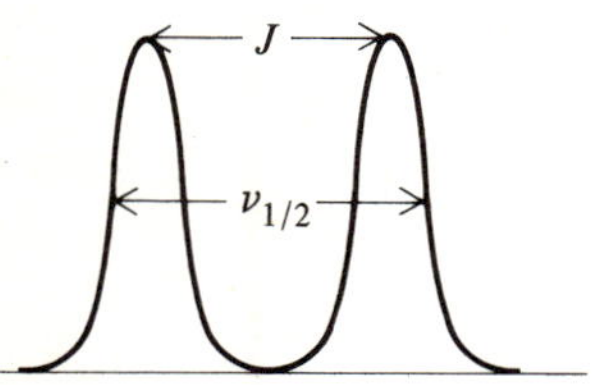


*FIGURE 9.9*
*Use of the total linewidth for measuring coupling constants.*

Similar simplifications will affect magnetic resonance spectra whenever some weak coupling is present. For instance, the esr spectra of all monoradicals are very simple and can be decomposed in terms of the various coupling constants to the hydrogen spins ($a_H$), nitrogen spins ($a_N$), and so on in the molecule. One such simplification is that in many cases complicated spectra can be broken down into simpler components from which all of the necessary information can be extracted. This procedure is referred to as subspectral analysis. Still other simplifications occur as a consequence of

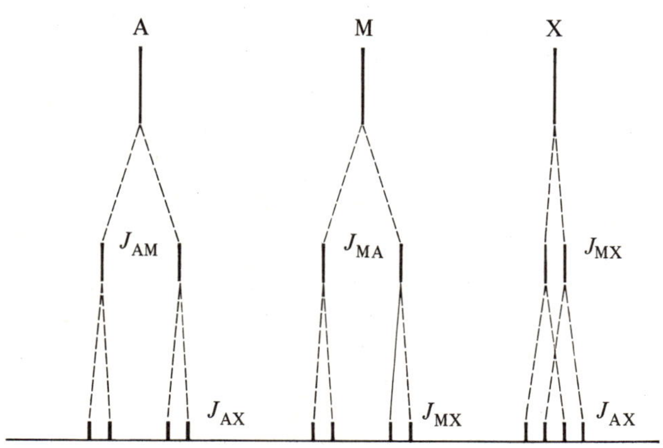


*FIGURE 9.10*
*Coupling in an AMX spectrum.*

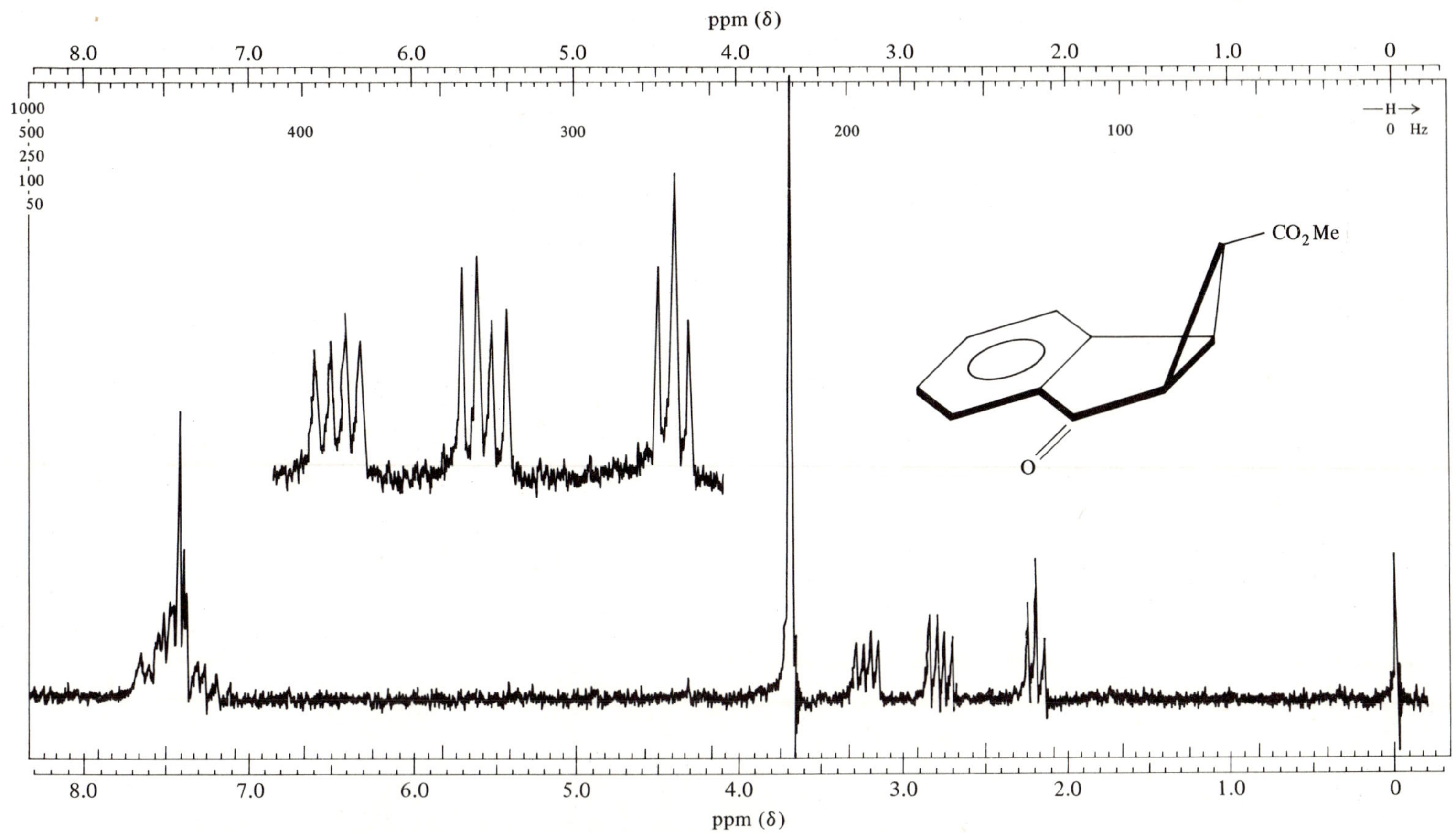


*FIGURE 9.11*
*Example of an AMX system displayed by the cyclopropyl protons of the compound shown.*

weak coupling; we shall return to them after examining simplifications resulting from magnetic equivalence.

### 9.3.2 *Simplification Resulting from Magnetic Equivalence*

We shall indicate the simplifications resulting from magnetic equivalence by a formal analysis of the energy diagrams pertaining to the various spin-interaction systems.

For one nucleus of spin $\frac{1}{2}$ there are two energy states $\alpha$ and $\beta$, which can be connected with the corresponding transition. We shall use horizontal lines connected with a vertical bar as a symbol for the actual transition:

For two nuclei with a spin of $\frac{1}{2}$, the energy diagram is:

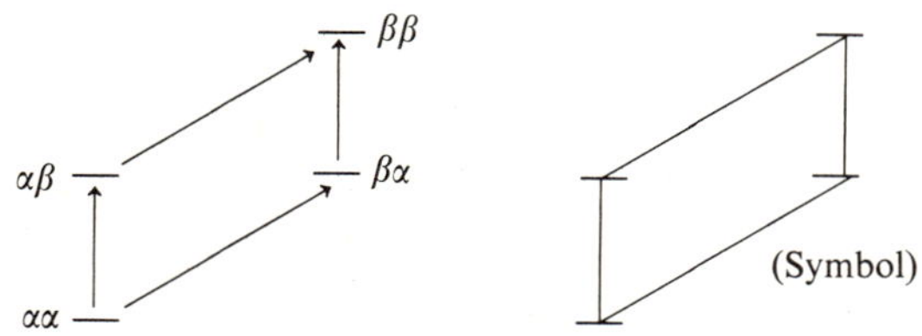


An alternate way of looking at this diagram for two nuclei is as the resultant of the symbolic multiplication:

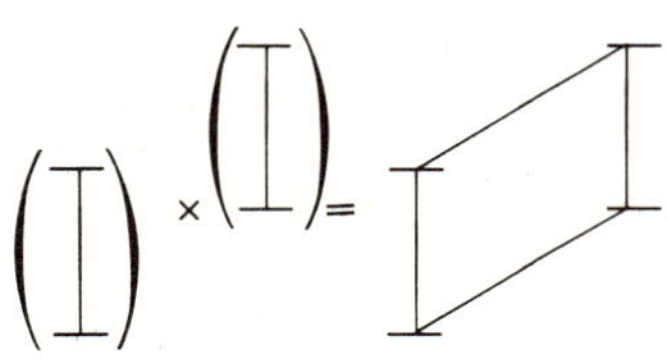

This is the most general AB case of two interacting nuclei with a spin of $\frac{1}{2}$. It is simplified by magnetic equivalence into an $A_2$ system (where the dashed lines represent forbidden transitions):

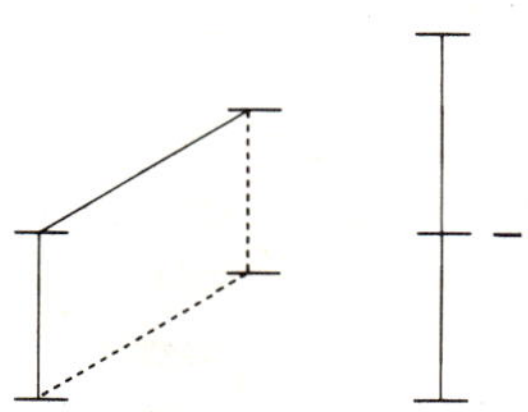

the two dotted transitions occur between states of different symmetry and are thus forbidden. Furthermore, the two allowed transitions have the same energy, which accounts for the observation of a single line in the $A_2$ case.

Let us now discuss the various possibilities for three nuclei with spins of $\frac{1}{2}$. In the $AB_2$ case we effect the following symbolic multiplication:

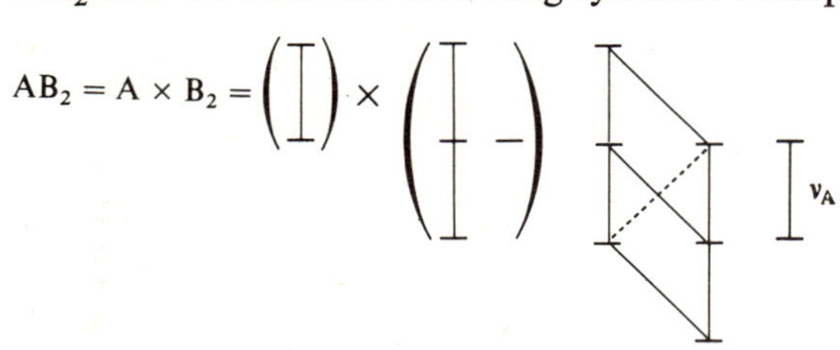

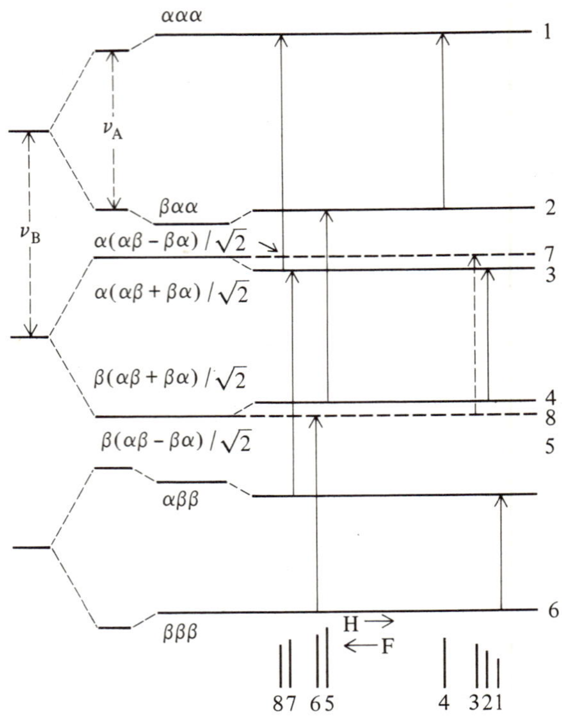


*FIGURE 9.12*
*Energy-level diagram of an $AB_2$ spin system for which $\nu_B > \nu_A$. The spin functions noted against the different energy levels are the respective spin-product functions. The various transitions are also shown schematically with heights proportional to single-resonance intensities. The lines 1 to 4 are "A" transitions and the lines 5 to 8 are "B" transitions. Line 3 is the transition between antisymmetric states 7 and 8 (dashed lines).*

predicting the appearance of one unperturbed A line exactly at the chemical shift $\nu_A$. The rest of the spectrum has to be calculated from quantum mechanics, as in the AB case, but some simplification is possible because of the magnetic equivalence in the $B_2$ part. The corresponding energy-level diagram and spectrum are shown in Figure 9.12.

In the ABX case the change from an AB case (strong coupling) to an AX case (weak coupling) can be represented by:

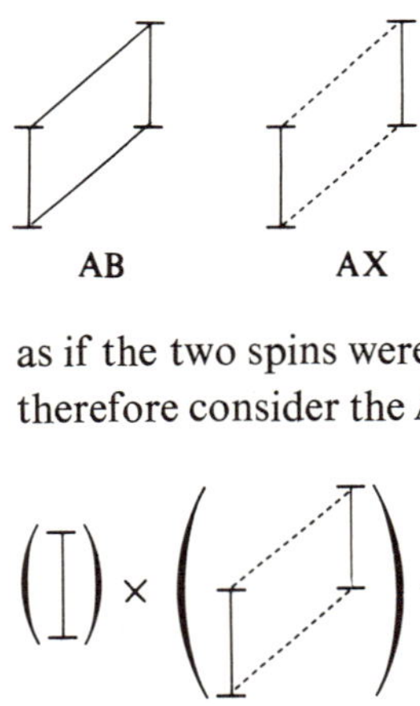


as if the two spins were becoming increasingly "insulated" from one another. Let us therefore consider the ABX case as the result of the symbolic product ABX = A × BX:

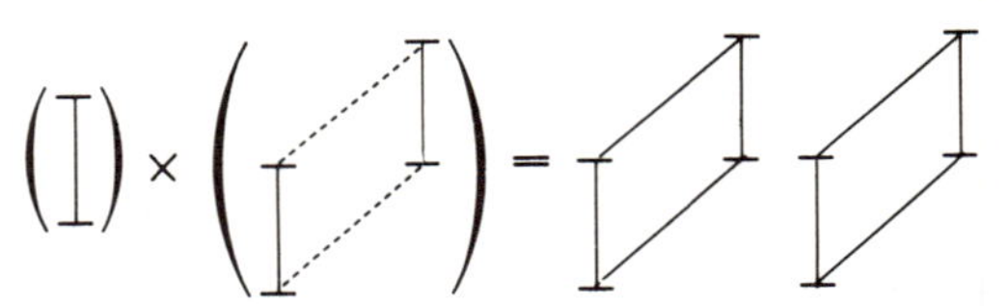

that is, the ABX system is found to contain two AB systems. In other words, for each orientation of the X spin, either α or β, there is one and only one AB system for the interaction between the two remaining spins A and B, unaffected by the presence of X. The actual appearance of the AB *part* of an ABX spectrum (easily recognizable by inspection of the actual spectrum) is then the superposition of two AB spectra (Figure 9.13)—hence the name subspectral analysis for this general method.

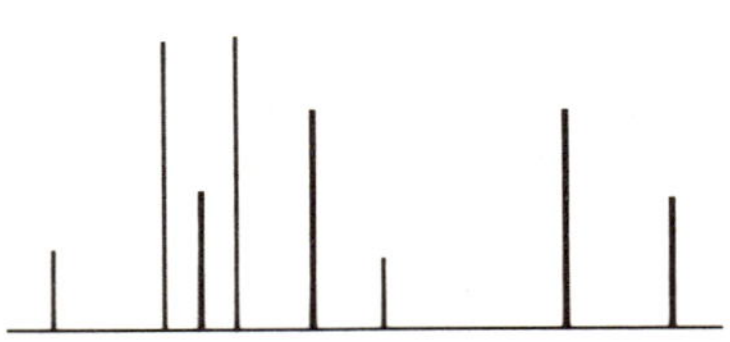

*FIGURE 9.13*
*The AB part of an ABX system. (The two ab subspectra are shown in different line thicknesses.)*

For each subspectrum one can define similar quantities $\sqrt{Ls}$ as above, which are then no longer *real* chemical shifts but are *related* to the two chemical shifts and coupling constants as follows:

$\Delta_1 = \sqrt{L_1 s_1}$ (for the first subspectrum)
$\bar{\nu}_1$ = center of gravity (for the first subspectrum)
$\Delta_2 = \sqrt{L_2 s_2}$ (for the other subspectrum)
$\bar{\nu}_2$ = center of gravity (for the other subspectrum)

Then:

$$\nu_A = \tfrac{1}{2}(\bar{\nu}_1 + \bar{\nu}_2) + \tfrac{1}{4}(\Delta_1 + \Delta_2)$$
$$\nu_B = \tfrac{1}{2}(\bar{\nu}_1 + \bar{\nu}_2) - \tfrac{1}{4}(\Delta_1 + \Delta_2)$$
$$J_{AX} = (\bar{\nu}_1 - \bar{\nu}_2) + \tfrac{1}{2}(\Delta_1 - \Delta_2)$$
$$J_{BX} = (\bar{\nu}_1 - \bar{\nu}_2) - \tfrac{1}{2}(\Delta_1 - \Delta_2)$$

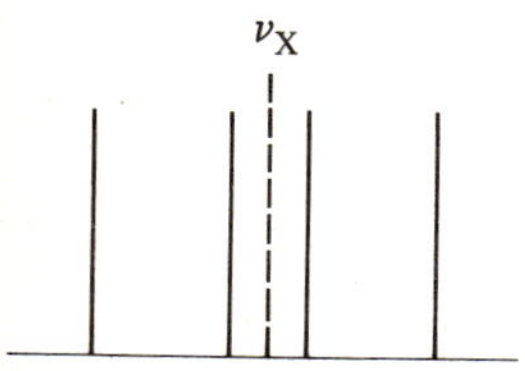


FIGURE 9.14
The X part of an ABX spectrum.

$J_{AB}$ is identical to the "coupling constant $J_{ab}$" which can be extracted from either *ab* subspectrum, since it occurs in both. In order to complete the analysis, only the value of $\nu_X$ has to be determined. It is given very simply by the center of gravity of the four symmetrical lines to be observed for the X nucleus (Figure 9.14).

We have seen, by a formal manipulation of symbols, how the ABX system can be shown to contain two simpler *ab* subspectra. Let us now analyze in detail this occurrence by constructing the AB part of the ABX system "from scratch":

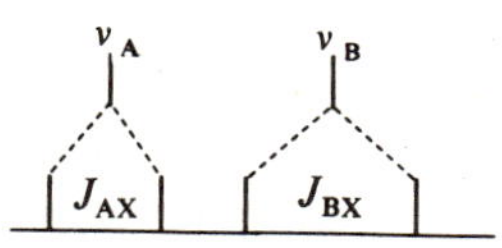


We thus split each of the A and B resonances into two, because of the weak coupling with the X proton. If $J_{AX}$ and $J_{BX}$ have the same sign, we obtain the arrangement shown:

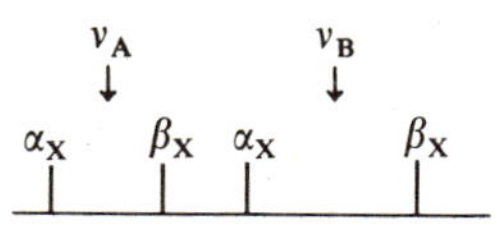


$J_{AX} J_{BX} > 0$

If $J_{AX}$ and $J_{BX}$ are of opposite signs:

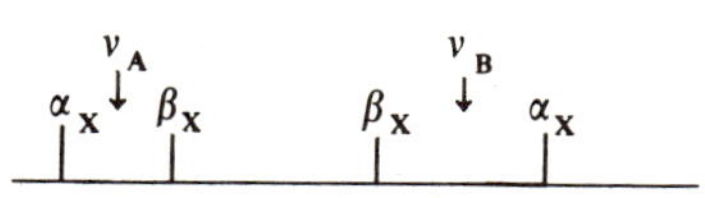


$J_{AX} J_{BX} < 0$

Then we construct *ab* subspectra, *one for each orientation of the X nuclear spin* (Figure 9.15) and obtain the two *ab* subspectra pointed out previously.

A note of caution is in order here. It is only legitimate to proceed in the order chosen, that is, first splitting the A and B resonances by their couplings to the weakly coupled nucleus X, and only *then* "building" the resulting *ab* subspectra. The converse procedure, that is, starting with the AB system and splitting each of the four lines into two because of the coupling $J_{AX}$ or $J_{BX}$ is only correct for the AMX system weakly coupled throughout, and is not applicable here because it does not produce the right mixing of the energy states. For the same reason it is not possible to construct the X part of the ABX spectrum by splitting the X resonance with the coupling constants to the A and to the B nucleus.

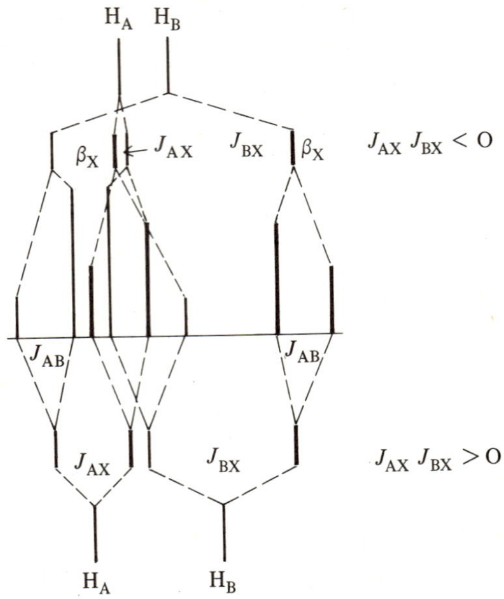


*FIGURE 9.15*
*Coupling in the AB part of an ABX system.*

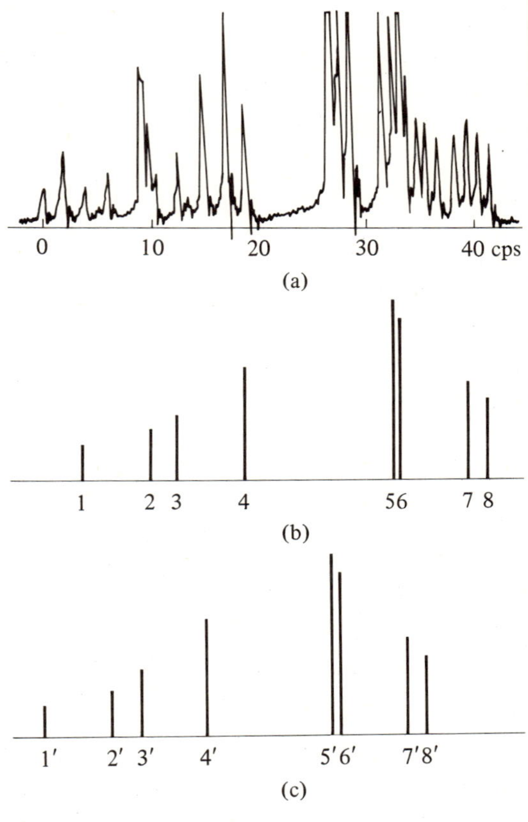


*FIGURE 9.16*
*The proton nmr spectrum of pyridine at 60 MHz. (a) The ABB′ part of the ABB′XX′ spectrum. (b) The $ab_2$ subspectrum corresponding to $F_z(XX') = +1$, where $J/\delta = J_{AB}/\delta_{ab} = J_{AB}/\delta_{AB} + \frac{1}{2}(J_{BX} + J_{BX'} - 2J_{AX})$. (c) The $ab_2$ subspectrum corresponding to $F_z(XX') = -1$, where $J/\delta = J_{AB}/\delta_{ab} = J_{AB}/\delta_{AB} - \frac{1}{2}(J_{BX} + J_{BX'} - 2J_{AX})$.*

The ABC case results formally from the multiplication:

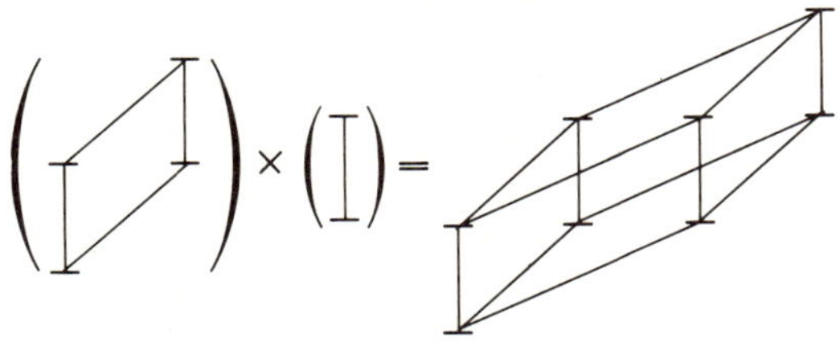

Fifteen transitions can occur, corresponding to the 12 edges and the 3 diagonals of the above cubic energy diagram. Here no simplification appears, and the ABC case can be approached only by direct calculation with a computer, a rather involved procedure as compared to the ABX case.

This task becomes all the more formidable when the number of coupled particles increases. Simplifications such as subspectral analysis, however, often offer a convenient means of extracting *all* of the information available in the spectrum from only *part* of it. An illustrative example of the breakdown of a complex experimental spectrum into its subspectral elements is indicated in Figure 9.16.

### *9.3.3 The Four-Spin AA′BB′ and $A_2B_2$ Spin Systems*

Let us consider finally some examples of four-spin systems. Interaction patterns of the type ABCX or ABKX can be treated by the method of subspectral analysis and decomposed into subspectra of the ABC and ABK types, respectively. Such analysis will usually yield all the desired information in terms of chemical shifts and coupling constants. The strongly coupled ABCD type is best treated by digital computation using trial-and-error methods. Symmetry, however, again proves to be a powerful help in strongly coupled but symmetrical systems such as AA′BB′ or $A_2B_2$. Let us write the symmetrized spin functions for the former case (Table 9.1). Setting up such a table is only a matter of careful bookkeeping.

*TABLE 9.1*

*Spin Functions for the AA′BB′ Case*

| Symmetry | A nuclei | B nuclei | *m* value | designation |
|---|---|---|---|---|
| $s$ | $\alpha\alpha$ | $\alpha\alpha$ | 2 | $s_2$ |
| $s$ | $\frac{1}{\sqrt{2}}(\alpha\beta + \beta\alpha)$ | $\alpha\alpha$ | 1 | $1s_1$ |
| $s$ | $\alpha\alpha$ | $\frac{1}{\sqrt{2}}(\alpha\beta + \beta\alpha)$ | 1 | $2s_1$ |
| $s$ | $\beta\beta$ | $\alpha\alpha$ | 0 | $1s_0$ |
| $s$ | $\alpha\alpha$ | $\beta\beta$ | 0 | $2s_0$ |
| $s$ | $\frac{1}{\sqrt{2}}(\alpha\beta + \beta\alpha)$ | $\frac{1}{\sqrt{2}}(\alpha\beta + \beta\alpha)$ | 0 | $3s_0$ |
| $s$ | $\frac{1}{\sqrt{2}}(\alpha\beta - \beta\alpha)$ | $\frac{1}{\sqrt{2}}(\alpha\beta - \beta\alpha)$ | 0 | $4s_0$ |
| $s$ | $\frac{1}{\sqrt{2}}(\alpha\beta + \beta\alpha)$ | $\beta\beta$ | −1 | $1s_{-1}$ |
| $s$ | $\beta\beta$ | $\frac{1}{\sqrt{2}}(\alpha\beta + \beta\alpha)$ | −1 | $2s_{-1}$ |
| $s$ | $\beta\beta$ | $\beta\beta$ | −2 | $s_{-2}$ |
| $a$ | $\frac{1}{\sqrt{2}}(\alpha\beta - \beta\alpha)$ | $\alpha\alpha$ | 1 | $1a_1$ |

*(Continued)*

*TABLE 9.1 (Continued)*

| Symmetry | A nuclei | B nuclei | $m$ value | designation |
|---|---|---|---|---|
| $a$ | $\alpha\alpha$ | $\frac{1}{\sqrt{2}}(\alpha\beta - \beta\alpha)$ | 1 | $2a_1$ |
| $a$ | $\frac{1}{\sqrt{2}}(\alpha\beta + \beta\alpha)$ | $\frac{1}{\sqrt{2}}(\alpha\beta - \beta\alpha)$ | 0 | $1a_0$ |
| $a$ | $\frac{1}{\sqrt{2}}(\alpha\beta - \beta\alpha)$ | $\frac{1}{\sqrt{2}}(\alpha\beta + \beta\alpha)$ | 0 | $2a_0$ |
| $a$ | $\frac{1}{\sqrt{2}}(\alpha\beta - \beta\alpha)$ | $\frac{1}{\sqrt{2}}\beta\beta$ | $-1$ | $1a_{-1}$ |
| $a$ | $\beta\beta$ | $\frac{1}{\sqrt{2}}(\alpha\beta - \beta\alpha)$ | $-1$ | $2a_{-1}$ |

Application of the selection rules, according to which the only transitions allowed are between states with the same symmetry and separated by $\Delta m = \pm 1$, allows the listing of the transitions for the AA′BB′ case (Table 9.2). A total of 28

*TABLE 9.2*

*Allowed Transitions for the AA′BB′ Case*

| | |
|---|---|
| $2a_{-1} \rightarrow 2a_0$ | $s_{-2} \rightarrow 2s_{-1}$ |
| $1a_{-1} \rightarrow 2a_0$ | $s_{-2} \rightarrow 1s_{-1}$ |
| $1a_{-1} \rightarrow 1a_0$ | $2s_{-1} \rightarrow 4s_0$ |
| $2a_{-1} \rightarrow 1a_0$ | $2s_{-1} \rightarrow 3s_0$ |
| $2a_0 \rightarrow 2a_1$ | $2s_{-1} \rightarrow 2s_0$ |
| $2a_0 \rightarrow 1a_1$ | $2s_{-1} \rightarrow 1s_0$ |
| $1a_0 \rightarrow 2a_1$ | $1s_{-1} \rightarrow 4s_0$ |
| $1a_0 \rightarrow 1a_1$ | $1s_{-1} \rightarrow 3s_0$ |
| | $1s_{-1} \rightarrow 2s_0$ |
| | $1s_{-1} \rightarrow 1s_0$ |
| | $4s_0 \rightarrow 2s_1$ |
| | $4s_0 \rightarrow 1s_1$ |
| | $3s_0 \rightarrow 2s_1$ |
| | $3s_0 \rightarrow 1s_1$ |
| | $2s_0 \rightarrow 2s_1$ |
| | $2s_0 \rightarrow 1s_1$ |
| | $1s_0 \rightarrow 2s_1$ |
| | $1s_0 \rightarrow 1s_1$ |
| | $2s_1 \rightarrow s_2$ |
| | $1s_1 \rightarrow s_2$ |

allowed transitions is expected. Furthermore, the spectrum will be symmetrical about the midpoint $(\nu_A + \nu_B)/2$. Inspection of Table 9.2 shows that the A part of the spectrum is identical to the B part of the spectrum. As for the $A_2B_2$ system, it has a total of only 18 allowed transitions, because of the magnetic equivalence within the A and B groups of nuclei. To calculate the energies for the various transitions in such a system it is convenient to consider the various "spin particles" formed by identical protons, in this instance particles with a total spin number of 1, each having $I_z = -1$, 0, and +1.

In summary, in this section we have been concerned with a description of the conceptual tools used in the calculation of nmr spectra. There is, however, another useful technique which is helpful in the ordering of the energy levels for diagrams of the type we have discussed. This is "spin tickling," a double-resonance method that we shall now discuss.

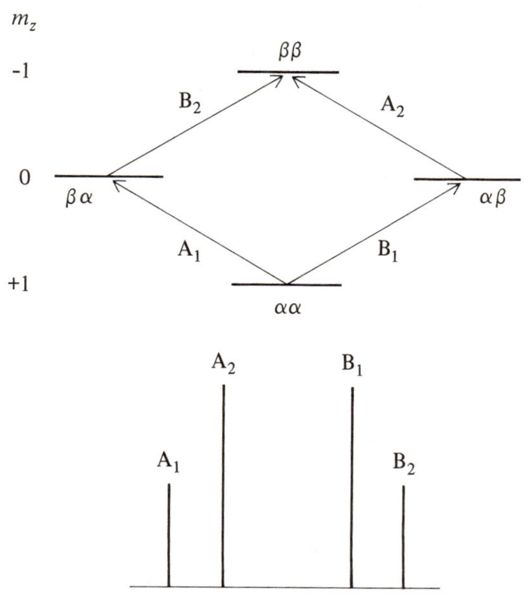


*FIGURE 9.17*
*Energy diagram and spectrum of an AB system.*

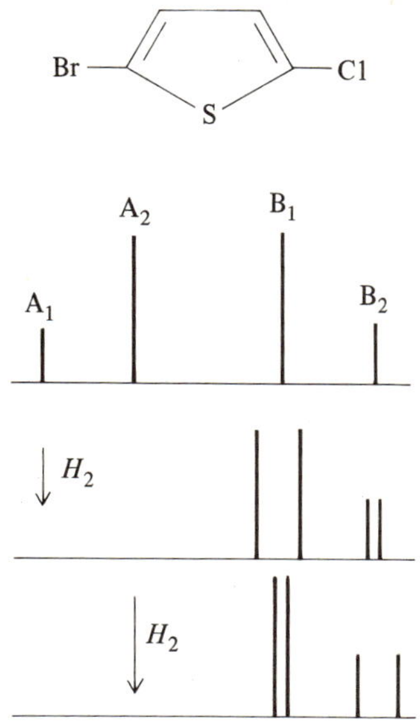


*FIGURE 9.18*
*Spin tickling of 2-chloro-5-bromothiophene.*

## 9.4 *Spin Tickling*

To obtain the desired coupling constants and chemical shifts in cases of strong coupling, typified by an AB system with $|\nu_A - \nu_B| < 5|J_{AB}|$ the technique of "spin tickling" can be used. The tickling experiment is a double-resonance experiment in which the RF field $H_2$ has a weak amplitude, so that $\gamma H_2/2\pi$ is of the order of the linewidths in the spectrum. The energy diagram for the AB system is shown again in Figure 9.17. Looking at the energy diagram it is seen that both transitions $B_1$ and $B_2$ have an energy level in common with $A_1$, but $A_2$ has none. Hence, by "tickling" the $A_1$ transition it is to be expected that the perturbation of $A_1$ is going to affect lines $B_1$ and $B_2$, but not $A_2$. This is indeed what is observed experimentally.

Transitions such as $A_1$ and $B_2$, or $B_1$ and $A_2$, are related in a *progressive* fashion, since the end result is that the total spin number changes by two units. Transitions such as $A_1$ and $B_1$, or $A_2$ and $B_2$, are termed *regressive*, since the total spin number is left unchanged by their succession. It can be shown, and it is found experimentally, that "tickling" of a line splits the lines corresponding to the transitions having an energy level in common with the transition irradiated. Furthermore, transitions of a regressive character are split into sharp doublets whereas the progressive transitions give only poorly resolved doublets! The example of 2-chloro-5-bromothiophene[2] is shown in Figure 9.18. In general, this technique is an elegant means for determining the ordering of energy levels.

## 9.5 *ILLUSTRATIVE EXAMPLES*

Figure 9.19 shows the *proton* spectrum of trideuterioacetophenone, $C_6H_5COCD_3$. Note the low-field absorption characteristic of aromatic protons and the quintet at 2.40 ppm. This quintet arises because of incomplete deuteration and the presence of a small amount of hydrogen, that is, $CD_2H$, which couples with the deuteriums. It is a quintet because deuterium has a spin of 1, hence $2nI + 1 = (2)(2)(1) + 1 = 5$. The observed coupling $J_{HD} = 2{\cdot}2$ Hz is unusually large and accounts for the well-resolved quintet. In most instances replacement of hydrogen by deuterium results only in a slight line broadening of the proton resonance and one does not observe any coupling

---

[2] R. Freeman and W. A. Anderson, *J. Chem. Phys.*, *37*, 2053 (1962).

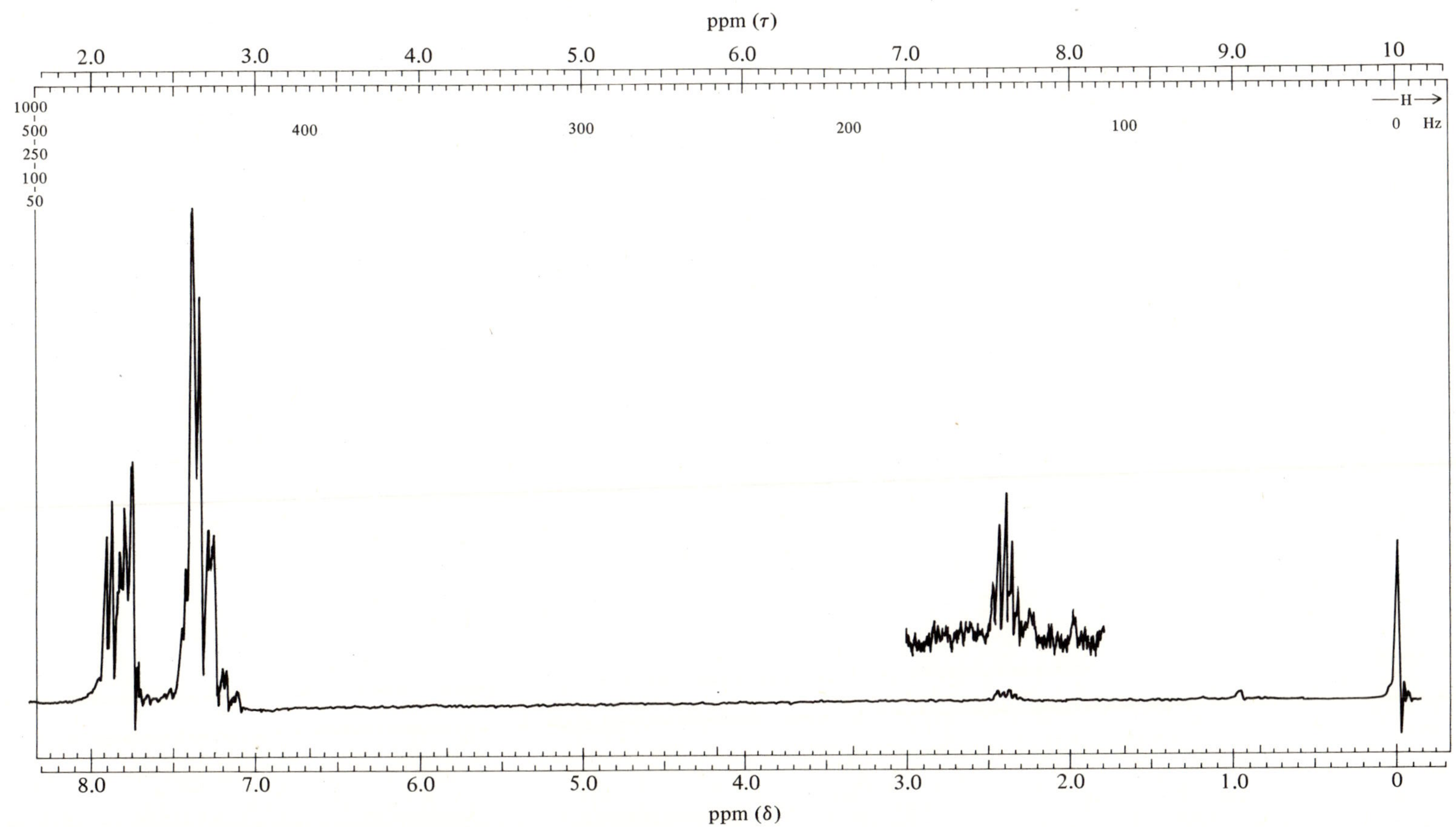


*FIGURE 9.19*
*Trideuterioacetophenone, 60-MHz proton nmr spectrum.*

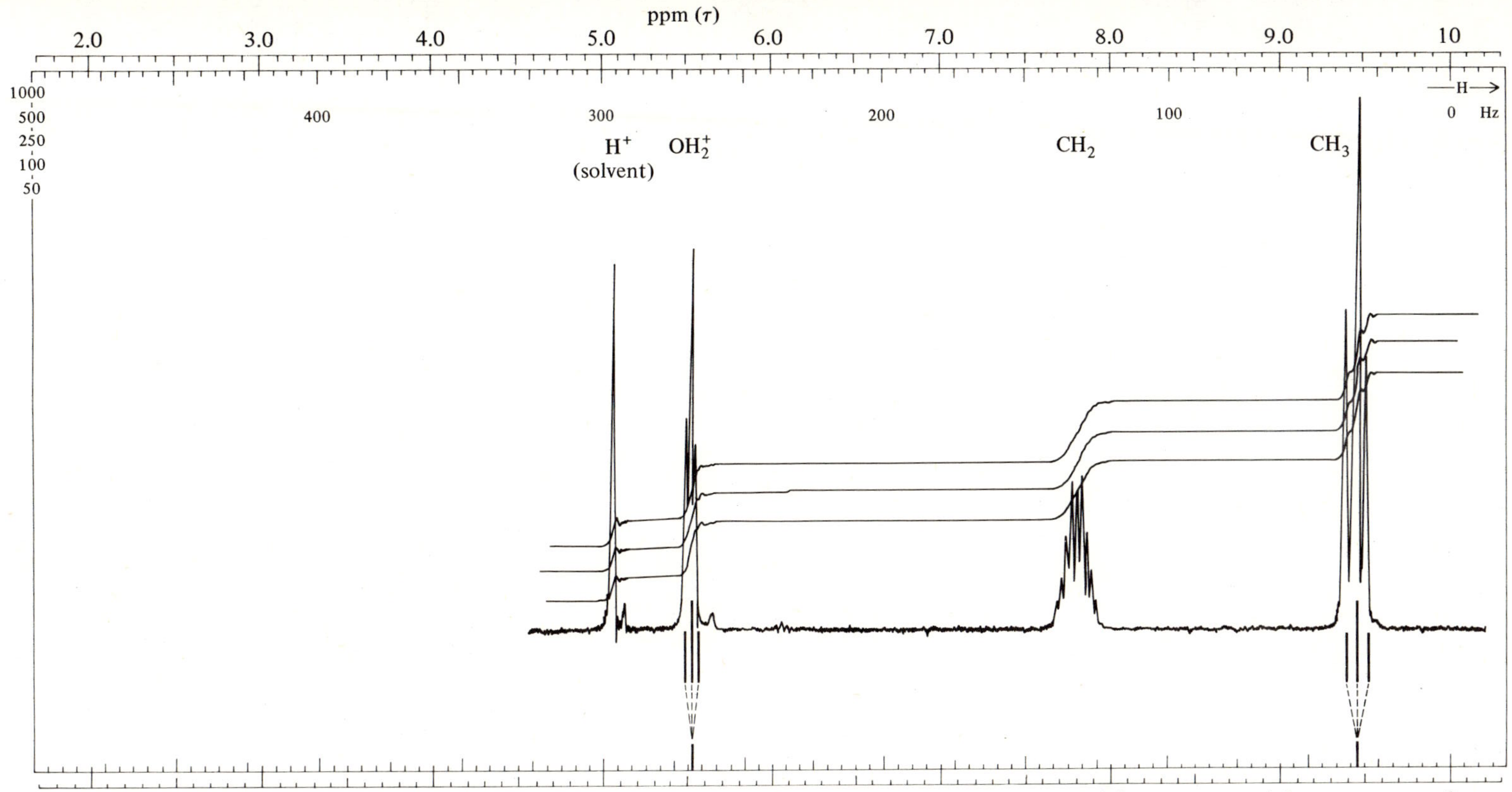


*FIGURE 9.20*
*Protonated ethanol in* $SbF_5-HFSO_3-SO_2$, *60-MHz proton nmr spectrum.*

due to deuterium. This example also illustrates how one can determine the interproton coupling of *equivalent* protons such as the $CH_3$ group of acetophenone. Replacement of some of the hydrogens by deuterium has destroyed the magnetic equivalence of the protons in the $CH_3$ group and hence the interproton coupling can be observed and calculated by means of the gyromagnetic ratio as $\gamma_H/\gamma_D \times J_{HD} = J_{HH} = (6.52)\ (2.2) = 14.4$ Hz.

The spectrum shown in Figure 9.20 is an excellent example of simple first-order splitting. The upfield $CH_3$ resonance is split into a triplet with $^3J = 7.3$ Hz by the two adjacent methylene protons. The downfield triplet is due to the $OH_2^+$ resonance split into a triplet by the two adjacent methylene protons with a coupling of 3.3 Hz. These same coupling interactions must of course affect the methylene protons, giving the observed splitting pattern. The exact coupling interaction of the methylene resonance is illustrated diagrammatically in Figure 9.21.

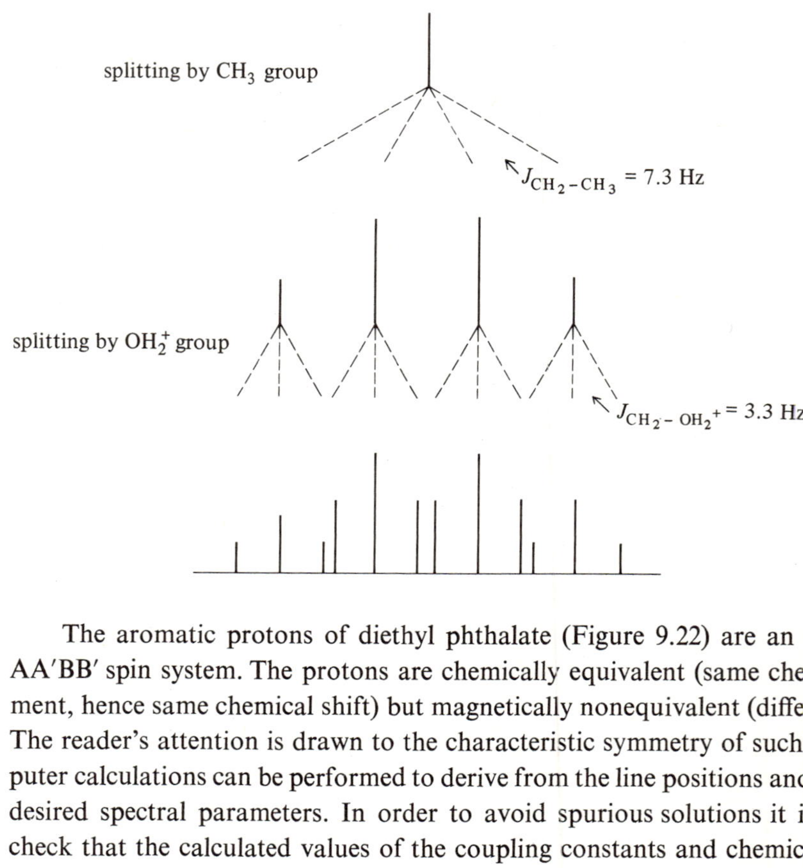


*FIGURE 9.21*
*Coupling pattern of the methylene protons in the spectrum of* $CH_3CH_2OH_2^+$.

The aromatic protons of diethyl phthalate (Figure 9.22) are an example of an AA′BB′ spin system. The protons are chemically equivalent (same chemical environment, hence same chemical shift) but magnetically nonequivalent (different coupling). The reader's attention is drawn to the characteristic symmetry of such spectra. Computer calculations can be performed to derive from the line positions and intensities the desired spectral parameters. In order to avoid spurious solutions it is important to check that the calculated values of the coupling constants and chemical shifts fit the experimental spectra at two different spectrometer frequencies, in this instance 60 and 100 MHz. The computer calculated values are

| | 60 MHz | 100 MHz |
|---|---|---|
| $\delta_A$ (Hz) | 457.54 | 762.69 |
| $\delta_B$ (Hz) | 445.87 | 743.42 |
| $J_{AA'}$ (para coupling) | 0.58 | 0.52 |
| $J_{AB}$ (ortho coupling) | 7.73 | 7.73 |
| $J_{AB'}$ (meta coupling) | 1.28 | 1.29 |
| $J_{BB'}$ (ortho coupling) | 7.50 | 7.51 |

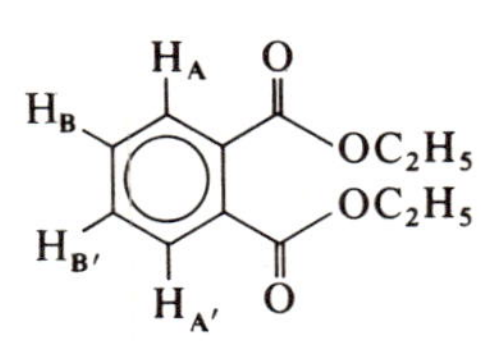

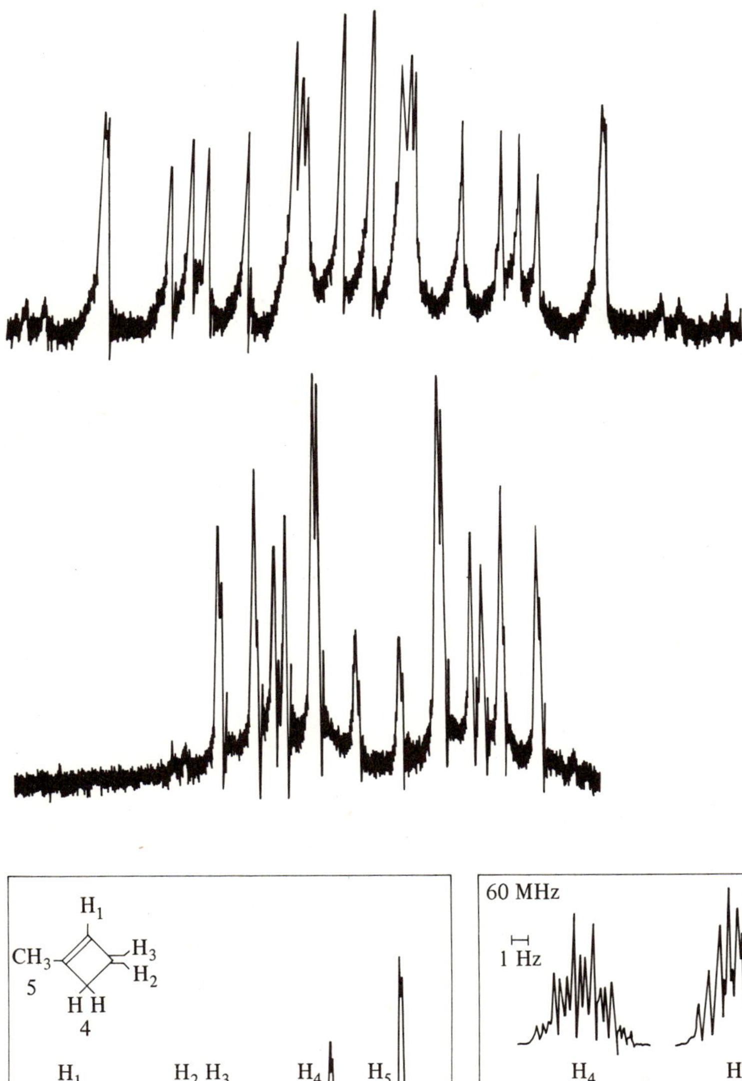

*FIGURE 9.22*
*The aromatic region of the nmr spectrum of diethyl phthalate at 60 MHz (top) and 100 MHz (bottom). The scales are different.*

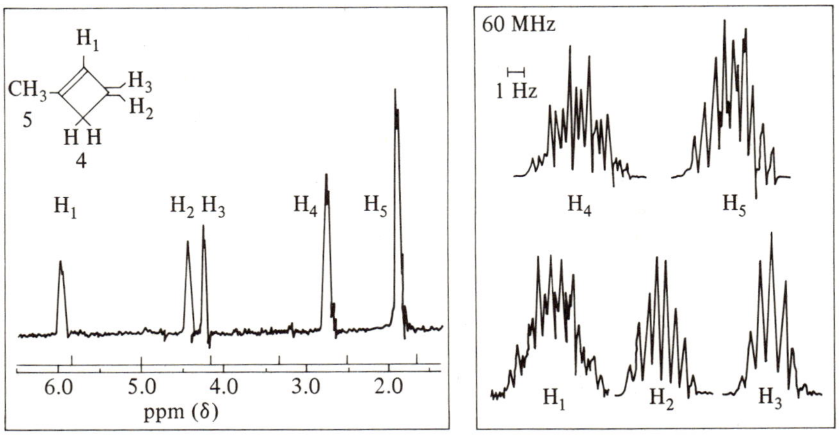


*FIGURE 9.23*
*Nuclear magnetic resonance spectrum (60 MHz) of 1-methyl-3-methylenecyclobutene (left), and under expanded scale and high-resolution conditions (right).*

The well individualized multiplets for each of the five different types of protons point to a first order spectrum (Figure 9.23). The simplest way to analyze this spectrum is by equating the distance $S$ between the outermost components of each multiplet—all clearly visible—to the sum of all the coupling constants involving the corresponding proton. For instance, $J_{1,4} + J_{2,4} + J_{3,4} + 3J_{4,5} = 6.5$ Hz for the $H_4$ protons and so on for all the five different protons. Setting these five equations equal to the observed values of $S$ (6.4, 4.4, 4.4, 6.5, and 5.4 Hz for $H_1$ through $H_5$) and solving for the five unknown coupling constants yields $J_{1,2} = 0$, $J_{1,3} = 0.77$, $J_{1,4} = 0.50$, $J_{1,5} = 1.55$, $J_{2,4} = 1.28$, $J_{2,5} = 0.60$, $J_{3,4} = 0.83$, $J_{3,5} = 0.65$, and $J_{4,5} = 1.31$ Hz.

It is also possible—but more time consuming—to reconstitute the appearance of each of these multiplets in terms of, say, $H_4$ being a doublet ($H_1$) of doublets ($H_2$) of doublets ($H_3$) of quartets ($H_5$) using trial values of the coupling constants until a fit is obtained.

## 9.6 *PROBLEMS*

SAMPLE PROBLEM

Ordering of energy levels and signs of coupling constants from "spin tickling" and double-resonance experiments, or a game of cubes.

Unsymmetrical three-spin systems have eight energy levels and at high magnetic fields generate 12 strong transitions. These figures suggest an interesting analogy with the topology of a cube. The corners of the cube may be taken to represent the eight energy levels, so that when the cube is oriented with one body diagonal vertical, the energy levels are grouped 1, 3, 3, 1 corresponding to the states $m = -\frac{3}{2}, -\frac{1}{2}, +\frac{1}{2}, +\frac{3}{2}$. The three sets of four parallel edges represent the A, B, and C quartets of transition lines. The three body diagonals represent "combination" lines which are very weak in the AMX case but undistinguishable from the other 12 lines in the strongly coupled ABC case. Six double-quantum transitions, observable at rather higher RF field strengths than normally employed, correspond to one face diagonal on each of the six faces. This analogy has been found very useful for keeping a clear picture of the energy-level diagram, and especially for predicting double-irradiation behavior.

If lines are numbered according to the increasing frequency $A_1$ through $C_4$, the four possible "cubes," when the relative signs of the coupling constants change, are shown below:

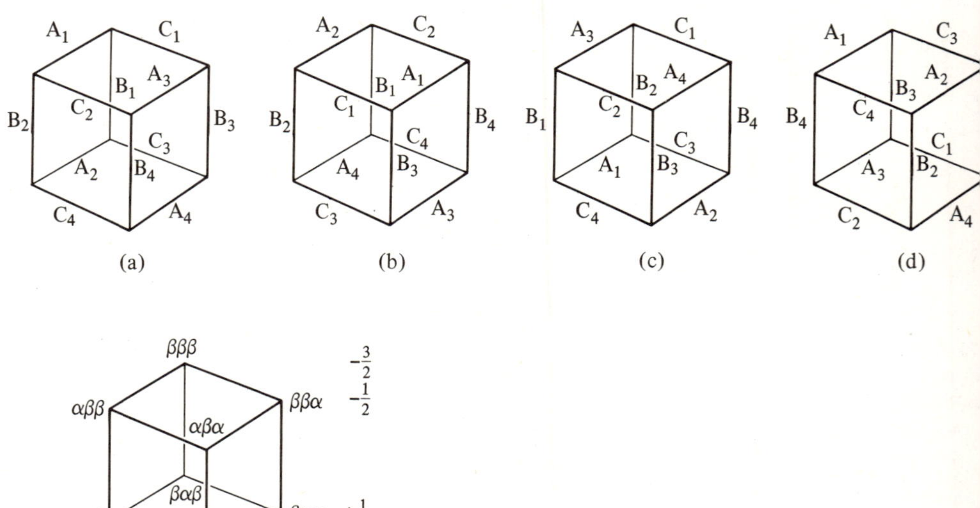

(a) (b) (c) (d)

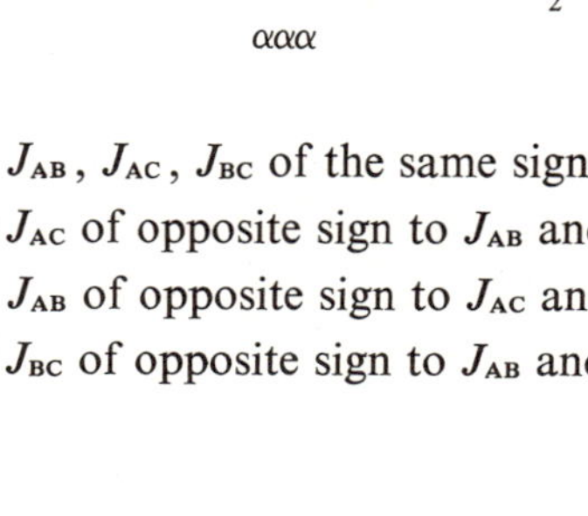


(a) $J_{AB}$, $J_{AC}$, $J_{BC}$ of the same sign.

(b) $J_{AC}$ of opposite sign to $J_{AB}$ and $J_{BC}$.

(c) $J_{AB}$ of opposite sign to $J_{AC}$ and $J_{BC}$.

(d) $J_{BC}$ of opposite sign to $J_{AB}$ and $J_{AC}$.

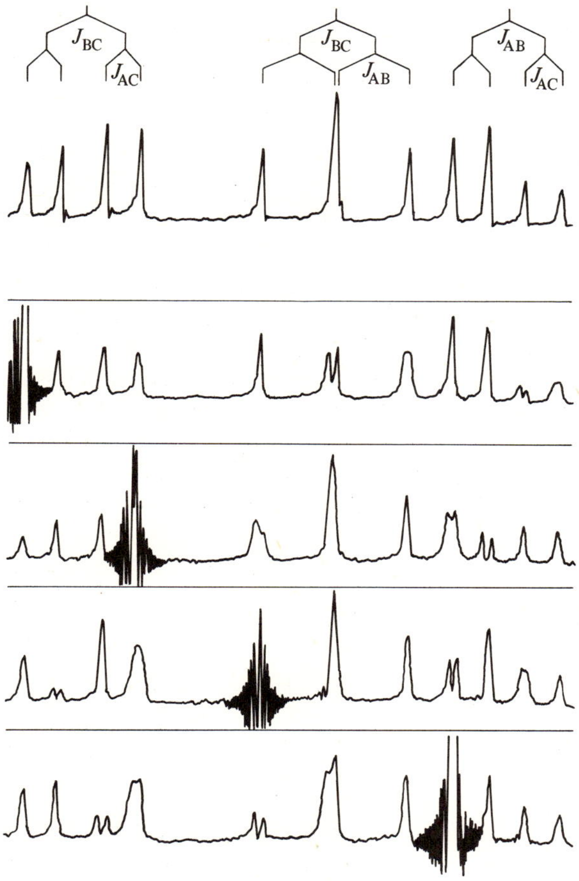


The spectra above correspond to the ABC part of the spectrum of a dibromopropionic acid obtained by sweeping the frequency of the RF field $H_1$, with an indication of the A, B, and C quartets. Lines are numbered $A_1$ through $C_4$ in order of increasing frequency, *right to left here*. The beat corresponds to the line being irradiated.

What are the relative signs of $J_{AB}$, $J_{AC}$, and $J_{BC}$? Why? What is the structure of the molecule?

In the preceding example "progressive" and "regressive" transitions were identified by the property of the latter to be split into sharp doublets, and of the former to be split into broad doublets when a line having an energy level in common with it is "tickled."

Another method which provides the same results is double-quantum absorption, which although forbidden at the normal RF $H_1$ fields used becomes observable at high values of $H_1$. Double-quantum transitions are the absorption of two energy quanta *at once* in a transition where the total spin number $m$ changes by $\pm 2$, instead of $\pm 1$. Use the energies for the AB system to demonstrate that the double-quantum (DQ) transition will occur at the center of gravity, therefore independent of $J_{AB}$, and will consist of a single line.

For an ABC system the eight spin states [the corners of the cube: see diagram (c)] have the following energies:

$\beta\beta\beta$ $\quad \frac{1}{2}(\nu_A + \nu_B + \nu_C)$

$\beta\beta\alpha$ $\quad \frac{1}{2}(\nu_A + \nu_B - \nu_C) - \frac{1}{2}(J_{AC} + J_{BC})$

$\beta\alpha\beta$ $\quad \frac{1}{2}(\nu_A + \nu_B + \nu_C) - \frac{1}{2}(J_{AB} + J_{BC})$

$\alpha\beta\beta$ $\quad \frac{1}{2}(-\nu_A + \nu_B + \nu_C) - \frac{1}{2}(J_{AB} + J_{AC})$

$\beta\alpha\alpha \quad -\frac{1}{2}(-\nu_A + \nu_B + \nu_C) - \frac{1}{2}(J_{AB} + J_{AC})$

$\alpha\beta\alpha \quad -\frac{1}{2}(\nu_A - \nu_B + \nu_C) - \frac{1}{2}(J_{AB} + J_{BC})$

$\alpha\alpha\beta \quad -\frac{1}{2}(\nu_A + \nu_B - \nu_C) - \frac{1}{2}(J_{AC} + J_{BC})$

$\alpha\alpha\alpha \quad -\frac{1}{2}(\nu_A + \nu_B + \nu_C)$

Give the number and frequencies of all the possible DQ transitions and explain how one could use them:

(a) to determine the relative signs of $J_{AB}$, $J_{BC}$, and $J_{AC}$;
(b) to recognize the "progressive" and the "regressive" single-quantum transitions.

ANSWER

Using a pencil, and your answer to the question about "tickling," show where they would appear on the top trace of the figure on page 129. For the example shown,[3] that of 2,3-dibromopropionic acid, $J_{AB}$ is of opposite sign to $J_{AC}$ and $J_{BC}$. Therefore, the proper assignment is $BrH_AH_BC{-}CH_CBr{-}COOH$, since $^2J$s are indeed negative.

Six DQ transitions are easily calculated from $2h\nu = \Delta E$, as:

$\frac{1}{2}(\nu_A + \nu_B) - \frac{1}{4}(J_{AC} + J_{BC})$

$\frac{1}{2}(\nu_A + \nu_B) + \frac{1}{4}(J_{AC} + J_{BC})$

$\frac{1}{2}(\nu_A + \nu_C) - \frac{1}{4}(J_{AB} + J_{BC})$

$\frac{1}{2}(\nu_A + \nu_C) + \frac{1}{4}(J_{AB} + J_{BC})$

$\frac{1}{2}(\nu_B + \nu_C) - \frac{1}{4}(J_{AB} + J_{AC})$

$\frac{1}{2}(\nu_B + \nu_C) + \frac{1}{4}(J_{AB} + J_{AC})$

These three pairs of lines will appear as doublets, at a resonance position equal to the average of those for nuclei $i$ and $j$, the splitting being the average of $J_{ik}$ and $J_{jk}$, where $i$, $j$, and $k$ are successively replaced by A, B, and C.

From the expressions above, and since the absolute magnitudes of $J_{AB}$, $J_{AC}$, and $J_{BC}$ are known from the regular spectrum, it will be an easy matter to find the relative signs. The progressive single-quantum transitions are placed symmetrically with

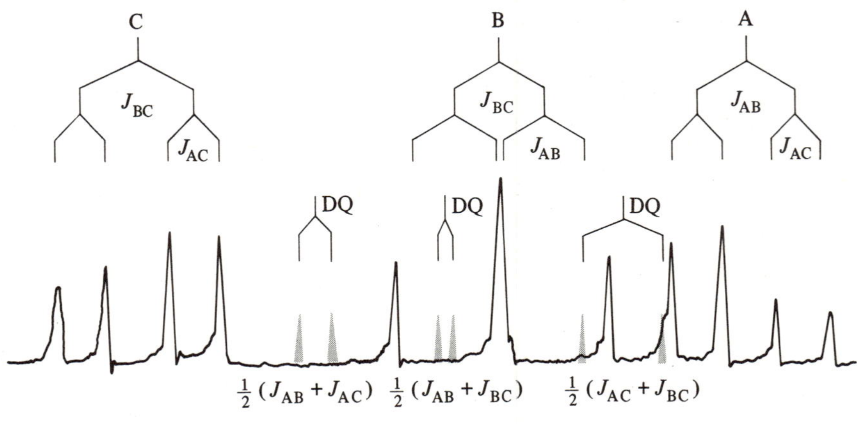


respect to the corresponding double-quantum transitions. Convince yourself of the truth of this statement in the AB case by constructing the energy diagram.

In this particular ABC case, note that the progressive $A_4$ and $C_1$ are symmetrical with respect to the corresponding (A,C) DQ transition.

*Experimentally*, the intensity of the DQ transitions increases when the separation of the corresponding SQ resonances decreases. In practice, the RF power available

[3] Ibid.

on some routine instruments is not sufficient; one needs the extra decibels built into more advanced spectrometers. There is a narrow margin of RF magnitudes between which the DQ spectra can be observed, since one must also take account of the saturation phenomena.

1. The proton magnetic resonance spectra of *cis*- and *trans*-2,3-epoxyhexane are shown in the accompanying figure. In each case the pattern observed for the epoxydic protons is reproduced, since it is known from model compounds that $^3J_{cis} \sim 4$ Hz and $^3J_{trans} \sim 2$ Hz in epoxides. Analyze these spectra and show which one belongs to which isomer.

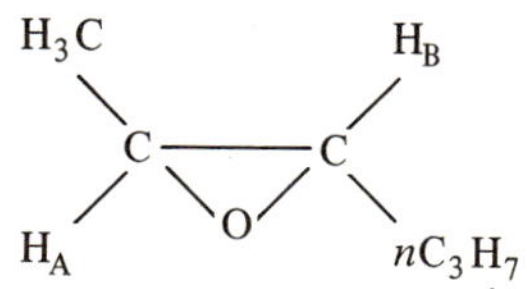

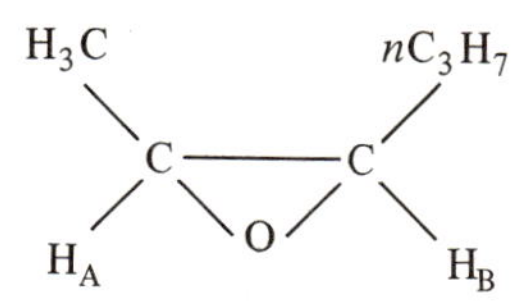

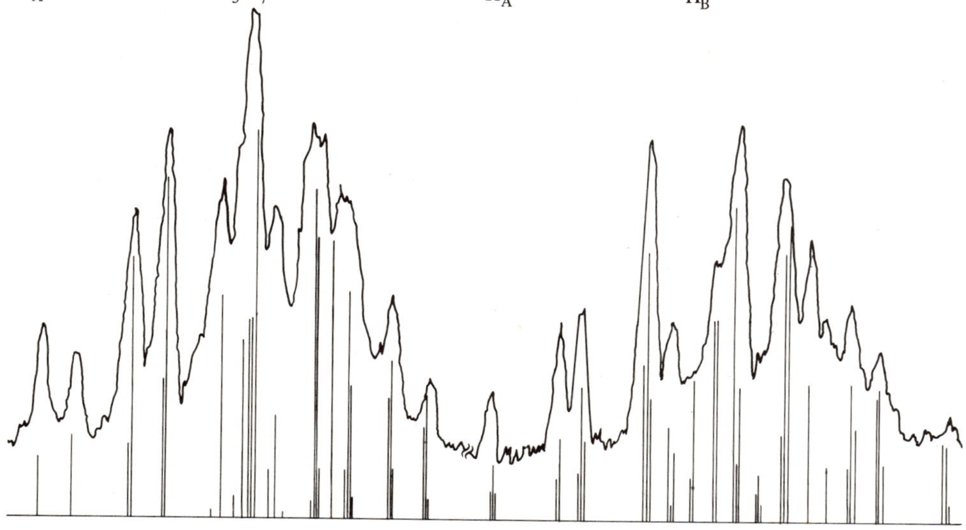

Isomer 1: AB protons  Isomer 2: AB protons

2. Find $J_{AB}$, $\nu_A$—$\nu_B$ (Hz), $\delta_A$(ppm), $\delta_B$(ppm).

60-MHz AB spectra

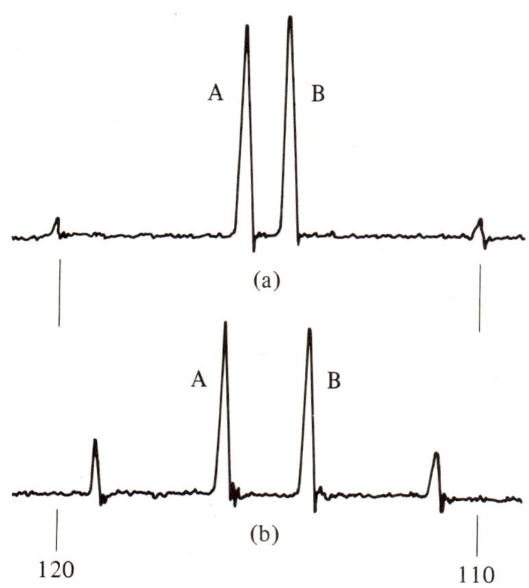

3. Determine $\nu_A$, $\nu_B$, $\nu_X$ and $J_{AB}$, $J_{AX}$, and $J_{BX}$ for both (a) and (b).

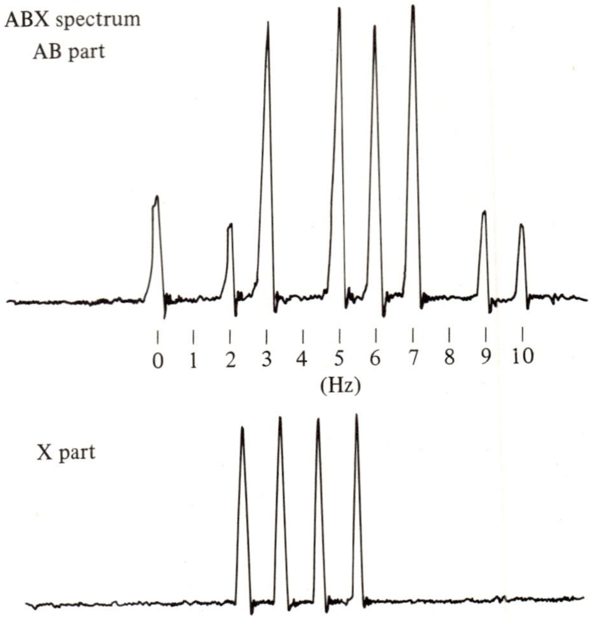


(a) $J_{AX}$ and $J_{BX}$ are of like signs

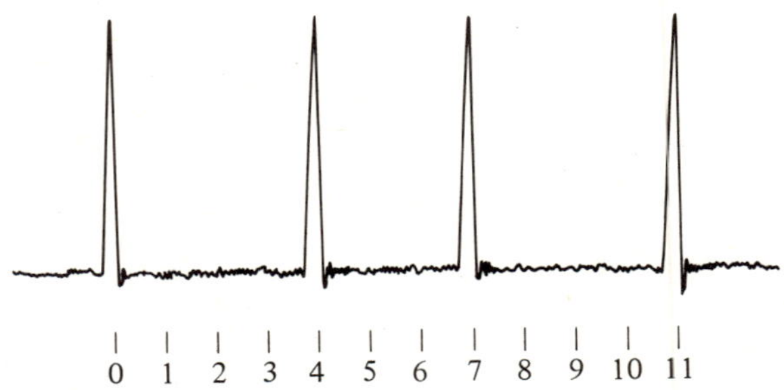


(b) $J_{AX}$ and $J_{BX}$ are of unlike signs

4. Analyze this spectrum in a similar manner as in Problem 3.

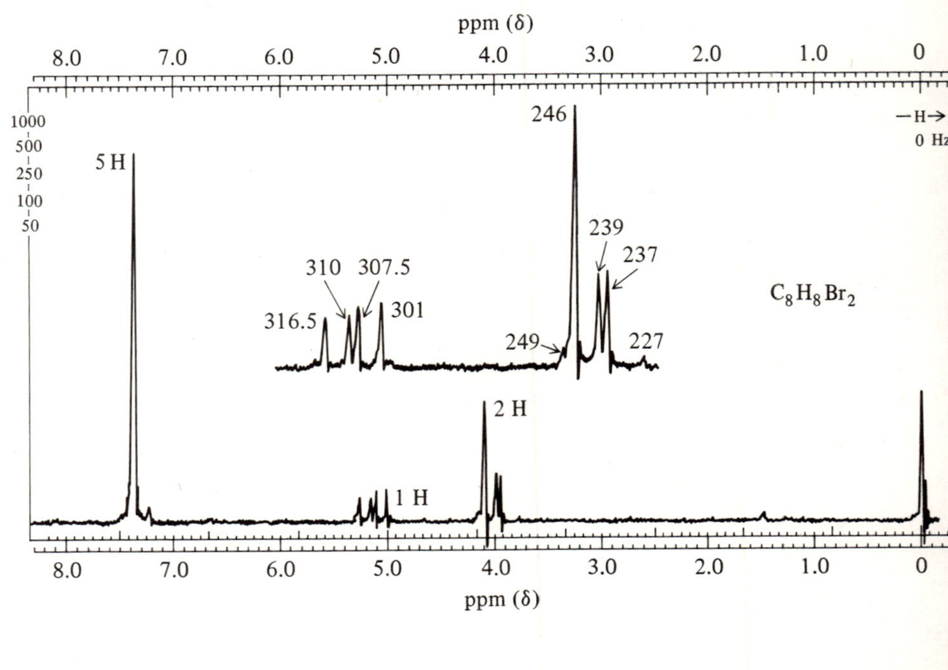

## 9.7 *SUGGESTED READINGS*

J. D. Roberts, *An Introduction to the Analysis of Spin-Spin Splitting in High-Resolution Nuclear Magnetic Resonance Spectra*, W. A. Benjamin, Inc., New York, 1961.

P. L. Corio, *Structure of High-Resolution NMR Spectra*, Academic Press, New York, 1966.

E. W. Garbisch, Jr., "Analysis of Complex NMR Spectra for the Organic Chemist," *J. Chem. Ed., 45*, 311, 402, 480 (1968).

P. Diehl, R. K. Harris, and R. G. Jones, "Sub-Spectral Analysis," in *Progress in Nuclear Magnetic Resonance Spectroscopy*, J. W. Emsley, J. Feeney, and L. H. Sutcliffe, eds., Pergamon Press, Oxford, 1967, vol. 3.

R. A. Hoffmann, "Analysis of High-Resolution NMR Spectra," *Pure Appl. Chem., 11*, 543 (1965).

E. Rahkamaa, "Sub-spectral Analysis of the 100 MHz PMR Spectrum of $^{15}$N– Pyrrole," *Mol. Phys., 19*, 727 (1970).

P. L. Corio and R. C. Hirst, "Magnetic Resonance Spectra of Multispin Systems," *J. Chem. Ed., 46*, 345 (1969).

R. J. Abraham, *The Analysis of High-Resolution NMR Spectra*, American Elsevier Publishing Co., Inc., New York, 1970.

R. G. Jones, "The Use of Symmetry in Nuclear Magnetic Resonance," in *NMR, Basic Principles and Progress*, P. Diehl, E. Fluck, and R. Kosfeld, eds., Springer-Verlag, Berlin, 1969, vol. 1, p. 97.

# CHAPTER 10 MASS SPECTROMETRY

The basic simplicity of mass spectrometry is apparent from its definition: the analysis of charged fragments arising from a molecule upon electron impact. The information so derived is then used to reconstruct the molecular formula. The various ions obtained are combined as in a jigsaw puzzle. The filiation of these ions from each other must also be established to understand the cleavages and rearrangements that occur within the mass spectrometer.

Most spectra are usually recorded with impact energies of about 70 eV (1 eV ~ 23 kcal/mole), sufficient energy to break the strongest chemical bonds. In most instances, therefore, the molecular ions decompose further. For instance, the $M^{\dagger}$ ion obtained from an A—B diatomic molecule by removal of one electron can in principle follow these four decomposition paths:

$$M^{\dagger} \equiv (A - B)^{\dagger} \rightarrow \begin{cases} A^{+} + B\cdot \\ A\cdot + B^{+} \\ A^{\dagger} + B \\ A + B^{\dagger} \end{cases}$$

These decompositions frequently occur so as to produce the most stable ions or radicals. Furthermore, many of the mechanisms of mass spectral fragmentations and rearrangements are thought to resemble those of the more familiar condensed-phase organic chemistry to a surprising extent. The mass spectrometrist relies heavily upon such analogies. It is an additional useful empirical observation that whenever different liquid-phase mechanisms are found under thermal or photochemical conditions, the mass spectral reactions tend to resemble the highly energetic photochemical rather than the ground-state thermal reactions. For instance, the molecule below is thermally stable but loses two carbon monoxide molecules under either photochemical or electron impact. Peaks occur in the mass spectrum at $m/e = 234,178$, and 56, suggestive of the loss of two CO units to give anthracene[1]:

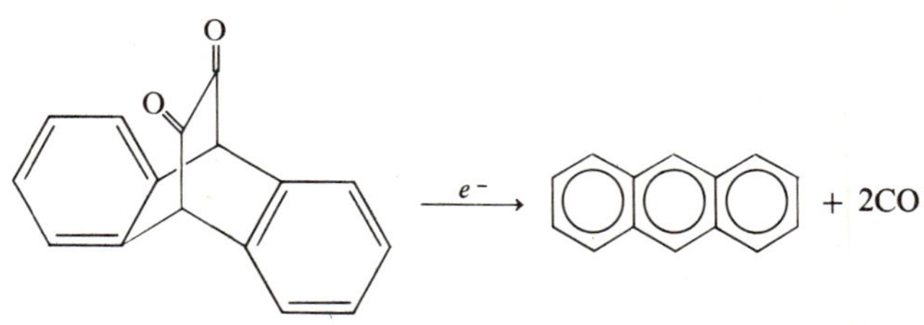

## 10.1 *Principles of the Mass Spectrometer*

The basic principles for the operation of a mass spectrometer are quite simple. The sample is vaporized in an inlet. The molecules in the gas phase diffuse into an ionization chamber in which they are subjected to electron bombardment. A high vacuum of about $10^{-7}$ torr is maintained in the ionization chamber as well as in the analyzing portions of the mass spectrometer. Electrons are emitted from a hot filament in a beam which crosses the molecular beam. Their energy—70 eV in the usual case—is acquired by maintaining the potential of the filament 70 V below that of the ionizing chamber into which the molecules are introduced. A schematic representation of a mass spectrometer is shown in Figure 10.1. The various ions which result from impact of the electron beam upon the molecules of the sample are then accelerated by passage of the beam (dashed line in Figure 10.1) through electrodes at a potential V. This

[1] J. Strating, B. Zwanenburg, A. Wagenaar, and A. C. Udding, *Tetrahedron Lett.*, 125 (1969).

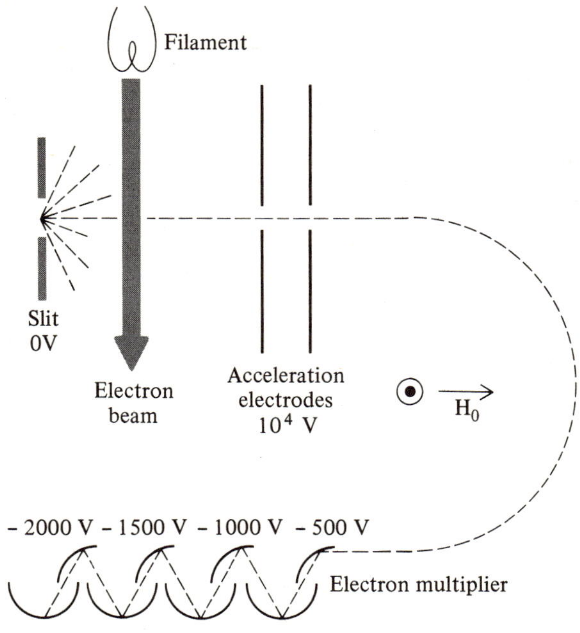


*FIGURE 10.1*
*Schematics of mass spectrometer.*

accelerating potential imparts kinetic energy to an ionic fragment of mass $m$ according to equation (1):

$$eV = \tfrac{1}{2}mv^2 \tag{1}$$

Application of a magnetic field $\mathbf{H}_0$ perpendicular to the velocity $\mathbf{v}$ of the particles will curve the trajectory of the beam into a circle centered on its axis. The resulting electromagnetic force $\mathbf{F} = e\mathbf{v} \times \mathbf{H}_0$ is equal to the centrifugal force,

$$evH_0 = \frac{mv^2}{r} \tag{2}$$

where $r$ is the radius of the orbit. From equations (1) and (2), equation (3) is readily obtained:

$$\frac{m}{e} = \frac{H_0^2 r^2}{2V} \tag{3}$$

From this equation it is seen that, at fixed values of $\mathbf{H}_0$ and $V$ only particles with a specific mass-to-charge ratio can be detected. However, by varying either the magnetic field $\mathbf{H}_0$ or the accelerating potential $V$ a spectrum of ions can be obtained.

Upon electron impact the molecule $M$ is ionized into negative and positive fragments. Most commercial instruments are capable of analyzing for both positive and negative ions. The former are more abundant, by one or two orders of magnitude, and the operation of a mass spectrometer is normally confined to the analysis of the positive ions arising from decompositions of the molecular ion $M^{+\cdot}$. Since in most commercial instruments the radius $r$ of the trajectory is fixed by the geometry of the instrument and the accelerating voltage $V$ is kept constant (typically at 100–1000 V), scanning of the magnetic field $H$ will result in the collection of ions with all the possible values of $m/e$ by the electron multiplier. The electron multiplier consists of a series of plates upon which different potentials are applied so that the positive ion striking the first plate will release a number of secondary electrons, say three or four, which in turn will strike the second plate and release more secondary electrons and so on. The result of this cascading process is a multiplication of the original charge by as much as $10^6$. This electrical current is then of sufficient magnitude to be recorded. A bar spectrum is obtained and the intensity of the peak at each integral mass unit is directly proportional to the abundance of the corresponding fragment. By convention intensities are represented as percentages of the most intense peak in the spectrum (the so-called *base* peak).

## 10.2 *The Molecular Peak*

In most instances the peak of greatest intensity *at the high-mass region* of the spectrum corresponds to the molecular peak $M^{+\cdot}$. It is always accompanied by satellites at $M + 1$ and $M + 2$, because chemical elements are mixtures of the various isotopes. To identify the molecular peak (sometimes referred to as the *parent* peak, for obvious reasons) the ionizing voltage can be reduced to 10–15 eV which further cuts down fragmentations of the molecular ion. The other peaks in the spectrum are greatly reduced and the molecular peak then stands out. Important exceptions are compounds with acidic hydrogens capable of the reaction $MH + M^+ \rightarrow MH^+ + M$, usually the case with many alcohols, amines, ethers, nitriles, sulfides, and glycols. These compounds have a high cross section for such an ion-molecule reaction, which in practice means that the very low pressure inside the mass spectrometer may not be sufficient to render this bimolecular reaction impossible. In such instances a further decrease in pressure reduces the extent of such bimolecular reactions and thus identifies the correct parent peak. Conversely, increasing the pressure increases the probability of ion molecule reactions in the ionizing chamber; this is of great diagnostic value for recognizing the compound types listed above. With modern high-resolution "double-focusing" instruments it is possible to measure the *exact* molecular weight for any peak in the spectrum. By resorting to a table of the distribution of isotopes in nature (such as Table 10.1) the elemental formula can be derived. In such "double-focusing" instruments magnetic analysis is preceded by an electrostatic analysis ensuring that the ion beam is monoenergetic, that is, that all particles of the same mass have the same speed. It is then a simple matter to distinguish, for example, between the following molecules:

| CO | $N_2$ | DCN | $H_2C{=}CH_2$ |
|---|---|---|---|
| 27.9949 | 28.0061 | 28.0187 | 28.0313 |

all of which would give the same peak at 28 mass units in a low-resolution spectrum. The methods and techniques involved in the determination of elemental composition from high-resolution data are well described in the specialized texts on mass spectrometry. From the elemental composition the degrees of unsaturation are easily derived. Given a compound of formula $C_cH_hN_nO_o$, its degree of unsaturation (number of rings or multiple bonds) is given by $c - (h/2) + (n/2) + 1$. For example, $CH_3CN{=}$

*TABLE 10.1*

*Natural Isotope Distribution for Some Elements*

| | | | | | | | |
|---|---|---|---|---|---|---|---|
| H | 99.984 | D | 0.016 | | | | |
| $^{12}C$ | 98.892 | $^{13}C$ | 1.108 | | | | |
| $^{14}N$ | 99.64 | $^{15}N$ | 0.36 | | | | |
| $^{16}O$ | 99.76 | $^{17}O$ | 0.04 | $^{18}O$ | 0.20 | | |
| $^{19}F$ | 100 | | | | | | |
| $^{32}S$ | 95.06 | $^{33}S$ | 0.74 | $^{34}S$ | 4.18 | $^{36}S$ | 0.014 |
| $^{28}Si$ | 92.28 | $^{29}Si$ | 4.67 | $^{30}Si$ | 3.05 | | |
| $^{31}P$ | 100 | | | | | | |
| $^{35}Cl$ | 75.4 | $^{37}Cl$ | 24.6 | | | | |
| $^{79}Br$ | 50.57 | $^{81}Br$ | 49.43 | | | | |
| $^{127}I$ | 100 | | | | | | |

$C_2H_3N$ has $2 - (3/2) + (1/2) + 1 = 2$ degrees of unsaturation (its triple bond). When only low-resolution spectra are available it is still possible to derive an elemental formula although with less accuracy. Next to the molecular peak the $M + 1$ and $M + 2$ peaks have intensities which mirror the elemental composition of the sample: for instance, methyl bromide has $M + 1$ and $M + 2$ peaks with 3.5 and 96 percent, respectively, of the intensity of the molecular peak at $m/e = 94$. Unlike bromine, however, most elements have only one heavily predominant isotope, so that generally the $M + 1$ and $M + 2$ peaks are only a small percentage of the molecular peak. If one considers only molecules containing the elements carbon, hydrogen, oxygen, and nitrogen, the relationships below can be used to determine the approximate number of carbons, oxygens, and nitrogens in the molecule, based upon the information in Table 10.1:

$$\frac{M+1}{M} \sim 1.1 \text{ percent c} + 0.36 \text{ percent n}$$

$$\frac{M+2}{M} \sim \frac{(1.1 \text{ percent})\, c^2}{200} + 0.20 \text{ percent o}$$

The computer can be used to generate tables of relative abundances of $M + 1$ and $M + 2$ peaks for all possible stoichiometric combinations of elements and it is straightforward to match the experimental data with such calculated output.

## 10.3 *Fragments and the Fragmentation Process*

The positive ions, either the molecular ion or fragment ions ("daughter" ions according to the familial nomenclature), can be divided into two classes: "odd-electron" ions containing an unpaired electron and designated $A^{\dot{+}}$, and "even-electron" ions designated $A^+$. By and large, odd-electron ions are relatively rare. The molecular ion *must* be an odd-electron ion and this is important in the recognition of the presence or absence of the molecular peak in a spectrum. Another important consequence is the so-called "nitrogen rule": an odd-electron ion will occur at an even $m/e$ number, unless it contains an odd number of nitrogen atoms.

Loss of neutral fragments is usually observed when small stable molecules can be expelled, the most common examples being CO, $N_2$, $H_2$, $H_2O$, $CH_4$, $CO_2$, $NH_3$, HCN, NO, $NO_2$, $H_2C{=}C{=}O$, $H_2C{=}CH_2$, and $HC{\equiv}CH$. The most reliable way to establish the origin of a given fragment or "daughter" ion is by means of the so-called metastable peaks. These are usually observed in a number of places throughout the spectrum at nonintegral masses with a low intensity (of the order of 1 percent) and appear as diffuse gaussian-shaped peaks. They correspond to decompositions which have occurred during the flight of the ion through the instrument: *after* nearly complete acceleration but *before* complete mass analysis. Decomposition of an ion affects the momentum and acceleration of the ion by orders of magnitude more than it affects its velocity. Hence the ion is accelerated with the original mass $m_1$ and is analyzed with the resultant mass $m_2$:

$$r = \frac{m_2}{eHv} = \frac{m_2}{eH}\left(\frac{2eV}{m_1}\right)^{1/2}$$

which can be set identical to the equation of the motion of a hypothetical ion with mass $m^*$:

$$r = \frac{1}{H}\left(\frac{2m^*V}{e}\right)^{1/2} \longrightarrow m^* = \frac{m_2^2}{m_1}$$

A more exact expression is

$$m^* = \frac{m_2^2}{\{m_1[1 + (m_1 - m_2)]/m_2\}\,\dfrac{E}{ev}}$$

where $E$ is the mean energy released in the fragmentation of the ion with mass $m_1$ to that with mass $m_2$. However, the approximate expression $m^* = m_2^2/m_1$ usually suffices to establish which are the two ions related by a given metastable $m^*$.

## 10.4 *Structure of Fragment Ions*

The determination of the exact structure of the various ions is perhaps the most difficult as well as ambiguous part of mass spectrometry. For the ionizing voltage provides sufficient energy to break, at least in principle, any bond and the only available information is the mass of the resulting fragment.

The most convenient technique for deriving the constitution of a particular fragment or deriving the structure of a functional group is isotopic labeling. Some of the more widely used reactions follow:

$$\text{ROH, RCO}_2\text{H, RNH}_2 \xrightarrow[\text{exchangeable hydrogens}]{D_2O} \text{ROD, RCOOD, RND}_2$$

$$-\text{CH}{=}\text{CH}- \xrightarrow[\text{unsaturation}]{D_2,\ Pd/C} \text{CHD}-\text{CHD}-$$

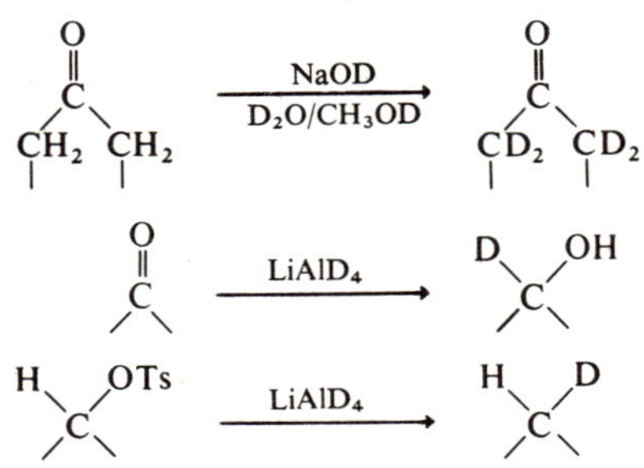


$$\text{RCOOH} \xrightarrow[D_2O]{CD_2N_2} \text{RCOOCD}_3$$

The elegant and classical experiments which have demonstrated the occurrence of the $C_7H_7^+$ tropylium ion in the fragmentation of benzyl derivatives $C_6H_5{-}CH_2X$ will be summarized as an example. It is observed that the mass spectra of toluene and of cycloheptatriene are virtually identical. It is then likely that the more stable aromatic tropylium ion, a $6\pi$-electron species, forms in the mass spectrometer from the benzyl cation $C_6H_5{-}CH_2^+$ generated initially. This would account for the considerable intensity of those peaks in the spectra of aromatic molecules whenever such a benzylic fragmentation is feasible. Isotopic labeling was used to show that this $C_7H_7^+$ ion is indeed the tropylium ion. Consider a molecule of the type $C_6H_5\underset{\alpha}{CH_2}{-}\underset{\beta}{CH_2}R$. Deuterium substitution of the $\beta$ carbon does not affect the position of the $C_7H_7^+$ peak, while $\alpha$ deuterium substitution increases its $m/e$ value by one unit per deuterium. Hence, fragmentation occurs at the $\alpha{-}\beta$ bond. The $C_7H_7^+$ ion can and does decompose further by expulsion of a neutral acetylene molecule H$-$C$\equiv$C$-$H and formation of a $C_5H_5^+$ ion. This sequence is corroborated by the observation of the corresponding metastable peak.

If the $C_6H_5{-}CH_2{-}$ fragment has been deuterium labeled, the deuterium will partition itself between the $C_2H_2$ and $C_5H_5^+$ fragments, in a manner dependent upon the initial structure of the $C_7H_7^+$ ion. Let us then assume that the acetylene fragment originates from two *adjacent* carbon atoms in $C_7H_7^+$. The different possibilities, corresponding either to a benzyl cation or to a tropylium cation formulation, are summarized below:

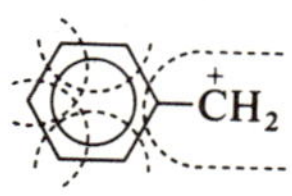


B

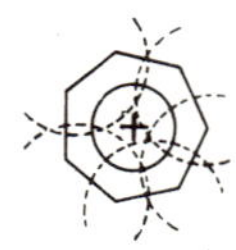

T

Making the further assumption that all such fragmentations are equally probable under the high-energy conditions within the mass spectrometer, a deuterium introduced in the *meta* position, for instance, would have a probability of $\frac{3}{5}$ of being *retained* in the $C_5H_5^+$ ion if $C_7H_7^+$ has the benzyl structure and a probability of $\frac{5}{7}$ if it has the tropylium structure. The calculated values for the various sites of isotopic substitution are indicated below; comparison with the experimental values leaves no doubt that the tropylium ion formulation is consistent with the experimental data.

Percent D retained by $C_5H_5^+$ fragment if $C_7H_7^+$ ion is:

| Benzyl | Tropylium | Exptl. |
|---|---|---|
| α-D ($\frac{4}{5}$) 80 percent D | ($\frac{5}{7}$) 71 percent D | 67 |
| *o*-D ($\frac{4}{5}$) 80 percent D | ($\frac{5}{7}$) 71 percent D | 68 |
| *m*-D ($\frac{3}{5}$) 60 percent D | ($\frac{5}{7}$) 71 percent D | 69 |
| *p*-D ($\frac{3}{5}$) 60 percent D | ($\frac{5}{7}$) 71 percent D | 70 |

## 10.5 *The McLafferty Rearrangement*

Since the McLafferty rearrangement is not confined to carbonyl-containing molecules, even though it has a strong probability of occurrence there, it will be stated here in general terms. Given the following arrangement of atoms:

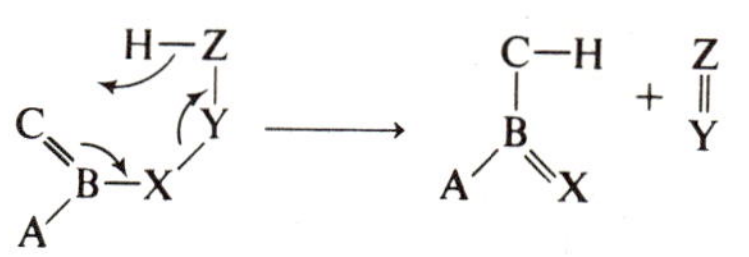


the McLafferty rearrangement[2] consists of a hydrogen transfer concomitant with an electronic reorganization resulting in the above fragments. Often, as in the case of ketones, the Z=Y fragment is an olefin, whereas the A—BX—CH moiety contains an oxygen heteroatom and is thus stabilized by resonance:

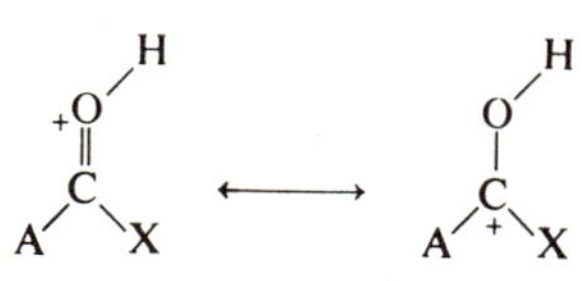


In all these rearrangements the migrating atom is almost always hydrogen, but it is not known whether the transfer occurs as a proton, radical (H·) or hydride shift. The atoms X, Y, and Z are most often carbon, oxygen, and nitrogen, whereas the doubly bonded atoms B and C can be two carbons, one carbon and an oxygen, one carbon and one nitrogen, and so on. The olefinic daughter ion competes with the heteroatom moiety

---

[2] F. W. McLafferty, *Anal. Chem.*, *31*, 82 (1959).

for the charge. For instance, it is found that in the mass spectral decomposition of ethyl esters of the type

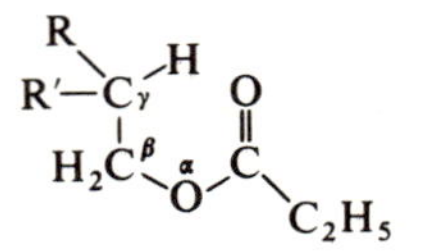

when R = $C_6H_5$ and R′ = H, the styrene formed corresponds to the most intense peak in the spectrum; when R = R′ = $CH_3$ the olefin peak is still rather strong, but it decreases markedly with R = $CH_3$ and R′ = H, and is unimportant with the ethyl esters for which R = R′ = H. Increased lability of the transferable hydrogen will facilitate the rearrangement, secondary hydrogens being superior to primary hydrogens. Djerassi[3] utilized isotopic labeling to demonstrate this effect:

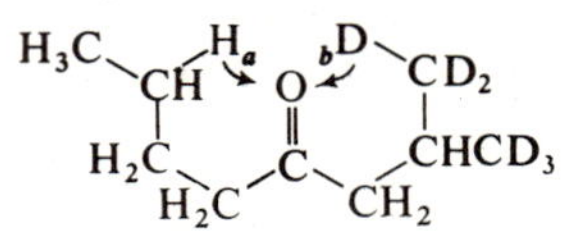

Path *a* is 10 times more favorable than path *b*. However, the isotope effect must also be taken into account. For $CDH_2CH_2CH_2COOCH_3$ and $CD_2HCH_2CH_2COOCH_3$ the isotope effect (ratio of hydrogen to deuterium transferred) is found to be 1.0:0.88. It seems that a critical H···O interatomic distance of less than 1.8 Å between the $\gamma$ hydrogen and the oxygen of the carbonyl is a requirement for ketones to undergo the McLafferty rearrangement, and numerous examples have been drawn from the steroid field.[4] However, the importance of the relative orientation of the C—H and the C—O bonds in the high-energy species undergoing the rearrangement has yet to be assessed.

## 10.6 *Ion Cyclotron Resonance Spectroscopy*

It starts almost as a fairy tale: After mass spectrometry and magnetic resonance were introduced to each other it was love at first sight, they married, and lived happily ever after; of this marriage, one beautiful child is already born and is called ion cyclotron resonance. . . .

The principles of this new spectroscopy are rather simple, and can be explained by reference to the equation of motion for a charged particle in a magnetic field **H**. Equating the centrifugal force $mv^2/r$ with the interaction $q\mathbf{v} \times \mathbf{H}$ of the charge $q$ traveling at a velocity **v** with the magnetic field **H** gives equation (4), identical to the equation of motion for an ion in a mass spectrometer:

$$\frac{mv^2}{r} = qvH \tag{4}$$

The charged particle travels in a circular orbit centered on the axis of the magnetic field **H**. It follows that the angular velocity $\omega_c = v/r = qH/m$ is characteristic of a given mass-to-charge ratio for a fixed value of the magnetic field $H_0$, whereas the radius of the orbit $r = mv/qH$ depends on the velocity of the charged particle. Hence, for a swarm of ions with a distribution of velocities **v** in a uniform magnetic field **H**, there will be a distribution of concentric circular orbits, but they will all be traversed at

[3] C. Djerassi *et al.*, *Tetrahedron*, *22*, 1391 (1966).
[4] C. Djerassi *et al.*, *J. Amer. Chem. Soc.*, *87*, 817 (1965).

the same angular velocity $\omega_c$. The idea then is to dispose a pair of parallel plates, and to establish between them an oscillating electrical field $E(t)$, which is applied normal to **H** and with a frequency $\omega/2\pi$. When $\omega/2\pi$ becomes equal to the frequency $\omega_c/2\pi$ characteristic of the cyclotronic motion of a given ionic species, energy is absorbed by the orbiting ions (Figure 10.2).

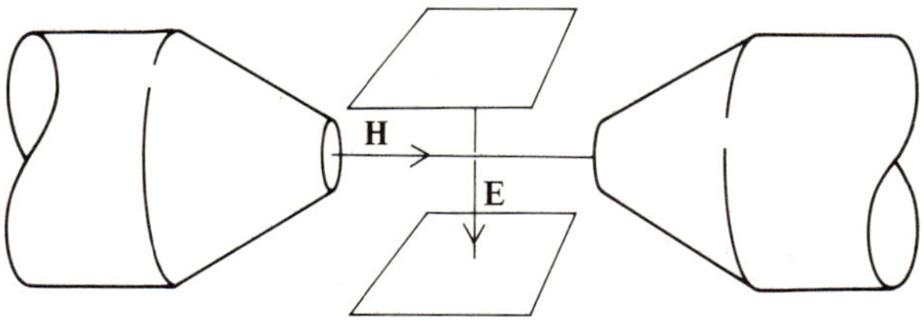


*FIGURE 10.2*
*The ion cyclotron resonance apparatus.*

The analogy with nmr is obvious and the detailed phenomenological equations are very similar to their nmr counterparts. It can be shown that the power absorbed from the oscillating electrical field corresponds to a Lorentzian line shape, as does the power absorbed in the nmr experiment by the system of spins subjected to a RF magnetic field. It is possible to calculate the icr (ion cyclotron resonance) line shapes in the presence of charge transfer and other chemical reactions: again, as in magnetic resonance, line broadenings are observed and can be related quantitatively to the rate of exchange. Double-resonance techniques are also possible. If the velocity of one type of ion is increased by the application of an electrical field $E_2(t)$ with a large amplitude and a frequency $\omega_2/2\pi$, there will be changes in the concentrations and in the velocities of all the other ions in chemical equilibrium with the ion irradiated. The ions are generated in a source similar to that in a normal mass spectrometer, then injected into the cyclotron equipped with the condenser plates used to induce resonance and eventually double resonance. For example, a reaction which can be initiated in the mass spectrometer ion source is

$$CD_4 + N_2 \longrightarrow CD_3^+ + CD_4^+ + N_2^+ + N_2D^+ + C_2D_5^+$$

In this reaction it is observed that the $N_2D^+$ product does *not* disappear when the ionizing potential is reduced below the appearance potential for $N_2^+$ production. Hence, one can presume that a reaction such as

$$CD_4^+ + N_2 \longrightarrow N_2D^+ + CD_3$$

is responsible for the production of $N_2D^+$. Ion cyclotron resonance can be used to transform these presumptions into a virtual certainty. Upon strong irradiation at the resonance frequency for the $CD_4^+$ ion ($\omega_2/2\pi = 425.5$ kHz), the $N_2D^+$ resonance ($\omega_1/2\pi = 285.5$ kHz) is sharpened markedly; the above reaction must therefore be an important pathway for production of $N_2D^+$. A second example comes from the study of the acetonitrile ion-molecule chemistry. One of the numerous reactions for which the mechanism can be identified securely by icr spectroscopy is

$$\underset{m/e\,=\,40}{CH_2CN^+} + CH_3CN \longrightarrow \underset{m/e\,=\,54}{C_3H_4N^+} + HCN$$

By using the labeled molecules $CH_3^{14}CN$, $CH_3C^{15}N$ and $CH_3CN$, it is possible to show that this reaction proceeds with retention of the neutral reactant nitrogen atom in the product ion, while the HCN eliminated has its hydrogen originating only from the ionic reactant. A cyclic transition state can be postulated to account for these observations:

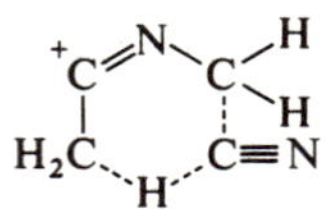

It is indeed found that there is a peak at *m/e* 81 corresponding to this complex between $CH_3CN$ and $CH_2CN^+$ in the single-resonance spectrum. However, irradiation of the *m/e*-81 resonance does *not* affect the *m/e*-54 resonance of the $C_3H_4N^+$ product. Hence the $C_4H_5N_2^+$ cannot be the intermediate leading to it. Furthermore, double-resonance experiments demonstrate the availability of two distinct paths between the *m/e*-40 reactant and the *m/e*-54 product:

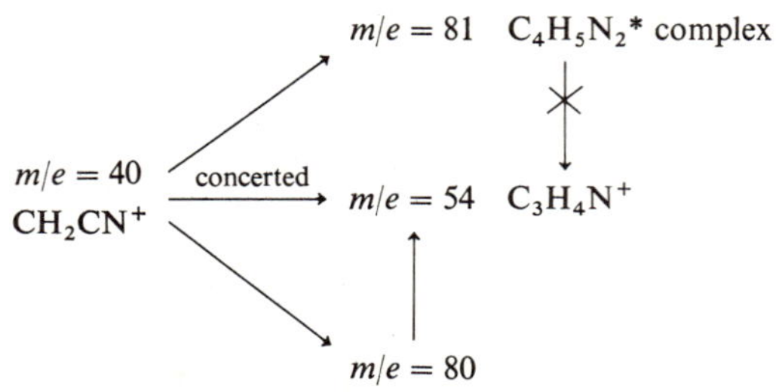


The second path is a two-step reaction via a *m/e*-80 ionic intermediate, which could conceivably be either $C_4H_4N_2^+$ or $C_5H_6N^+$. The former possibility can be rejected since use of trideuterioacetonitrile shifts this peak to *m/e* 86 and not to *m/e* 84. The $C_5H_6N^+$ intermediate is in turn formed by the reaction

$$\underset{\substack{\text{(complex)}\\ m/e\,=\,81}}{C_4H_5N_2^+} + CH_3CN \longrightarrow \underset{m/e\,=\,80}{C_5H_6N^+} + CN_2H_2$$

followed by

$$C_5H_6N^+ \longrightarrow \underset{m/e\,=\,54}{C_3H_4N^+} + C_2H_2$$

All these elementary steps can be demonstrated rigorously using icr spectroscopy by such combinations of isotope labeling and double resonance. Ion cyclotron resonance, although a relatively new technique still in its infancy, is clearly going to revolutionize the study of reactions in the gas phase.

## 10.7 *PROBLEMS*

SAMPLE PROBLEM

The mass spectrum of the β-diketone

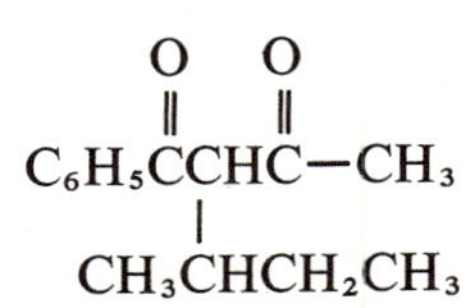

displays important peaks at $m/e = 176$, 161 (15 percent) and 147 (65 percent). Account for these fragmentations.

ANSWER

The McLafferty rearrangement:

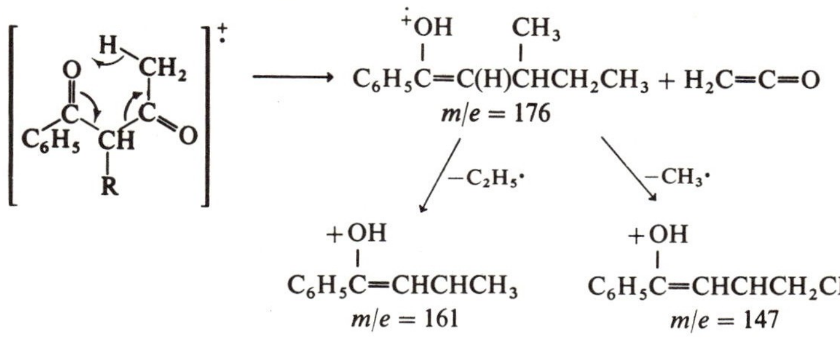


1. Substituted *n*-butylbenzenes decompose upon electron impact by two major routes resulting in peaks at $M = 42$ and $M = 43$ mass units. Indicate a likely mechanism for each pathway.
2. It is frequently observed that NO is lost by nitrobenzenes in the mass spectrometer. In *para*-methoxy nitrobenzenes, the relative intensities of the $(M{-}NO)^+$ ion are:

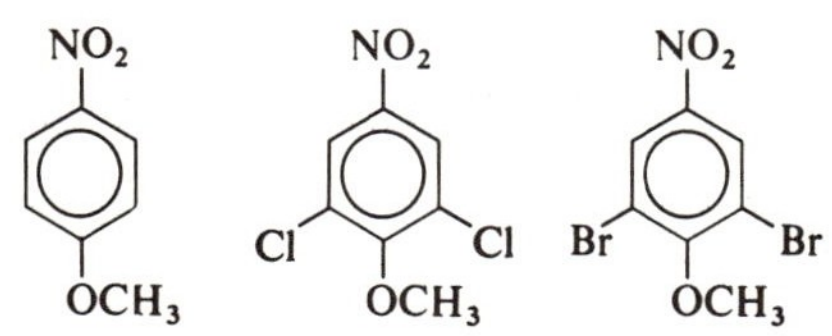


| | | | |
|---|---|---|---|
| $\frac{(M - NO)^+}{M^{+\cdot}}$ | 1.4 | 0.06 | 0.08 |

Account for these findings.
3. An important fragmentation of hexahelicene (below) leads to $m/e = 300$ directly from the molecular ion. Account for this observation.

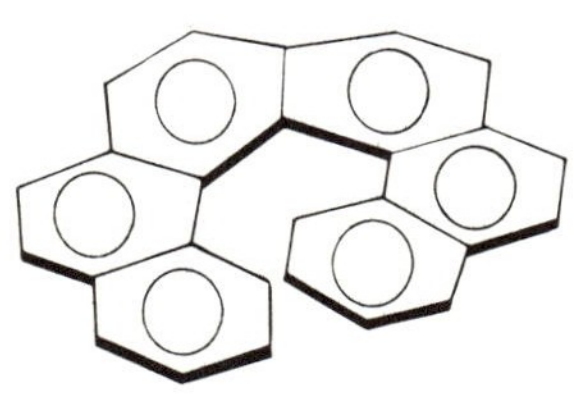

4. The most intense peak in the mass spectrum of cyclohexanone occurs at $m/e = 55$. It is absent from the spectrum of cyclobutanone. Instead, the base peak of cyclobutanone is observed at $m/e = 42$. In the spectrum of 3,3,5,5-cyclohexanone-$d_4$, the base peak is displaced to $m/e = 57$, and to $m/e = 56$ in the spectrum of 2,2,6,6-cyclohexanone-$d_4$. How do you reconcile these seemingly different observations?
5. The relative acetyl ion, $CH_3CO^+$, abundances in the mass spectra of acetophenone, *p*-amino acetophenone, and *p*-nitro acetophenone are 1.00, 0.26, and 4.5 respectively. Explain.
6. Pyridazine (below) shows strong mass spectral peaks at $m/e = 52$ and $m/e = 26$. Why are they important?

## 10.8 *SUGGESTED READINGS*


F. W. McLafferty, *Interpretation of Mass Spectra*, W. A. Benjamin, Inc., New York, 1966.

K. Biemann, W. J. McMurray, and P. V. Fennessey, "Ion Types, a Useful Concept in the Interpretation of High-Resolution Mass Spectra," *Tetrahedron Letts.*, 3997 (1966).

R. C. Dougherty, "A Perturbation MO Approach to the Interpretation of Organic Spectra. The Relationship Between Mass Spectrometric, Thermolytic, and Photolytic Fragmentation Reactions," *J. Amer. Chem. Soc.*, *90*, 5780, 5788 (1968).

K. Biemann, "Interpretation of High-Resolution Mass Spectra," *Advan. Mass Spectrometry*, *4*, 139 (1968).

J. D. Baldeschwieler, "Ion Cyclotron Resonance Spectroscopy," *Science*, *159*, 263 (1968).

J. M. S. Henis, "An Ion Cyclotron Resonance Study of Ion-Molecule Reactions in Methanol," *J. Amer. Chem. Soc.*, *90*, 844 (1968).

G. A. Gray, "Study of Ion-Molecule Reaction Mechanisms in Acetonitrile by Ion Cyclotron Resonance," *J. Amer. Chem. Soc.*, *90*, 2177, 6002 (1968).

G. Eadon, J. Diekman, and C. Djerassi, "Application of Ion Cyclotron Resonance to the Structure Elucidation of the $C_3H_6O^{\ddagger}$ Ion Formed in the Double McLafferty Rearrangement," *J. Amer. Chem. Soc.*, *97*, 3986 (1969).

M. Inoue and S. Wexler, "Isotopic Exchange in $CH_4-D_2$ and $CD_4-H_2$ Mixtures Studied by Ion Cyclotron Resonance Spectroscopy. The Mechanism of Self-Induced Labeling of Methane by Tritium," *J. Amer. Chem. Soc.*, *91*, 5730 (1969).

J. I. Brauman and K. C. Smyth, "Photodetachment Energies of Negative Ions by Ion Cyclotron Resonance Spectroscopy. Electron Affinities of Neutral Radicals," *J. Amer. Chem. Soc.*, *91*, 7778 (1969).

M. T. Bowers and D. D. Elleman, "Analysis of Ion-Molecule Reactions in Hydrogen-Methane Mixtures Using Ion Cyclotron Resonance," *J. Amer. Chem. Soc.*, *92*, 1847 (1970).

R. Venkataraghaven, R. J. Klimowski, and F. W. McLafferty, "On-Line Computers in Research. High-Resolution Mass Spectrometry," *Accounts Chem. Res.*, *3*, 158 (1970).

G. C. Goode, R. M. O'Malley, A. J. Ferrer-Correia, and K. R. Jennings, "Ion Cyclotron Resonance Mass Spectrometry," *Nature*, *227*, 1093 (1970); *Chem. Brit.*, *7*, 12 (1971).

W. T. Huntress, Jr., and D. D. Elleman, "An Ion Cyclotron Resonance Study of the Ion-Molecule Reactions in $CH_4$—$NH_3$ Mixtures," *J. Amer. Chem. Soc.*, *92*, 3565 (1970).

M. T. Bowers, D. D. Elleman, R. M. O'Malley, and K. R. Jennings, "Analysis of Ion-Molecule Reactions in Allene and Propyne by Ion Cyclotron Resonance," *J. Phys. Chem.*, *74*, 2583 (1970).

# CHAPTER 11 APPLICATION OF SPECTROSCOPY TO REACTION KINETICS

From the organic chemist's point of view the purpose of kinetic measurements is to deduce information about the mechanism of a particular reaction, and to establish relationships between reactivity and structure. The various spectroscopies provide convenient means for following the course of a reaction, and thus measuring its rate. In fact, spectroscopic techniques can be used in two different ways to obtain kinetic information:

1. as a mere analytical tool for monitoring the concentrations of various chemical species as a function of time; and
2. as a special probe into the transformation in a chemical reaction of the energy state for one molecule into the energy state of another, the effect upon the spectroscopic transition being dependent both upon its duration and upon the lifetimes of the interconverting molecular systems.

## 11.1 *Spectroscopy as an Analytical Tool*

Use of spectroscopy as an analytical tool for measuring concentrations is straightforward. A few examples from the optical spectroscopies (ir, uv, etc.) will illustrate the methods and techniques employed.

It is a very general experimental finding that the absorbance $A = \log_{10} I/I_0$, where $I_0$ is the intensity of the radiation to which the sample is submitted and $I$, its value after passage through the sample, is related to the concentration $c$ of absorbing molecules by

$$A = \log_{10}\frac{I}{I_0} = abc$$

where $b$ is the path length and the proportionality constant $a$ is the extinction coefficient for a particular chromophore; it is a constant if the wavelength is also held constant. This relationship is referred to as Beer's law. The absorbance is not always proportional to concentration, however, either because the radiation used is not strictly monochromatic, or sometimes because of molecular interactions in solution (self-association obviously tends to negate Beer's law).

Consider a first-order reaction where reactant R goes to product P:

$$\mathrm{R} \xrightarrow{k} \mathrm{P}$$

The corresponding rate expressions are

$$\ln\frac{C_0(\mathrm{R})}{C_t(\mathrm{R})} = kt \qquad \text{if disappearance of reactant is followed}$$

$$\ln\frac{C_\infty(\mathrm{P}) - C_0(\mathrm{P})}{C_\infty(\mathrm{P}) - C_t(\mathrm{P})} \qquad \text{if formation of product is followed}$$

where $C(\mathrm{R})$ and $C(\mathrm{P})$ represent concentrations of reactant R and product P at time zero, any time $t$, and at infinite time. Since in the expression of Beer's law the coefficients $a$ and $b$ are not functions of time, measurement of the absorbance (of either reactant or product) allows the determination of the rate $k$, given by the expression

$$k = \frac{1}{t} \ln \frac{A_\infty - A_0}{A_\infty - A_t}$$

(if the reaction is followed by the increase of an absorption band of the product), where $A$ represents the absorbance of the product at infinite time, time zero and time $t$.

For instance, the acid-catalyzed hydration of phenylacetylene to give acetophenone,

$$\mathrm{Ph{-}C{\equiv}C{-}H + H_2O \xrightarrow{H^+} Ph{-}\overset{\overset{\displaystyle O}{\|}}{C}{-}CH_3}$$

is conveniently followed spectrophotometrically by the appearance of a peak for acetophenone at 259 nm. This technique provides a convenient tool for following rates in the range of $10^{-2}$–$10^{-6}$ $\sec^{-1}$ (corresponding to reaction half-lives between 1 min and 200 hrs).

In a similar manner, ir spectroscopy can be used as an analytical tool to follow the concentration of a particular species. It may be a little more cumbersome to use than uv-visible spectra, since ir absorptions appear more complex and their extinction coefficients are not readily measured. Infrared techniques, however, are ideal for measuring rates of isotopic exchange since, for instance, the disappearance of a C—H stretching vibration will be accompanied by the appearance of a C—D band at approximately $1/\sqrt{2}$ the original frequency. For example, the rate of deuterio deprotonation of fluorene in methanol-$d_1$ ($CH_3OD$) is conveniently followed by the appearance of the C—D stretch at 2160 $cm^{-1}$. The analysis of the data and the rate expressions are identical to those described above for uv and visible.

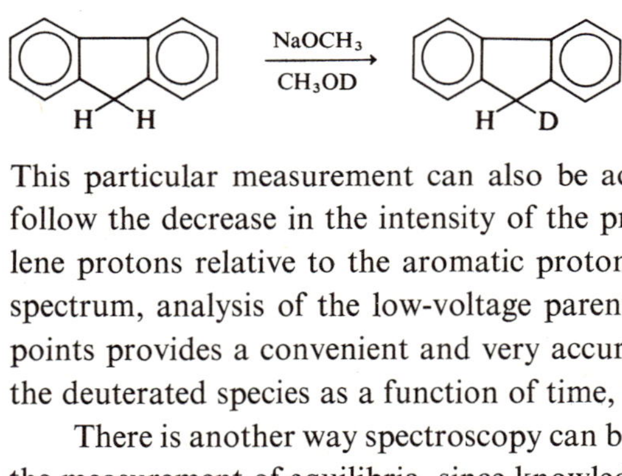


This particular measurement can also be accomplished by other methods. One can follow the decrease in the intensity of the proton magnetic resonance for the methylene protons relative to the aromatic protons as a function of time. Or, in the mass spectrum, analysis of the low-voltage parent peak for each of the individual kinetic points provides a convenient and very accurate measurement of the mole fraction of the deuterated species as a function of time, and hence the rate of exchange.

There is another way spectroscopy can be used as a simple analytical tool, namely, the measurement of equilibria, since knowledge of the concentrations of all individual species in an equilibrium constitutes a direct measure of the equilibrium constant.

For example, from a knowledge of the uv and visible absorption spectra of two individual hydrocarbons and their corresponding anions, one can obtain their equilibrium concentrations and therefore the difference in $pK_a$ values for the two hydrocarbons AH and BH:

$$\mathrm{A{-}H + M^+B^- \overset{K}{\rightleftharpoons} A^-M^+ + B{-}H}$$

Another example is the measurement of the equilibrium concentrations of the enol and keto forms of acetylacetone:

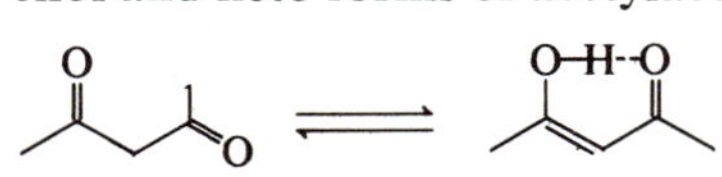

This can be accomplished either by ir or nmr. In the ir spectrum the relative intensities of the absorptions in the O—H stretching region and in the C—O stretching region indicate the equilibrium concentrations of each species. With nmr, either the proton or the $^{13}C$ spectrum can be used to determine the relative amounts of the keto and enol form and hence the equilibrium constant (see sample problem).

## 11.2 *Spectroscopy and Lifetimes of Molecular States*

The Heisenberg uncertainty principle states that the product of the uncertainty $\Delta E$ on the measurement of energy of a particular quantum by the uncertainty $\Delta t$ on the measurement of the time for the transition to occur is of the order of $h/2\pi$.

The average time spent in an excited (metastable) state is by definition short in comparison with that for the ground state, so $\Delta t$ can be taken as the lifetime of the excited state. Hence the linewidth $\Delta v$ for the transition

$$\Delta v = \frac{\Delta E}{h} = \frac{\Delta E\,\Delta t}{h\,\Delta t} \sim \frac{1}{2\pi\,\Delta t}$$

is proportional to the inverse of the lifetime of the excited state. Table 11.1 gives the minimum lifetimes for observation of *separate* species to be feasible with a particular type of observation.

*TABLE 11.1*

*Minimum Lifetimes for Observation of Separate Species*

| Type of observation | Lifetime (time scale in seconds) |
|---|---|
| Electron diffraction | $10^{-20}$ |
| Neutron diffraction | $10^{-18}$ |
| x-Ray diffraction | $10^{-18}$ |
| Ultraviolet | $10^{-15}$ |
| Visible | $10^{-14}$ |
| Infrared-Raman | $10^{-13}$ |
| Electron spin resonance | $10^{-4}$–$10^{-8}$ |
| Microwave | $10^{-4}$–$10^{-10}$ |
| Nuclear magnetic resonance | $10^{-1}$–$10^{-9}$ |
| Nuclear quadrupole resonance | $10^{-1}$–$10^{-8}$ |
| Mossbauer (iron) | $10^{-7}$ |
| Molecular beam | $10^{-6}$ |
| Physical isolation | $10^{+2}$ |

SOURCE: E. L. Muetterties, *Inorg. Chem.*, *4*, 769 (1965).

The optical spectroscopies (uv, visible, ir, and Raman) which obey the Franck–Condon principle have transitions which are essentially instantaneous in comparison with the rates of usual chemical reactions, since the electronic transitions occur much faster than the nuclear motions. Conversely, the magnetic resonance techniques are much slower, and give access to the lifetimes of excited states in the range of about $10^{0}$–$10^{-9}$ sec.

## 11.3 *Principles of the Magnetic Resonance Lineshape Methods*

In the simplest case two nuclei of spin $\frac{1}{2}$ can exchange between two sites A and B either intramolecularly or intermolecularly. If the chemical shifts for the two sites are identical, the spectrum will remain unaffected, as seen from the diagram of Figure 11.1 and consideration of the original magnetic resonance experiments (Chapter 1). The width of the line observed for the $A_2$ system—for example the amino protons of methylamine—is determined only by the lifetime of the excited state $\beta$. Figure 11.1 shows that the lifetime of the excited state is not affected by the horizontal lines connecting the ground states, provided chemical exchange is faster than the flipping of the nuclear spin. However, when the transition energies and hence the chemical shifts are

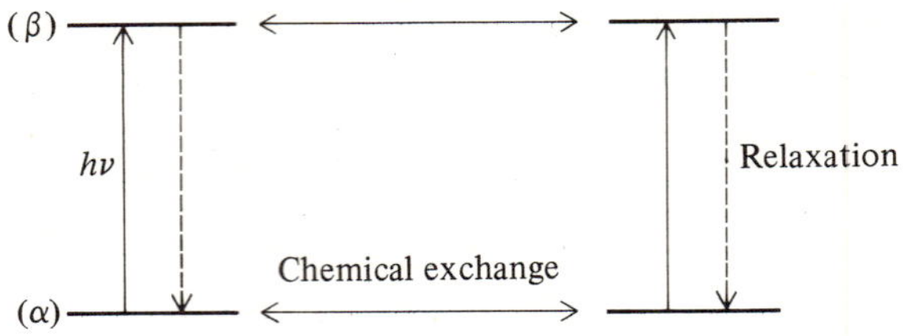


*FIGURE 11.1*
*Exchange in an $A_2$ system.*

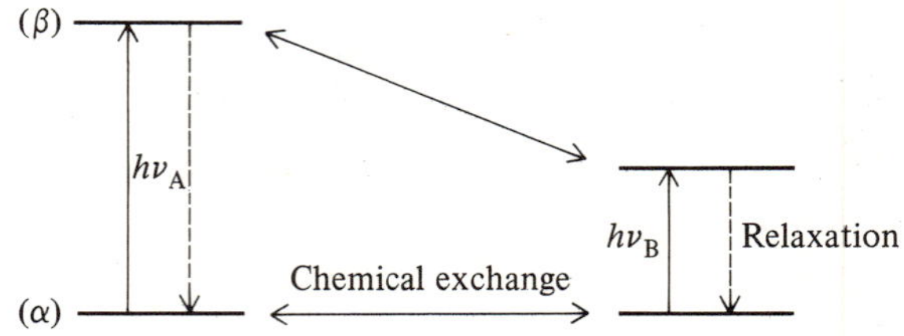


*FIGURE 11.2*
*Exchange in an AB system.*

different, then exchange-dependent linewidths are observed and the lineshape will be determined by the relative magnitudes of the exchange and of the relaxation rates (Figure 11.2).

All of the observations are consistent with the idea that the sudden jump of a nucleus from site A to site B is much faster than the duration of a transition during which it would flip its spin. The rate of exchange $k$ (i.e., the probability of a jump from A to B or vice versa) is related to the residence times $\tau$ at each site by

$$k = \frac{1}{2\tau}$$

There are various mathematical treatments that describe these exchange phenomena. The lineshape is obtained as a function of the populations of the different sites, their chemical shifts, and the probabilities for exchange between the various sites. In general, one handles these expressions with digital computers since analytical relationships are derived easily only for the simplest cases (AB exchange with or without coupling) and even these are already quite complex.

Qualitatively, in the ideal situation, the experimentalist can vary the rate of the reaction studied sufficiently so that he will obtain the magnetic resonance spectrum of the unreacting system in a given preferred instantaneous structure, at the bottom of a potential well. As the temperature is raised, the onset of exchange between the different potential wells available on the rearrangement path is evidenced by a general line broadening (due to the fact that the residence time in each site is decreasing). At still higher temperatures the exchange processes will be so rapid that the magnetic resonance spectrum sharpens to that of a composite spectrum corresponding to the weighted

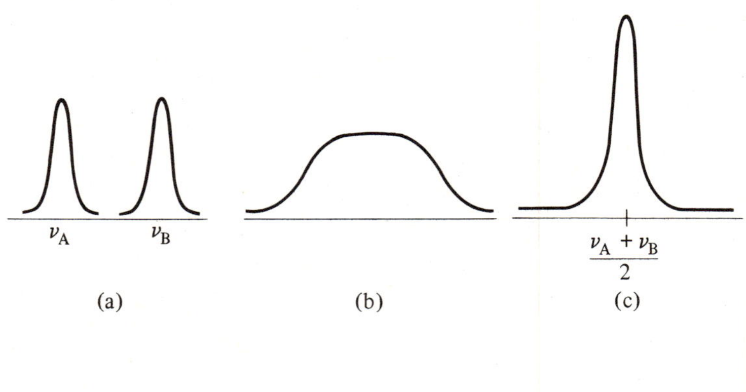


*FIGURE 11.3*
*An uncoupled AB system undergoing exchange. (a) Slow or no exchange. (b) Coalescence. (c) Fast exchange.*

average of the various constituents. For an uncoupled AB system undergoing exchange there are three phases of interest:

1. Slow exchange, represented by sharp resonance lines characteristic of the instantaneous structure (Figure 11.3a).
2. Coalescence, the point where the number of lines is reduced to one and the absorption line has a characteristic flat-topped shape (Figure 11.3b).
3. Fast exchange, in which the spectrum consists of one sharp line only (Figure 11.3c).

### *11.3.1 The Bullvalene Molecule*

A classical example of the determination of the rate of a fast reaction from the exchange-broadened magnetic resonance spectrum is the work of M. Saunders on the rearrangement of bullvalene. The molecule of bullvalene undergoes a succession of Cope rearrangements, each of which transforms it into the identical molecule (Figure 11.4). This isodynamic process can be followed nevertheless since it exchanges the various protons in the molecule between the four different sites cyclopropyl $H_1$, olefinic $H_2$, olefinic $H_3$, and another allylic site on a bridgehead, $H_4$ (Figure 11.5). If one follows a given proton, and submits the molecule to the succession of Cope rearrangements of the type shown, it is possible to move the proton from site to site and bring it back to its original position after it has occupied all the possible sites in the molecule.

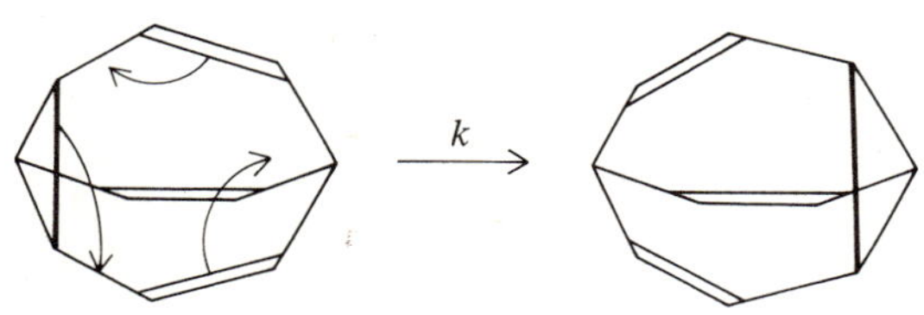


*FIGURE 11.4*
*Bullvalene undergoing a degenerate Cope rearrangement.*

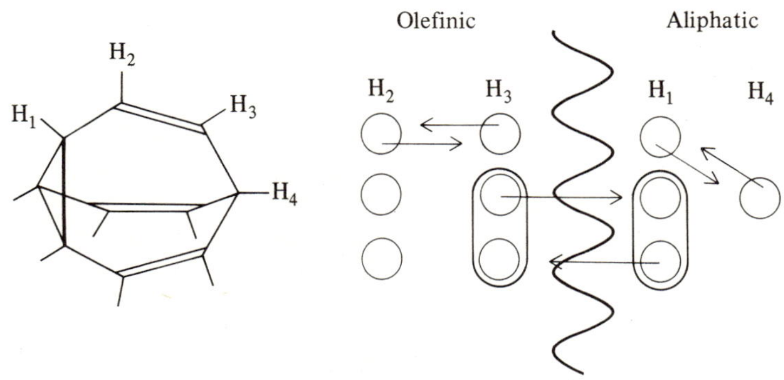


*FIGURE 11.5*
*Bullvalene and its rearrangement pattern.*

No Cope rearrangement, however, can interchange in a single step $H_1$ and $H_2$, or $H_3$ and $H_4$. The entire rearrangement pattern is schematized according to Saunders in Figure 11.6. One expects to see four separate groups of lines in the pmr spectrum at a temperature low enough so that bullvalene is "frozen" into its *instantaneous* nuclear configuration. Upon warming of the sample rearrangement starts to take place and the protons exchange between the various sites until at high-enough temperatures only one line is observable. A given proton can occupy all of the different sites if enough reaction steps occur. The predicted changes in the spectrum indeed occur and a sharp single line is observed above 90°C whereas there is much fine structure in the spectra recorded below −45°C.

These spectral changes with temperature can be matched with spectra calculated

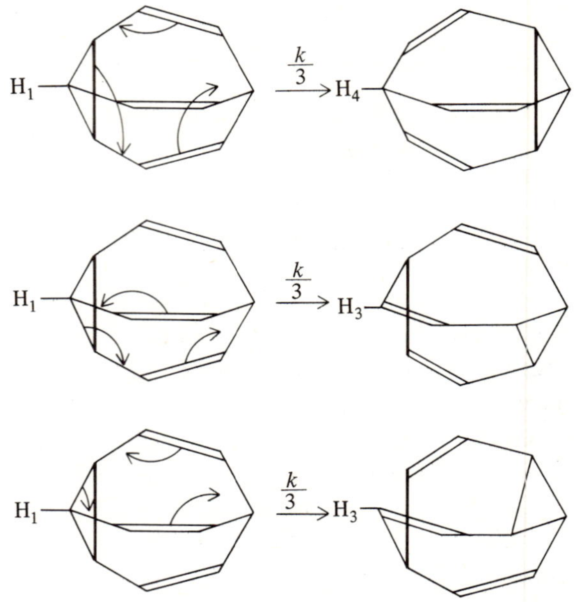


*FIGURE 11.6*
*Construction of the transition probability matrix (P).*

theoretically for various rates of exchange. For bullvalene the line shapes were calculated using the theory of Anderson[1] and Kubo.[2] With any theory of the exchange lineshape the input must consist of:

1. the chemical shifts for the various sites;
2. the relative populations of the various sites (for instance, 3:3:3:1 for sites 1, 2, 3 and 4, respectively, in bullvalene);
3. the probabilities for jumping from one site to another, expressed in matrix form ($P$) and as fractions of the rate $k$ for a single step.

$$(P) = \begin{vmatrix} \diagdown & 0 & 2k/3 & k/3 \\ 0 & \diagdown & k/3 & 0 \\ 2k/3 & k/3 & \diagdown & 0 \\ k & 0 & 0 & \diagdown \end{vmatrix}$$

For instance, the probability for a proton initially in site 1 of going to site 4 is $k/3$, as seen from the rearrangement pattern in Figure 11.6, since there are equal probabilities for that proton of "becoming" an $H_3$ proton when the two other bonds of the cyclopropane ring in Figure 11.4 are broken. Finally the input must also include the lineshape and the linewidth for each resonance in the absence of exchange at the temperature at which the rate is being calculated. It is often a valid approximation to use the lineshape and width of some standard, say TMS, at the same temperature. In the case of bullvalene, calculated lineshapes using the information summarized below were obtained and matched with the measured shape of a single coalesced line at a series of temperatures (Figure 11.7). The results obtained are summarized in Table 11.2. When the calculated spectra are compared with the experimental spectra the agreement is very good despite the neglect of spin-spin coupling in the mathematical treatment. [Inclusion of spin-spin coupling necessitates rather involved formalisms which are extremely costly in terms of computer time. Whereas exchange between four spins

[1] P. W. Anderson, *J. Phys. Soc. Jap.*, *9*, 316 (1954).
[2] R. Kubo, *J. Phys. Soc. Jap.*, *9*, 935 (1954).

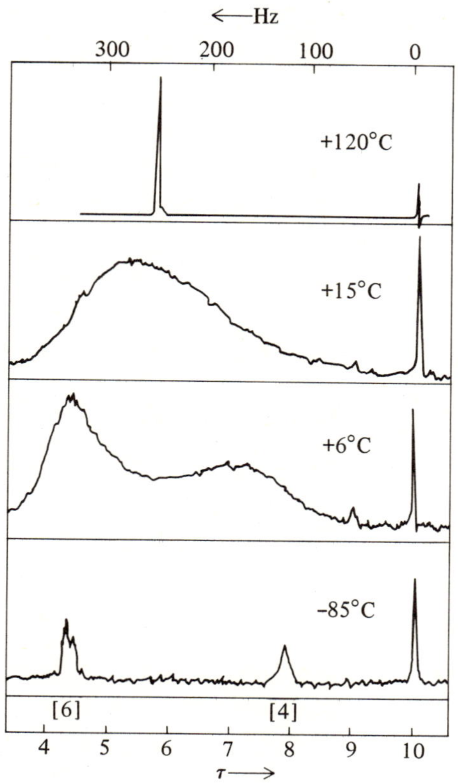


*FIGURE 11.7*
*The pmr spectrum of bullvalene at various temperatures (TMS internal reference). The low-field multiplet integrates for six hydrogens and corresponds to the $H_2$ and $H_3$ olefinic protons; $H_1$ and $H_4$ correspond to the relatively broad high-field band.*

*TABLE 11.2*

*Calculated Lineshape and Rates of Rearrangement of Bullvalene at Various Temperatures*

| $T$ (°C) | Linewidth at half-height (Hz) | $k$ (sec $^{-1}$) |
|---|---|---|
| 10.5 | 197 | 1,200 |
| 28.1 | 59.9 | 4,400 |
| 33.6 | 41.7 | 6,000 |
| 43.1 | 23.5 | 10,500 |
| 56.2 | 10.8 | 23,000 |
| 65.2 | 6.7 | 36,000 |
| 73.2 | 4.7 | 53,000 |
| 82.7 | 3.1 | 79,000 |

would require ~10 sec of computation for each rate constant if coupling is neglected, it can take ~1 hr if (strong) coupling has to be considered explicitly.]

### *11.3.2 Scope and Limitations of Magnetic Resonance Methods*

The extreme importance of magnetic resonance methods stems from the fact that they give access to rate processes in a range of considerable interest to the organic chemist—processes with energy barriers of 3–20 kcal/mole. This range encompasses rotational barriers, ring inversion barriers, atom-inversion barriers in molecules such as amines, pseudorotations as in certain pentacoordinate phosphorus compounds, intermolecular exchange reactions, and so on. These resonance methods allow the determination of rates of reactions much faster than those studied by conventional chemical means, yet slower than the very fast reactions for which Eigen has devised special relaxation methods of measurement (see ref. 5).

The limitations of these methods are essentially the gross approximations which are performed in the selection of the input data for the calculation of theoretical lineshapes. A first and severe limitation is that, except in the case of elementary spin systems such as the AB case, coupling is frequently neglected in order to keep computational time down to a manageable figure. The error involved is difficult to appreciate: As already stated, there is an extremely good apparent fit in the bullvalene case between the experimental lines and the calculated line shapes even though coupling is omitted. This does not ensure that the corresponding rate constants are identical to the true rate constants. This neglect of coupling is more legitimate if the spectrum shows only weak coupling (i.e., as one approaches the limit of exchange between several uncoupled sites). Another source of error, even more serious in practice, is that, due to solvent effects, the chemical shifts for the various sites in a molecule undergoing exchange are not temperature independent. Yet it is a common practice to "read" these chemical shifts from low-temperature spectra, in which the exchange is "frozen," and then to assume that the same chemical shifts apply throughout the entire temperature range studied. A classical example of the pitfalls in this approach is afforded by dimethyl formamide, $(CH_3)_2NCHO$. In this molecule two nonequivalent methyl resonances are seen at low temperatures, indicating partial double-bond character for the carbon-to-nitrogen bond:

From exchange broadening of these methyl resonances at higher temperatures the energy barrier can, in principle, be evaluated. It was first determined as 7 kcal/mole in 1956 and then quickly jumped to 22–23 kcal/mole. In subsequent years, other investigators again addressed themselves to the problem and the magnitude of the energy barrier continued to oscillate, as shown in Figure 11.8. It is now (hopefully) stabilized at about 20.5 kcal/mole. The reason for this wide oscillation is that, at any given temperature, both the rate constant $k$ for the internal rotation and the difference in chemical shift $\Delta\nu$ between the two methyl groups vary considerably with the concentration of dimethyl formamide (Figure 11.9). This variation is accounted for by the possibility of self-association of dimethylformamide. If a monomer-dimer equilibrium

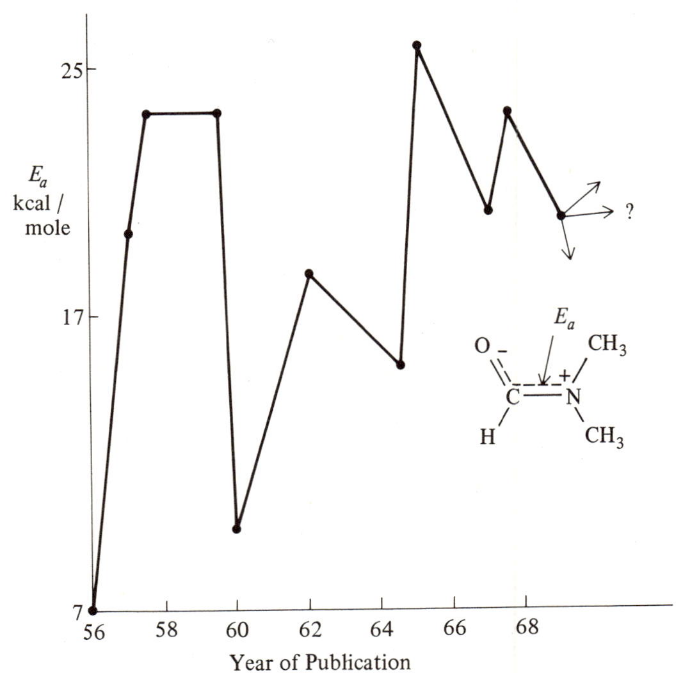


*FIGURE 11.8*
*Energy barrier for rotation in dimethyl formamide.*

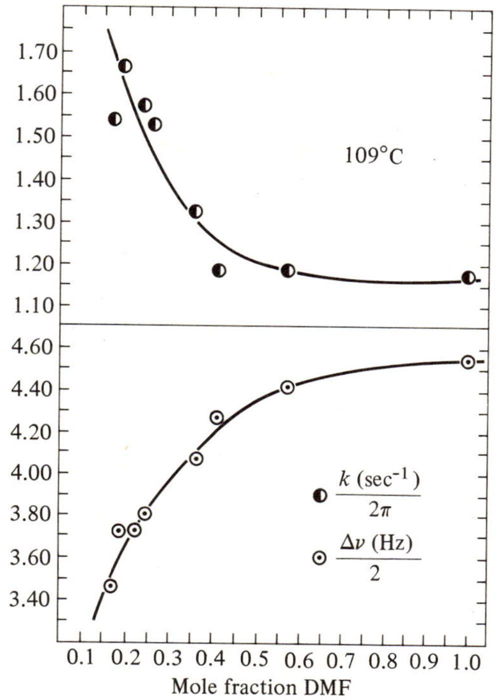


*FIGURE 11.9*
*Concentration dependence of the rate for internal rotation and of the limiting chemical shift: DMF in* $CCl_4$.

is assumed, then the thermodynamic parameters for the formation of the complex and the chemical shifts in the monomer and in the dimer are as follows[3,4]:

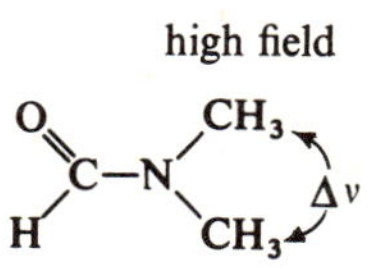


$$\begin{array}{lcl} 2\ \text{monomer} & \rightleftharpoons & \text{dimer} \\ M+M & & D \end{array}$$

$\Delta H^o \sim -6$ kcal/mole
$\Delta S^o \sim -14$ e.u.
$\Delta\nu_M = 0.1$ ppm
$\Delta\nu_D \sim 0.2$ ppm

This self-association was ignored or not properly taken into account by the earlier investigations of the hindered rotation in dimethyl formamide, hence the very dramatic errors. Part of these errors, however, were also due to recourse to approximate analytical expressions rather than digital computer simulation of the complete exchange-broadened line shapes. These approximate expressions should be used only to generate trial values for the total lineshape analysis using a computer. In a molecule such as dimethyl formamide one should also be wary of other possible origins for temperature-dependent linewidths than chemical exchange. For instance, quadrupolar relaxation of the $^{14}N$ nucleus washes out in part the $^{14}N$ couplings to the various kinds of protons in the molecule and this phenomenon is temperature dependent. A reliable study of the hindered rotation in this molecule then necessitates prior decoupling of the $^{14}N$ nucleus.

---

[3] M. Rabinovitz and A. Pines, *J. Amer. Chem. Soc.*, *91*, 1585 (1969).
[4] R. C. Neuman, Jr., and L. B. Young, *J. Phys. Chem.*, *69*, 2570 (1965).

Another parameter mandatory as input for the calculation of an exchange-broadened line shape is the line shape in the absence of exchange, and the corresponding linewidth: Often, this parameter is really difficult to procure, and substituting an ideal Lorentzian line shape and the linewidth of TMS at the same temperature may not be entirely satisfactory.

A further practical difficulty inherent in these resonance techniques for rate determinations is the measurement of temperature within the nmr sample tube at the exact location of the RF sensor. If the temperature cannot be measured to better than say $\pm 2°C$, an already difficult undertaking, then the energy barrier cannot be determined to better than about $\pm 1$ kcal/mole in terms of free energy, and therefore usually worse for either the enthalpy or the entropy. A final word of caution: The excellent linearity of an Arrhenius plot is the necessary accompaniment to a careful study but does not necessarily vindicate approximations made beforehand.

## 11.4 *Principles of the Magnetization Transfer Methods*

Another way to deduce rates from magnetic resonance spectra is an extension of the relaxation methods derived by M. Eigen and his coworkers.[5] A system at chemical equilibrium is pushed off equilibrium by some abrupt perturbation. The return to equilibrium occurs with a time constant which gives very simply the rate of reaction. The equations describing such relaxation behavior are

$$t\frac{dx}{dt} + x = 0 \rightarrow x = x_0 \exp\frac{-t}{\tau}$$

where $x$ is a property of the system under study, $x_0$ is its equilibrium value, and $\tau$ is the relaxation time.

For instance, in the case of exchange between two sites A and B, one can observe signal A and its magnetization $M^A$, and then watch its decay when B is irradiated with a strong secondary RF field. The A nuclei at site A will "tell" the effect of irradiation at site B since they exchange between the two sites A and B. If the linewidth of the exchange-broadened line is defined as

$$\frac{1}{\tau_{1A}} = \frac{1}{\tau_A} + \frac{1}{T_{1A}}$$

where $1/T_{1A}$ is the bandwidth in the absence of exchange, and $1/\tau_A$ is the residence time at site A, then

$$M^A = M^A(t \rightarrow \infty)\left[\exp\left(\frac{-t}{\tau_{1A}}\right) + 1\right]$$

so that the characteristic time for decay of the signal at A will be used to provide the value of the residence time $\tau_A$, and hence of the rate $k$ for exchange since $k = 1/(\tau_A + \tau_B) = 1/2\tau_A$. This experiment can be performed in alternate ways, for instance, by monitoring the recovery at site A when irradiation at site B stops. It is possible to identify nmr absorption bands for chemical groups which undergo slow reversible chemical exchange with water (*viz.* OH and NH protons) by this method. Even the

[5] M. Eigen, *Angew. Chem. Int. Ed.*, *3*, 1 (1964).

trace amounts of water in commercial solvents such as deuteriochloroform are sufficient. Taking for instance phenol, $C_6H_5OH$, in $CDCl_3$ solution, irradiation of the *water* resonance at about 1.5 ppm causes a marked decrease in the intensity of the *phenolic* OH at 5–6 ppm.[6]

## 11.5 *PROBLEMS*

SAMPLE PROBLEM

1. Deduce the molecular structure of the compound $C_5H_8O_2$ from the magnetic resonance spectrum, which was obtained at 15.1 MHz, and make any pertinent remarks.

ANSWER

The scale of chemical shifts, together with the frequency of measurement, rule out a $^1H$ spectrum and indicate that it is a carbon-13 spectrum. From the molecular formula, one calculates two units of unsaturation. The spectrum is seen to consist of a high-field quartet, a doublet in the middle of the spectrum and a singlet at low field (other minor lines look as if they were attributable to an impurity). The chemical groups: —C—H (doublet), $—CH_3$ (quartet), and —C=O (singlet) are the simplest consistent with the molecular formula and the spectrum. Acetylacetone comes to mind as fulfilling the description:

It is predominantly enolized and fast proton transfer renders the two methyl signals equivalent, together with the two oxygen-bonded carbons. The minor lines are due to the keto form: $^{13}C=O$ at low field, a $CH_3$ quartet overlapping that for the enol, and a triplet for $^{13}CH_2$ barely visible at slightly lower fields. The equilibrium constant for β-diketone ⇌ enol is deduced to be ~4 from the relative intensities of the carbonyl singlets.

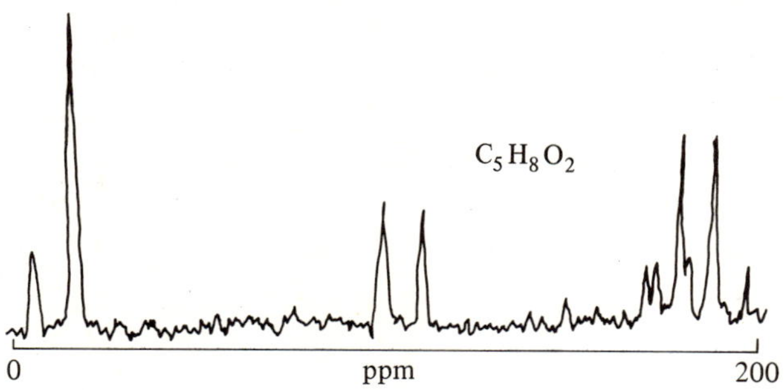


SAMPLE PROBLEM

2. Lineshapes at various temperatures are given for an uncoupled two-site system undergoing chemical exchange. The difference in the chemical shifts $\delta\nu_\infty$ of the two sites is 21.7 Hz, and the linewidth in the absence of exchange is 1.1 Hz. A slow-exchange curve is also obtained with a doublet separation = 20.5 Hz. Calculate the rate of exchange $k$ corresponding to each of these curves using the (very) approximate expressions:

$$k = \frac{\pi}{\sqrt{2}}\,\delta\nu_\infty \qquad \text{(coalescence)}$$

[6] J. Feeney and A. Heinrich, *Chem. Comm.*, 295 (1966).

$$k = \frac{\pi}{2}\frac{\delta v_\infty}{B} \qquad B = \text{exchange line broadening (fast exchange)}$$

$$k = \frac{\pi}{2}(\delta v_\infty^2 - \delta v_{exp}^2)^{1/2} \qquad \text{(slow exchange)}$$

ANSWER

Fast-exchange lineshapes, of course, always consist of one line with variable width; subtracting from the experimental linewidth, at half-height, the "natural" linewidth in the absence of exchange gives the following values of $B$, and hence $k$, from the above expression: The exact values are obtained from a total lineshape analysis,

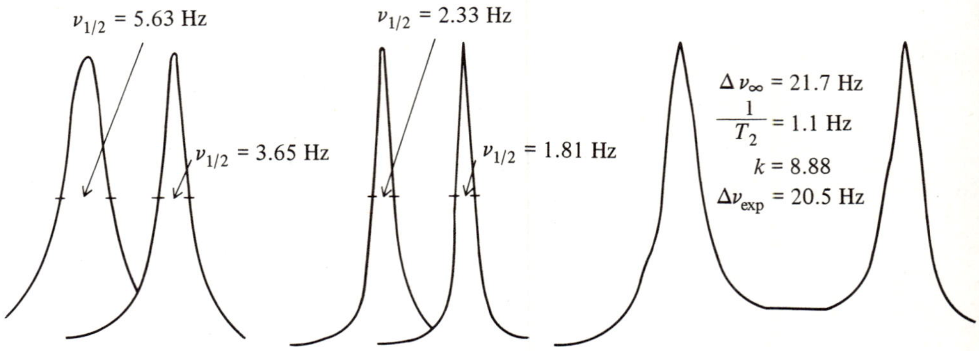


| $\nu_{1/2}$ (Hz) | B (Hz) | $k$(sec$^{-1}$) (approx.) | $k$(sec$^{-1}$) (exact) |
|---|---|---|---|
| 5.63 | 4.53 | 165 | 171 |
| 3.65 | 3.55 | 273 | 300 |
| 2.33 | 1.23 | 600 | 640 |
| 1.81 | 0.71 | 1035 | 1220 |

using a digital computer; the agreement between the approximate and the exact values is fair, within 10 percent. As a further check these fast-exchange rate constants are greater than the rate at coalescence $k = 48.3\ \text{sec}^{-1}$. Likewise, the slow-exchange rate is calculated to be $k = 15.6\ \text{sec}^{-1}$, as compared with the exact value of $8.88\ \text{sec}^{-1}$.

SAMPLE PROBLEM

3. The $^{31}$P nmr spectrum of $(CH_3)_2NPF_4$ is observed to be temperature-dependent. While it consists of a triplet of triplets at $-100°C$, it simplifies to a 1:4:6:4:1 quintet above $-50°C$. This behavior is readily attributed to exchange between the apical and equatorial fluorines in the trigonal bipyramid appropriate for this compound:

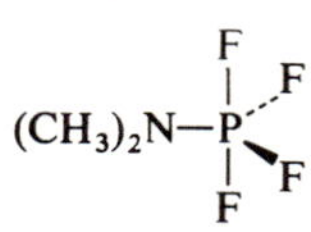


The central line of the spectrum remains sharp throughout the temperature range in which these spectral changes occur. Other lines are broadened markedly at some temperatures. Show that this information can be used either to prove or to disprove the Berry mechanism,[7] where in an isodynamic process the two apical fluorines exchange simultaneously with the two equatorial fluorines.

ANSWER

Only exchange between different sites produces line broadenings. The spectral assignment is as shown.[8] It can be written readily, taking advantage of the equivalence

[7] R. S. Berry, *J. Chem. Phys.*, *32*, 933 (1960).
[8] G. M. Whitesides and H. L. Mitchell, *J. Amer. Chem. Soc.*, *91*, 5384 (1969).

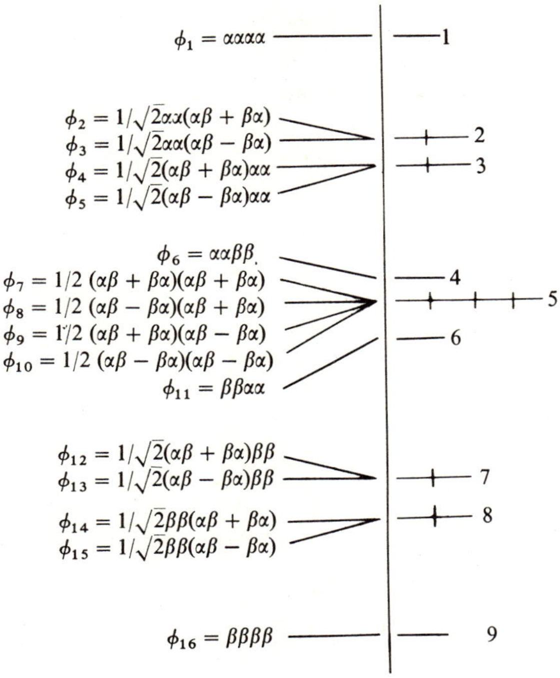

*Spectral assignments for the* $^{31}$P *spectrum of* $(CH_3)_2NPF_4$

between each of the apical and each of the equatorial fluorine nuclei. While, for instance, line 3 is exchanged into line 2 by the Berry mechanism, or line 4 into line 6, line 5 (the central line) is exchanged into itself, leading to no observable change. A non-Berry mechanism would exchange line 5 with, for example, line 4 or line 6.

1. The syn-anti isomerization of isophorone oxime (1), which is readily followed by pmr, is reversible and proceeds according to

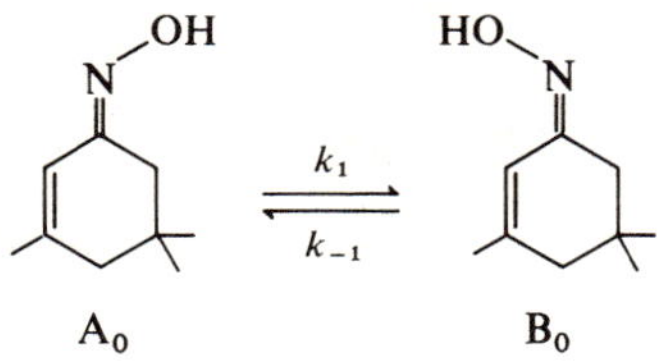

$A_0$      $B_0$

*1a*      *1s*

$$\frac{d(1a)}{dt} = -k_1(1a) + k_{-1}(1s)$$

$$\frac{d(1s)}{dt} = k_1(1a) - k_{-1}(1s)$$

The following results were obtained at 93°C:

| Time (min) | Percent **1a** |
|---|---|
| 0 | 89.4 |
| 32 | 81.9 |
| 45 | 80.3 |
| 58 | 79.8 |
| 71 | 77.5 |
| 84 | 76.5 |
| 110 | 74.4 |

At that temperature, the equilibrium constant $K = k_1/k_{-1} = 1.575$. Integrate the differential equations to obtain an expression for the velocity constant of the forward and of the reverse reactions in terms of the equilibrium constant $K$ and of the initial concentrations $A_0$ and $B_0$. Evaluate $k_1$ and $k_{-1}$.

2. The ring inversion of cyclohexane is a first-order reaction, for which the following rates are obtained:

| Temperature (°C) | $k$ (sec$^{-1}$) |
|---|---|
| −24 | 2490 |
| −34.4 | 1150 |
| −42.5 | 566 |
| −61.4 | 64 |
| −77.9 | 6.4 |
| −81.8 | 4.0 |

Plot $\log_{10} k$ against the reciprocal of absolute temperature, and use the least-squares slope to calculate the Arrhenius activation energy $E_a$ and the Eyring activation parameters $\Delta H^*$ and $\Delta S^*$.

3. In the heptamethylbenzeneonium ion a degenerate rearrangement occurs in which each carbon in turn becomes $sp^3$ hybridized and thus bears two methyl groups:

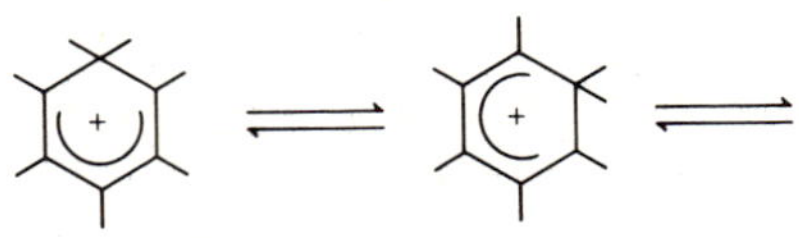


Two mechanisms have been proposed. Mechanism A is a succession of 1,2-methyl shifts, whereas mechanism B is a random process in which a methyl group is removed from position 1 and has then equal probabilities ($\frac{1}{6}$) of attaching itself to any carbon. It is found that upon irradiation at the chemical shift for $CH_3$ at positions 2 and 6, the intensity of the signal for the methyl group at position 1 is affected; additional irradiation at positions 3(5) and 4 produces no further decrease in the intensity of signal 1. Do these observations distinguish between the two mechanisms? How?

## 11.6 *SUGGESTED READINGS*


M. Saunders, "Measurement of Rates of Fast Reactions Using Magnetic Resonance," in *Magnetic Resonance in Biological Systems*, A. Ehrenberg, B. G. Malmström, and T. Vanngard, eds., Pergamon Press, Oxford, 1967.

F. A. L. Anet and A. J. R. Bourn, "Nuclear Magnetic Resonance Line-Shape and Double-Resonance Studies of Ring Inversion in Cyclohexane-$d_{11}$," *J. Amer. Chem. Soc.*, *89*, 760 (1967).

S. Forsén and R. A. Hoffman, "Study of Moderately Rapid Chemical Exchange Reactions by Means of Nuclear Magnetic Double Resonance," *J. Chem. Phys.*, *39*, 2982 (1963).

S. Forsén and R. A. Hoffman, "Exchange Rates by Nuclear Magnetic Multiple Resonance. III. Exchange Reactions in Systems with Several Nonequivalent Sites," *J. Chem. Phys.*, *40*, 1189 (1964).

G. Binsch, "The Study of Intramolecular Rate Processes by Dynamic Nuclear Magnetic Resonance," in *Topics in Stereochemistry*, E. L. Eliel and N. L. Allinger, eds., Wiley-Interscience Publishers, Inc., New York, 1968, vol. 3.

N. S. Ham and T. Mole, "The Application of NMR to Organometallic Exchange Reactions," in *Progress in Nuclear Magnetic Resonance Spectroscopy*, J. W. Emsley, J. Feeney, and L. H. Sutcliffe, eds., Pergamon Press, Oxford, 1969, vol. 4, p. 91.

D. A. Kleier, G. Binsch, A. Steigel and J. Sauer, "Dynamic Nuclear Magnetic Resonance Studies of Complex Spin Systems. Degenerate Valence Isomerization of a 3,4-Diazanorcaradiene," *J. Amer. Chem. Soc.*, *92*, 3787 (1970).

B. G. Cox, F. G. Riddell, and D. A. R. Williams, "A New Approach to Rate Measurements by Use of Nuclear Magnetic Resonance Spectroscopy. Digital Analysis of Lineshapes," *J. Chem. Soc.* (*B*), 859 (1970).

D. S. Kabakoff and E. Namanworth, "NMDR Studies of the Dimethylcyclopropylcarbinyl Cation. Measurement of the Rotation Barrier," *J. Amer. Chem. Soc.*, *92*, 3234 (1970).

# III

# DEDUCTION OF ORGANIC STRUCTURES FROM SPECTRA

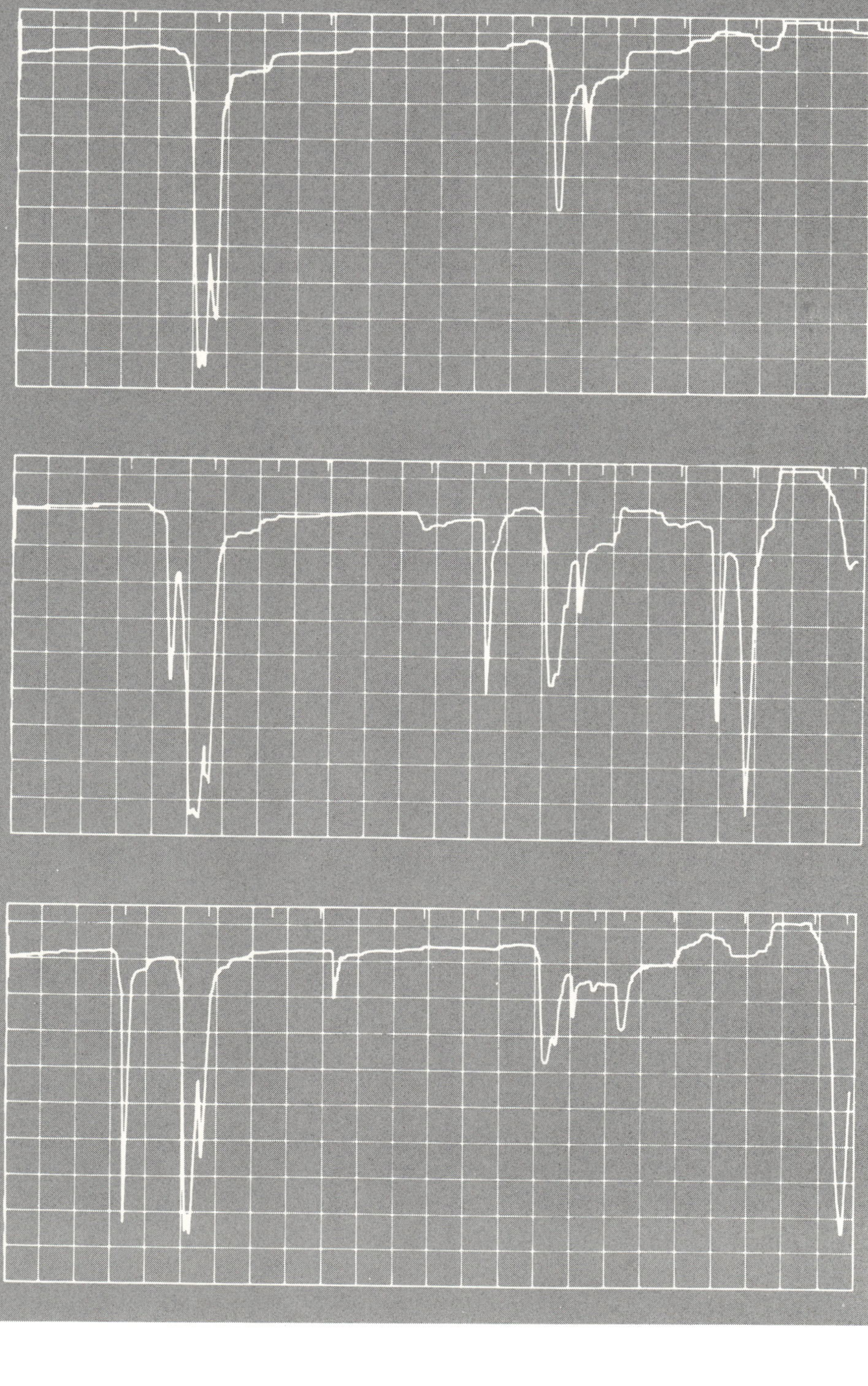

# CHAPTER 12 SATURATED HYDROCARBONS AND THEIR CONFORMATION

## 12.1 *General Characteristics*

Saturated hydrocarbons offer a challenge to the chemist who seeks to determine their structure; in this respect, they are worthy of their old appellation of "paraffins." Of course hydrocarbons cannot be said to be inert with respect to physical methods as they are towards chemical reactions. Still, it is somewhat difficult to differentiate between isomeric hydrocarbons even by modern spectroscopic methods. In general, the structural elements upon which these distinctions base themselves are (a) the molecular weight; (b) the acyclic or cyclic nature and the chain length or the ring size of the molecule; and (c) the presence and the nature of ramifications. We shall give examples here of how to answer these various questions. A further question of interest, that of conformation, is discussed in section 12.2.

### *12.1.1 Mass Spectra*

At the present time mass spectrometry is one of the most powerful tools for studying hydrocarbons. The appearance of the mass spectrum is determined by the fact that most carbon–carbon bonds have the same energy, and that the most substituted carbonium ions are the most stable (tertiary $R_3C^+$ > secondary $R_2CH^+$ > primary $RCH_2^+$). Hence, due to fragmentation, the intensity of the parent peak ($M^{+\cdot}$ molecular ion) in the mass spectrum of a hydrocarbon is usually small and decreases with increasing molecular weight. Furthermore, in long-chain hydrocarbons in particular, the spectrum has a regular appearance, 14 mass units apart, corresponding to loss of successive methylene $CH_2$ units. The intensity of the parent peak is also greater in the

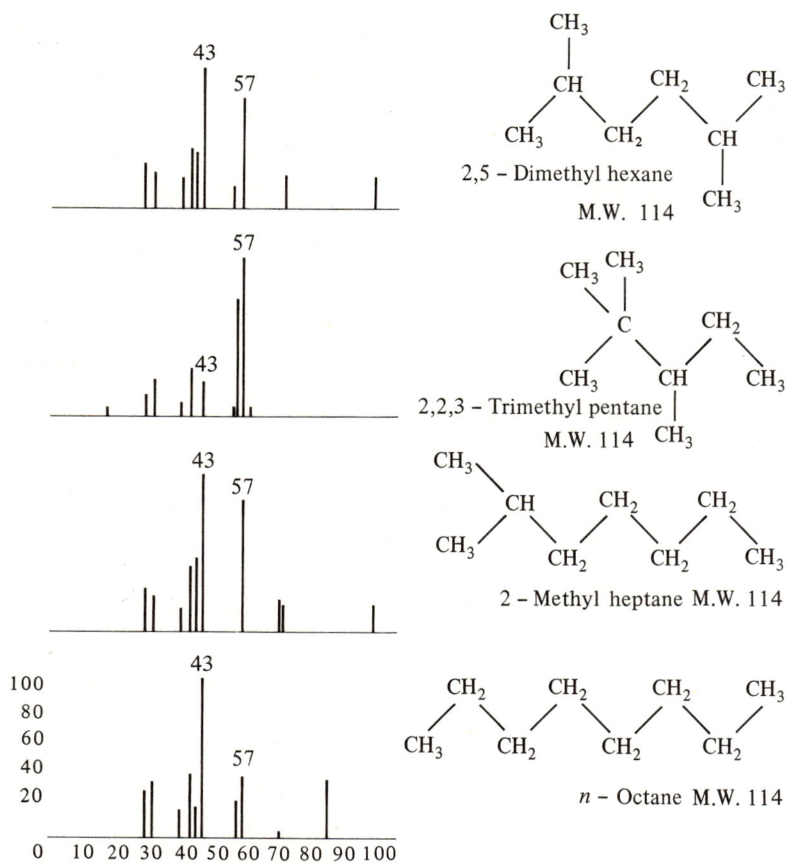


*FIGURE 12.1*
*Mass spectra of some $C_8H_{18}$ hydrocarbons.*

straight-chain isomers, and decreases with branching. In general, cleavage is favored at the branched carbons, since it will lead to the more stable ions. Often the most intense peaks in the spectrum are observed at $m/e = 43$ corresponding to the stable isopropyl cation $(CH_3)_2CH^+$ and at $m/e = 57$ for the $t$-butyl cation $(CH_3)_3C^+$. For cycloalkanes, as well as for aromatics, the parent peak is more readily apparent and intense, and again cleavage predominates at points of branching. Illustrative examples are shown in Figure 12.1; the isopropyl and the tertiary butyl peaks at $m/e = 43$ and 57 predominate not only for the hydrocarbons in which they correspond to cleavage of a chain, but also in the spectrum of $n$-octane, thus emphasizing that the stability of these ions is so great as to trigger rearrangements leading to them.

### 12.1.2 *Infrared Spectra*

The ir spectra of saturated hydrocarbons display two major areas of absorption: C—H stretching between 2800 and 3100 cm$^{-1}$ and bending, usually below 1500 cm$^{-1}$. The various kinds of stretching and bending modes for a $CH_2$ methylene group are summarized below:

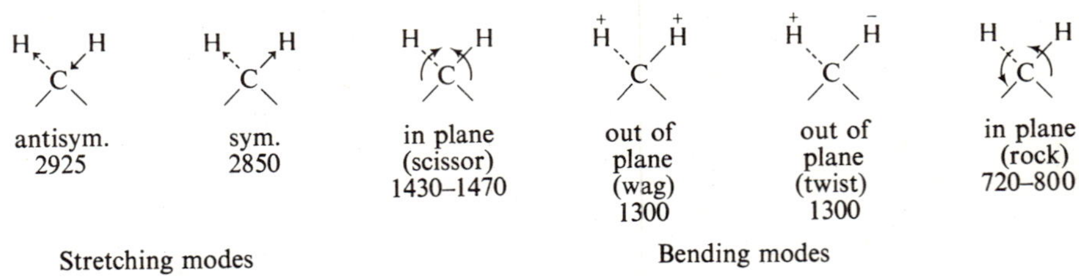


Methyl groups can be characterized by the presence of $M$-15 peaks in the mass spectrum, together with observation in the ir of absorptions for the stretching modes at 2960 cm$^{-1}$ (antisym.) and 2870 cm$^{-1}$ (sym.), while the bending modes give rise to strong bands at 1460 cm$^{-1}$ (antisym.) and 1380 cm$^{-1}$ (sym.). Geminal methyl groups on the same carbon will, of course, lead to singlet proton resonances if that carbon atom is tetrasubstituted; more generally, the ir spectrum of such compounds displays a doublet for the symmetrical bending vibration at 1380 cm$^{-1}$. As a further check, there are skeletal vibrations at 1195 and 1215 cm$^{-1}$. Likewise, isopropyl groups are characterized by similar skeletal vibrations at 1145 and 1170 cm$^{-1}$ while $t$-butyl groups have these at 1210 and 1255 cm$^{-1}$. $t$-Butyl groups are further characterized by intense $C_4H_9^+$ peaks in the mass spectrum, and a proton nmr singlet for nine hydrogens which cannot be missed and whose position depends upon the presence and nature of electronegative substituents.

In general, the statements made above are also valid for cycloalkanes. However, in addition, one should note fragmentations of the ring to eliminate $C_2H_4$ and $C_2H_5$ in the mass spectrum. The scissoring vibration between 1400 and 1500 cm$^{-1}$ for the methylene groups in cycloalkanes is often of special value, since it is influenced by the ring strain, occurring at 1450 cm$^{-1}$ in cyclohexane and at 1445 cm$^{-1}$ in cyclopentane versus 1405 cm$^{-1}$ for a methylene group in the $\alpha$ position to a carbonyl. This brings up the question of the ring size. A cyclopropane ring, in general, is easily characterized both by the high location of the asymmetric $CH_2$ stretching mode (3050 cm$^{-1}$) in the ir, and by the high-field position of the cyclopropyl proton resonances (0.2–1.0 ppm, in the absence of electronegative substituents on the same carbon).

### 12.1.3 *Carbon-13 Nuclear Magnetic Resonance Spectra*[1]

Beyond the identification of simple alkyl groups such as methyl, ethyl, $i$-propyl, and $t$-butyl, proton nmr is not always useful in the identification of hydrocarbons. This is particularly the case for high-molecular-weight alkanes or polybridged systems

[1] The authors are grateful to Professor D. M. Grant for helpful discussion and experimental data prior to publication.

where most or all of the protons resonate in a narrow range and hence do not convey much information. However, a powerful new technique—$^{13}C$ spectroscopy—is being increasingly applied in this area.

Since the range of chemical shifts for the $^{13}C$ nuclei is about 400 ppm compared with only about 30 ppm for $^{1}H$, it is obvious that $^{13}C$ shifts are much more sensitive to slight changes in the electronic environment and hence structure. The application of this technique is perhaps best described by some examples. Figures 12.2 and 12.3 show both the proton and carbon-13 spectra of *cis-syn-cis*-perhydrophenanthrene (1) and *trans-syn-cis*-perhydrophenanthrene (2) respectively.

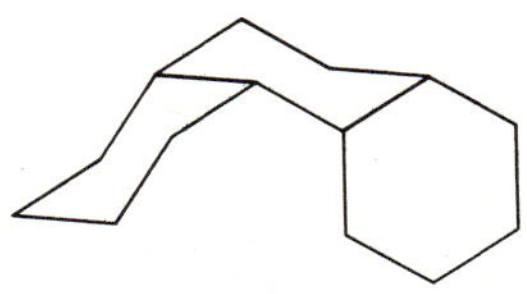

1

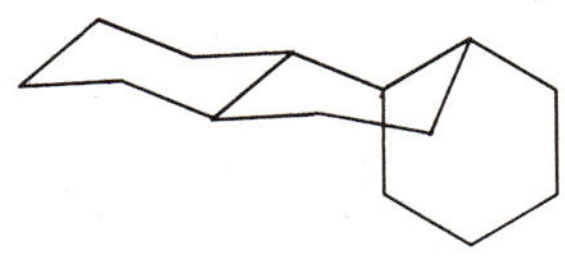

2

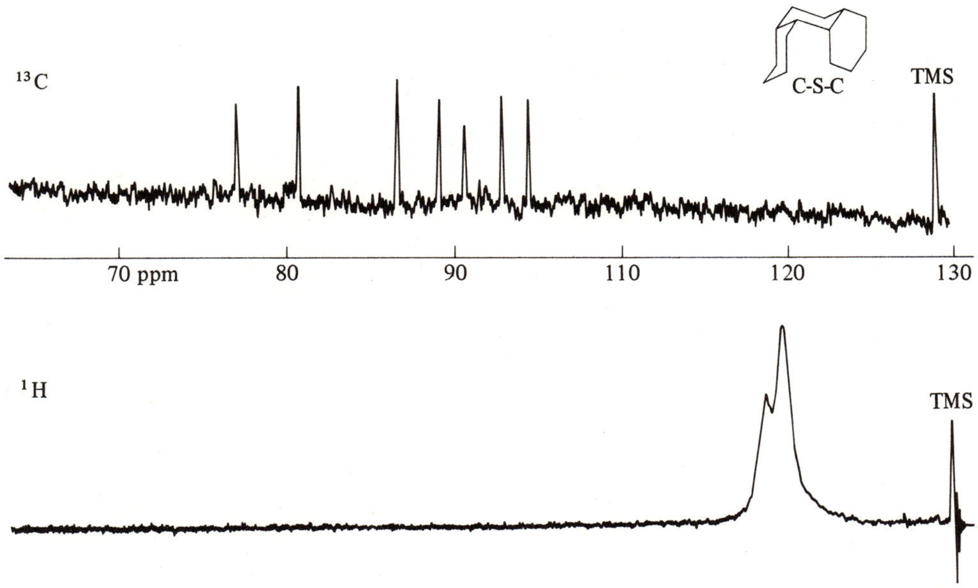


*FIGURE 12.2*
*$^{1}H$ and $^{13}C$ Spectrum of cis-syn-cis-perhydrophenanthrene.*

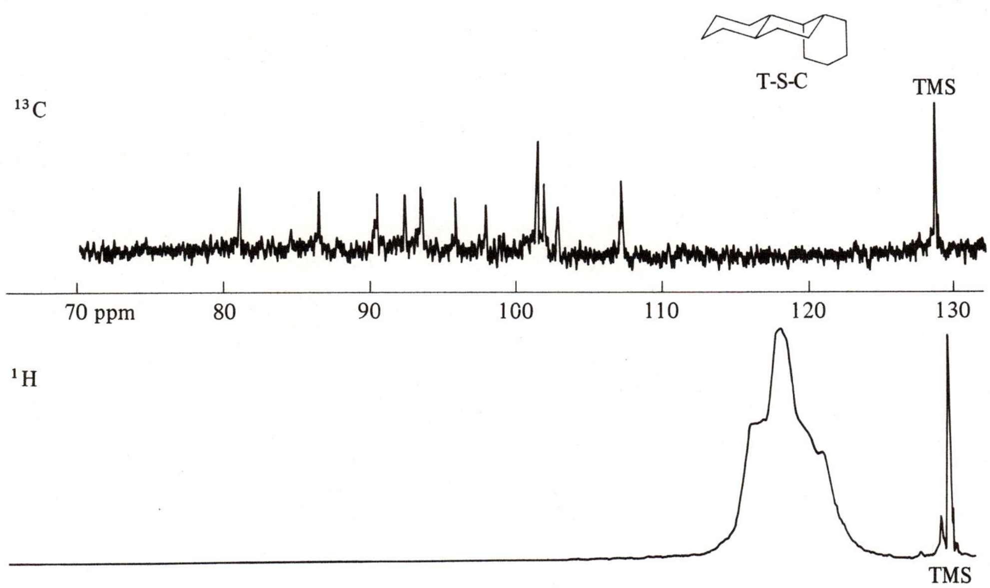


*FIGURE 12.3*
$^{1}H$ and $^{13}C$ *Spectrum of trans syn-cis-perhydrophenanthrene.*

As seen from these figures all of the protons resonate in a very narrow range and, outside of suggesting a hydrocarbon, contain no useful structural information. In contrast the $^{13}C$ spectra are considerably more detailed and hence reveal greater structural information. The spectrum of the *cis-syn-cis* isomer (Figure 12.2) shows seven resonance lines. This isomer is a mixture of conformations including the form:

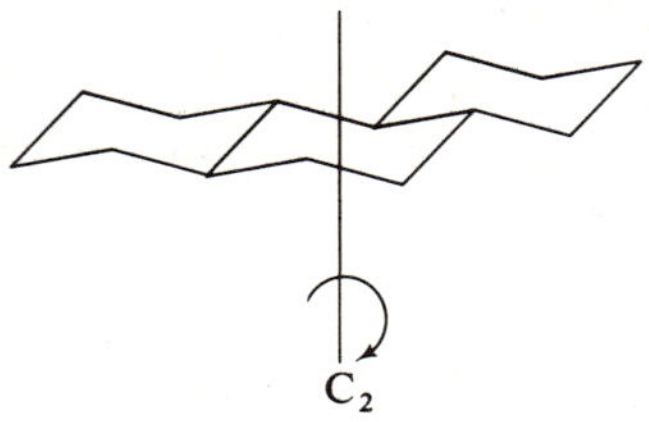


which possesses a symmetry element—$C_2$—hence kinetic averaging renders two sets of carbons equivalent, resulting in the observed seven-line spectrum. On the other hand the *trans-syn-cis* isomer does not possess a symmetry element in any of its conformations; hence, in principle, there should be 14 nonequivalent carbon atoms and 14 resonance signals. The fact that only 11 lines are observed is due to accidental equivalence of the chemical shifts for some of the carbons as indicated by the difference in intensities. Another example of a complex molecule with a detailed $^{13}C$ spectrum is that of cholesterol[2] shown in Figure 12.4.

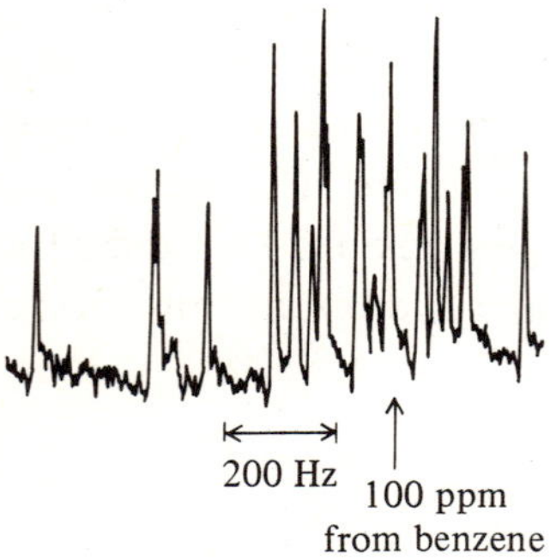


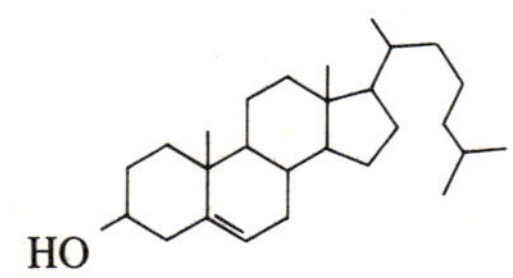


*FIGURE 12.4*
*Cholesterol (1.1 M in $CS_2$); 1000 scans. The 2 vinyl carbons are 1000 Hz downfield.*

Of course, to get the full information from these $^{13}C$ spectra one would have to know which resonance line corresponds to which carbon or, conversely, where different types of carbons resonate. There are two approaches to this problem. One is via additivity relationships that assign chemical shifts to carbons of a certain type such as primary, secondary, tertiary, bridgehead, and so on, and are derived both from empirical observation of key compounds and from theoretical considerations of $^{13}C$ chemical shifts. This approach is currently in its infancy; for example, it is at present not possible to assign all of the observed lines to specific carbons in any of the compounds shown in Figures 12.2, 12.3, and 12.4.

The second approach is strictly experimental. All of the spectra shown were obtained on samples with natural abundance only (i.e., ~ 1.1 percent) of $^{13}C$, by a technique known as "random-noise decoupling." This technique consists of irradiating the sample with a strong secondary RF field of *wide-enough range* to completely decouple *all protons simultaneously* while observing the $^{13}C$ spectrum. If, on the other hand, decoupling is done over a narrow frequency range the protons can be decoupled selectively from the various individual carbons in a particular molecule. Then, if the proton spectrum is known and the various proton resonances do not overlap, selective decoupling provides a means of assigning the $^{13}C$ spectrum. This is illustrated in Figure 12.5, which shows both the proton and the $^{13}C$ spectrum of *exo*-5-methoxy-2,3-diazabicyclo(2.2.1)hept-2-ene. In this spectrum, by irradiating at the frequency of proton 1 while observing the $^{13}C$ range, the resonance for carbon 1 appears. Similarly irradiation of the methyl protons of the methoxy group leads to the appearance of the corresponding carbon in the $^{13}C$ range, and so on for all the protons and carbons in the molecule. This second technique is, of course, applicable only to molecules that have distinct and assigned proton resonances and cannot be used, for instance, to make $^{13}C$ assignments for the molecules shown in Figures 12.2, 12.3, and 12.4.

It is clear from this brief discussion that as the technique of $^{13}C$ spectroscopy is

[2] F. J. Weigert, M. Jautelat, and J. D. Roberts, *Proc. Nat. Acad. Sci., U.S., 60*, 1152 (1968).

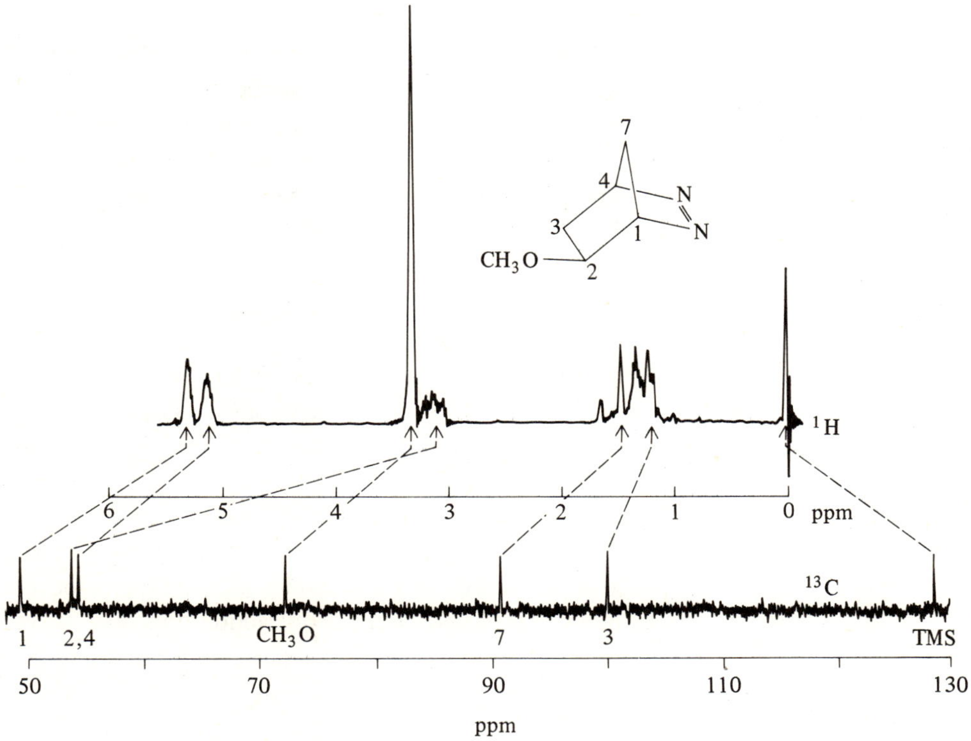


*FIGURE 12.5*
$^1H$ *and* $^{13}C$ *spectrum of methoxy diazabicyclo(2.2.1)heptene.*

developed and applied to, for instance, polybridged hydrocarbons it will become a useful and powerful tool for structure identification.

A brief summary of important $^{13}C$ chemical shifts is presented in the following chart (J. B. Leane, Perkin–Elmer, Ltd., Beaconsfield, England):

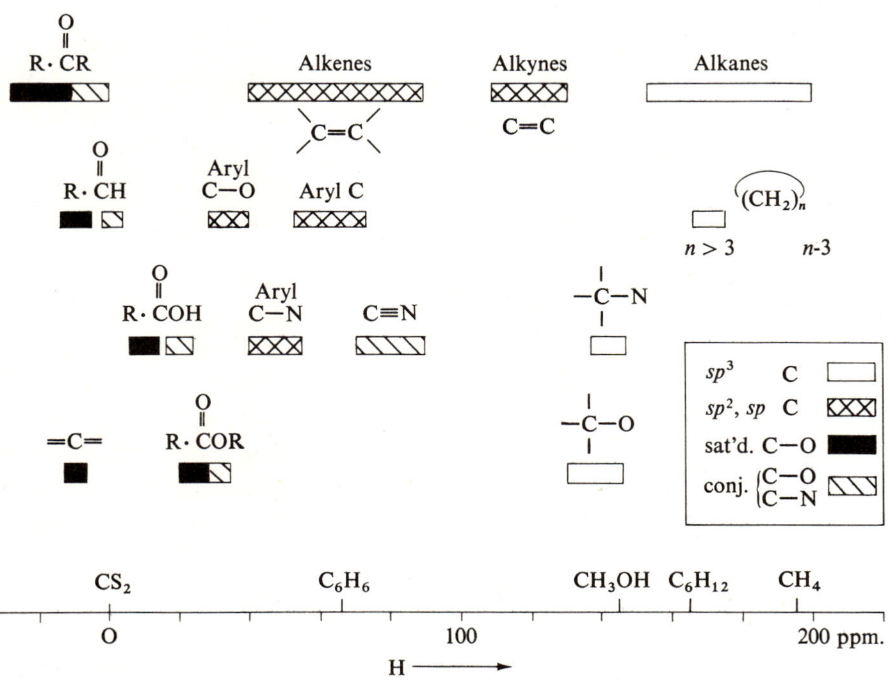


## 12.2 *Conformation of Hydrocarbons*

The ground state of a molecule is the stablest configuration of its atoms in space and, by definition, the conformation of lowest energy. One can break down the total energy of any molecule into four parts: nuclear-nuclear repulsion, interelectronic

*12.2.1 Methods and Techniques*

repulsion, nuclear-electron attraction, and the kinetic energy of the electrons. These terms are very difficult to evaluate, however, even for the simplest organic systems. A much less rigorous but operationally more easily understood formalism has been developed over the years. In this semiempirical scheme, molecular geometries are defined by four types of structural parameters; bond angles $\theta$, torsional angle $\phi$, bond length $r$, and nonbonded distances $d$. Strain energy (S.E.), in turn, is a function of these four parameters:

$$\text{S.E.} = E_r(r) + E_\theta(\theta) + E_\phi(\phi) + E_{nb}(d), \tag{1}$$

where $E_r$ and $E_\theta$ are energies for bond-length and bond-angle deformations, respectively, $E_\phi$ is the energy for torsional eclipsing, and $E_{nb}$ is the nonbonded interaction energy. According to this scheme molecules adopt conformations that have a minimum of strain energy.

In practice there are two basic approaches to the analysis of molecular conformation: (a) experimental, and (b) calculations. The experimental approach comprises x-ray and electron diffraction and microwave and nmr spectroscopy. Of these techniques x-ray and electron diffraction and microwave spectroscopy are the best and most accurate methods of determining bond length as well as bond and torsional angles, thereby yielding the exact geometries of the conformations of lowest energy. On the other hand, nmr is particularly well suited for determining the energy difference as well as the activation energy for interconversion of two (or more) conformations as well as some structural information about particular conformations.

With the advent of high-speed computers calculations can be considered as another technique of conformational analysis. Their advantage over the physical methods consists in their general applicability and in their ability to yield geometries and relative energies not only of the conformations of minimum energy but also of possible intermediates and transition states. Their disadvantages lie in the artificiality and in the large and somewhat arbitrary parametrization required at present for reasonable application. Calculation of molecular conformation requires the evaluation of the individual terms of the strain energy [equation (1)]. In brief, this is accomplished as follows: The $E_r$ and $E_\theta$ terms treat deformations in bond length and angles assuming harmonic restoring forces such that Hooke's law holds:

$$E_r = \sum \tfrac{1}{2}k_r(r - r^0)^2 = \sum \tfrac{1}{2}k_r(\Delta r)^2 \tag{2}$$

$$E_\theta = \sum \tfrac{1}{2}k_\theta(\theta - \theta^0)^2 = \sum \tfrac{1}{2}k_\theta(\Delta\theta)^2 \tag{3}$$

Here summation is taken over all bond lengths and angles, respectively, and $r^0$ and $\theta^0$ represent "strain-free" lengths and angles such as 1.54 Å for C—C single bonds or 109.4° for tetrahedral angles, and $k_r$ and $k_\theta$ are stretching- and bending-force constants, respectively, usually derived from vibrational (ir and Raman) spectroscopy. The torsional term $E_\phi$ is evaluated by means of a periodic function:

$$E_\phi = \tfrac{1}{6}\sum V_0(1 + \cos 3\phi) \tag{4}$$

where $V_0$ is one-third of the experimental value for a threefold rotational barrier (such as in ethane)—of the order of 1 kcal/mole—and the sum is taken over all dihedral angles $\phi$, about all C—C bonds. Finally the nonbonded term $E_{nb}$, commonly referred to as van der Waals interactions, is evaluated by either of two well-known intermolecular potential functions: the Buckingham [equation (5)] or Lennard–Jones potential [equation (6)]

$$E_{nb} = ae^{-b/d} - \frac{c}{d^6} \tag{5}$$

$$E_{nb} = \frac{a'}{d^{12}} - \frac{b'}{d^6} \tag{6}$$

where $a$, $b$, and $c$ are constants and $d$ corresponds to the distance between any two nonbonded atoms. The interaction is taken over all such nonbonded pairs of atoms.

The major purpose of this chapter is to examine the conformations of hydrocarbons which have been investigated extensively. Let us classify hydrocarbons into two categories, acyclic and cyclic molecules, and examine each in turn.

### 12.2.2 Acyclic Alkanes

Once it was recognized that internal rotation about single bonds in polyatomic molecules such as normal hydrocarbons was not entirely free, it became clear that such molecules can assume several possible conformations differing from one another by rotation around the various single bonds. The simplest hydrocarbon for which restricted rotation around a carbon-carbon single bond exists is ethane. A plot of the potential energy as a function of the dihedral angle $\phi$ (Figure 12.6) for ethane over a range of $2\pi$ reveals three equal-energy minima and three equal-energy maxima corresponding to the two extreme conformations of ethane: staggered and eclipsed. Any other point in the curve corresponds to some conformation of ethane in between these two extremes. Although the exact origin of this rotational barrier is still not well understood its magnitude can be determined easily and has been shown to be about 2.8 kcal/mole.

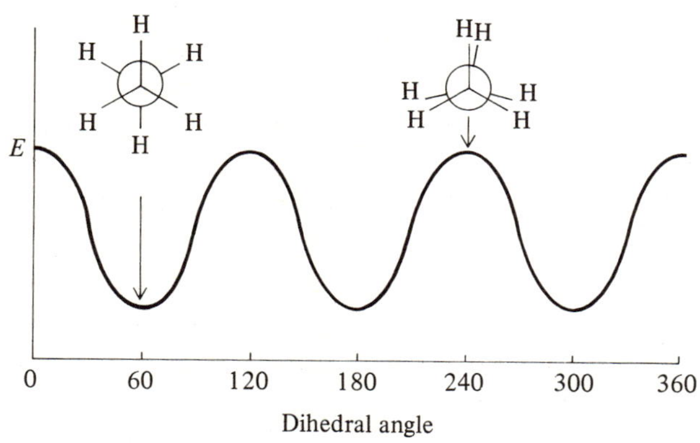


*FIGURE 12.6*
*Plot of potential energy versus dihedral angle for ethane.*

The next molecule of interest in this series is $n$-butane. For $n$-butane the energy minima and maxima are no longer equal because there are two possible eclipsed conformations, defined by methyl-hydrogen or methyl-methyl "contacts" (Figure 12.7). It is interesting to inquire how the molecule proceeds to decrease the repulsive gauche interaction between the two methyl groups to minimize the energy of this conformation relative to the "strain-free" anti form. The perfectly staggered conformation would have dihedral angles of $\phi = 0°$ for the anti conformation. However, both calculations[3] and electron diffraction data[4] indicate a displacement of 2–6° around the central C—C bond and of 5–10° around the two terminal C—C bonds. The calculated energy difference between the two staggered conformations (anti and

[3] L. S. Bartell *et al.*, *J. Chem. Phys.*, *32*, 827 (1960); N. L. Allinger *et al.*, *J. Amer. Chem. Soc.*, *89*, 4345 (1967).

[4] L. S. Bartell *et al.*, *J. Chem. Phys.*, *39*, 3097 (1963); R. A. Bonham and L. S. Bartell, *J. Amer. Chem. Soc.*, *81*, 3491 (1959).

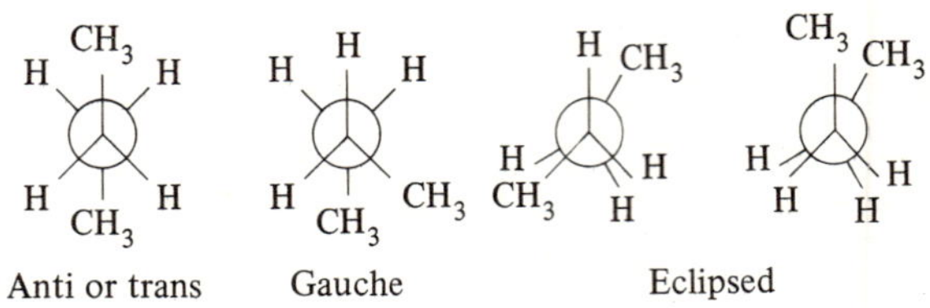


*FIGURE 12.7*
*Conformation of n-butane.*

gauche) is found to be 0.5–1.0 kcal/mole, depending upon the assumptions in the various calculations and the experimental uncertainty in the physical determination. The rotational barrier is found experimentally to be about 3.5 kcal/mole.

It is clear, of course, that as the number of C—C single bonds becomes larger with increasing chain length the number of possible conformations becomes enormous. Even if the assumption is made that there are only three possible rotational states around *each internal* C—C bond (anti with $\phi = 0°$ and two gauche states with $\phi = 120°$) *n*-decane, for instance, would have 2187 possible conformations!

### *12.2.3 Conformation of Cycloalkanes*

The basic aims of conformational analysis of the cycloalkanes as pointed out by Hendrickson[5] are threefold: a knowledge of (1) the detailed geometry and relative energy of each ring form which represents an energy minimum in the overall energy profile of conformational change; (2) the modes of conformational interconversions between these forms and their associated activation energies; and (3) the strain energies associated with placing substituents on each of the various geometrically distinguishable sites of substitution of each of the rings. However, up to now the only ring system for which all three aims have been accurately and completely accomplished experimentally is cyclohexane. Indeed, the bulk of the experimental work in conformational analysis has been done on cyclohexane or fused-ring systems containing six-membered rings. The other medium ring systems ($C_7$–$C_{12}$) possess, in general, a lower degree of symmetry, are therefore more complex conformationally, and consequently more difficult to investigate experimentally. A number of systems have nevertheless been investigated and calculations have been carried out on most medium-ring cycloalkanes. Let us examine each of these ring systems separately.

#### CYCLOHEXANE

Cyclohexane is unique among cycloalkanes in that its behavior can be rationalized assuming only three basic conformations, commonly referred to as the chair, the boat, and the twist-boat. The calculated as well as the experimental conformational energies for these three conformations and the transition states for their interconversion are summarized in Table 12.1.

*TABLE 12.1*

*Conformational Energy of Cyclohexanes*

| | ΔE (kcal/mole) | |
|---|---|---|
| Conformation | Calculated[a] | Experimental |
| Chair | 0.0[b] | 0.0[b] |
| Twist-boat | 5.0–5.6 | 4.8–5.9[c] |
| Boat | 6.4–6.9 | — |
| Transition state (half-chair) | 11.0–12.0 | 9.0–11.5[d] |

[a] N. L. Allinger *et al., J. Amer. Chem. Soc., 89,* 4345 (1967); *ibid, 90,* 1199 (1968); J. B. Hendrickson, *J. Amer. Chem. Soc., 89,* 7036 (1967).
[b] The arbitrary zero of energy in this system.
[c] W. S. Johnson *et al., J. Amer. Chem. Soc., 83,* 606 (1961); N. L. Allinger *et al., J. Amer. Chem. Soc., 82,* 2393 (1960); J. L. Margrave *et al., J. Amer. Chem. Soc., 85,* 546 (1963).
[d] F. R. Jensen *et al., Tetrahedron Lett.,* 4523 (1966); F. A. L. Anet *et al., J. Amer. Chem. Soc., 89,* 760 (1967); J. D. Roberts, *Chem. Brit.,* 529 (1966).

[5] J. B. Hendrickson, *J. Amer. Chem. Soc., 89,* 7047 (1967).

The calculated values were obtained by means of the strain-energy calculations outlined in Section 12.2.1 and vary within the indicated range depending on the assumptions made to obtain the needed parameters. Since the energy of the simple unsubstituted twist-boat cyclohexane is experimentally unaccessible it was derived from various substituted isomers such as the *cis*- and *trans*-1,3-di-*t*-butyl isomers or from the heats of combustion of *trans-syn-trans*- (1) and *trans-anti-trans*- (2) perhydroanthracene, again resulting in a range of values.

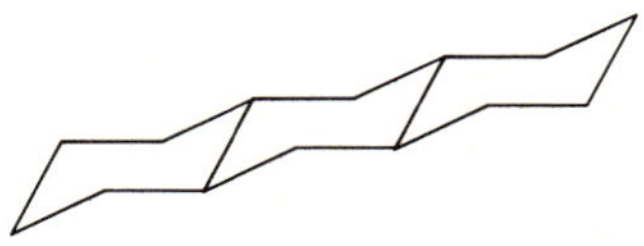

1

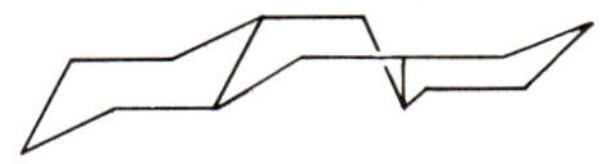

2

The experimental activation energy for interconversion of the two chair forms also varies over a range depending upon the exact technique and physical method used to evaluate it. The most common technique is dynamic nmr spectroscopy over a range of temperatures, some low enough that the kinetic process can be stopped and the conformation frozen. A particularly nice example of this technique is the work of Roberts and coworkers on the interconversion of 1,1-difluorocyclohexane. Figure 12.8 shows the $^{19}F$ nmr spectra of 1,1-difluorocyclohexane at $-100°C$ where two different chemical shifts are observed for the axial and equatorial fluorine in the "frozen" conformation. Figure 12.9 shows the temperature dependence of the spectrum between $-100°C$ and $+30°C$ as well as the calculated spectra and the corresponding inversion frequencies (or rather lifetimes $\tau = 1/2k$). The calculated spectra and inversion frequencies were obtained by the methods discussed in Chapter 11.

The calculated and experimental[6] values for the geometry of the chair form of cyclohexane are in excellent agreement, with endocyclic C—C—C angles of 111.5°, torsional (dihedral) angles of 55.9° and C—C bond lengths of 1.52 Å. Because of the high degree of symmetry in cyclohexane there are only two distinct substituent positions: axial and equatorial. The conformational free-energy differences for a wide variety of substituents in these two positions have been investigated extensively. These free-energy differences are commonly referred to as *A* values and are given in Table 12.2 for some of the most important substituents. For hydrocarbons the equatorial isomer is favored over the axial isomer because of repulsive 1,3-interactions in the latter.

*TABLE 12.2*

A *Values for Various Substituents on Cyclohexane*

| Group | *A* Value | Group | *A* Value |
|---|---|---|---|
| F | 0.25 | $CH_3$ | 1.7 |
| Cl | 0.4 | $C_2H_5$ | 1.8 |
| Br | 0.5 | $i$-$C_3H_7$ | 2.1 |
| I | 0.4 | $n$-$C_3H_7$ | 2.1 |
| OH | 0.7 | $(CH_3)_3C$ | ~12.0 |
| $OCH_3$ | 0.7 | $C_6H_5$ | 3.1 |
| $OC_2H_5$ | 0.9 | $C_6H_{11}$ | 2.15 |
| $OCOCH_3$ | 0.7 | $CO_2H$ | 1.2 |
| SH | 0.9 | $CO_2CH_3$ | 1.1 |
| $SCH_3$ | 0.8 | $CO_2C_2H_5$ | 1.2 |
| $N(CH_3)_2$ | 2.1 | CN | 0.2 |

[6] M. Davis and O. Hassel, *Acta Chem. Scand.*, *17*, 1181 (1963); H. R. Buys and H. J. Geise, *Tetrahedron Lett.*, 2991 (1970).

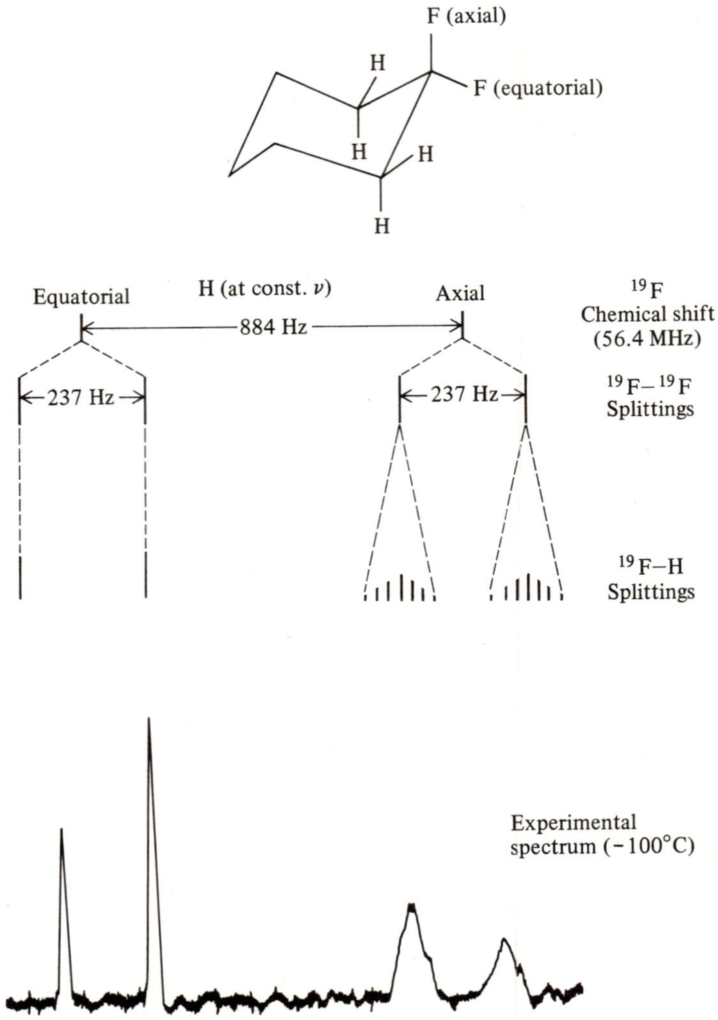


*FIGURE 12.8*
*Chemical shift and spin-spin splitting in $^{19}$F spectra of 1,1-difluorocyclohexane at −100°C.*

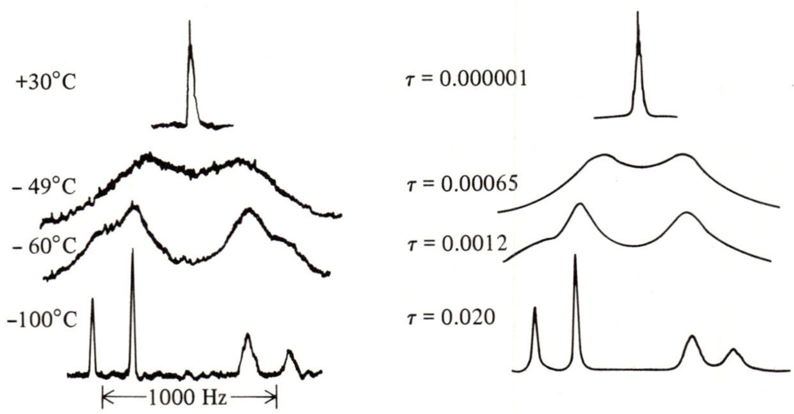


*FIGURE 12.9*
*Experimental and calculated $^{19}$F spectra of 1,1-difluorocyclohexane as a function of inversion frequency. Each of the calculated spectra is labeled with the appropriate τ, the mean lifetime in seconds before inversion occurs.*

The possible modes of conformational interconversion for cyclohexane have been summarized in an elegant manner by Hendrickson.[7] This representation corresponds to a three-dimensional map which for cyclohexane is represented best by a globe (Figure 12.10). In this representation an equatorially substituted cyclohexane in the chair conformation corresponds to the north pole with zero conformational energy. Such a conformation can change symmetrically (a transition in which a single symmetry element, such as a plane or axis, is retained throughout a transition from one form or family to another) by following a longitudinal path, over a barrier (‡), to the boat (B) or twist-boat (TB) form at the equator, where it may pseudorotate freely among the B/TB forms (around the globe) or proceed over another symmetrical barrier to the axially substituted chair conformation at the south pole. Pseudorotation (defined by Hendrickson as the passage of a ring with a plane of symmetry through an atom into one with an axis of symmetry bisecting the bonds adjacent to that atom and vice versa) converts one twist-boat form into another, the boat form being the transition state. Such pseudorotations usually have low energy barriers—in the case of cyclohexane, about 1 kcal/mole. The boat forms therefore represent saddle points in the overall energy profile in being maxima on the horizontal pseudorotation itinerary but minima on the longitudinal pathways between the two barriers leading to the two chair rings.

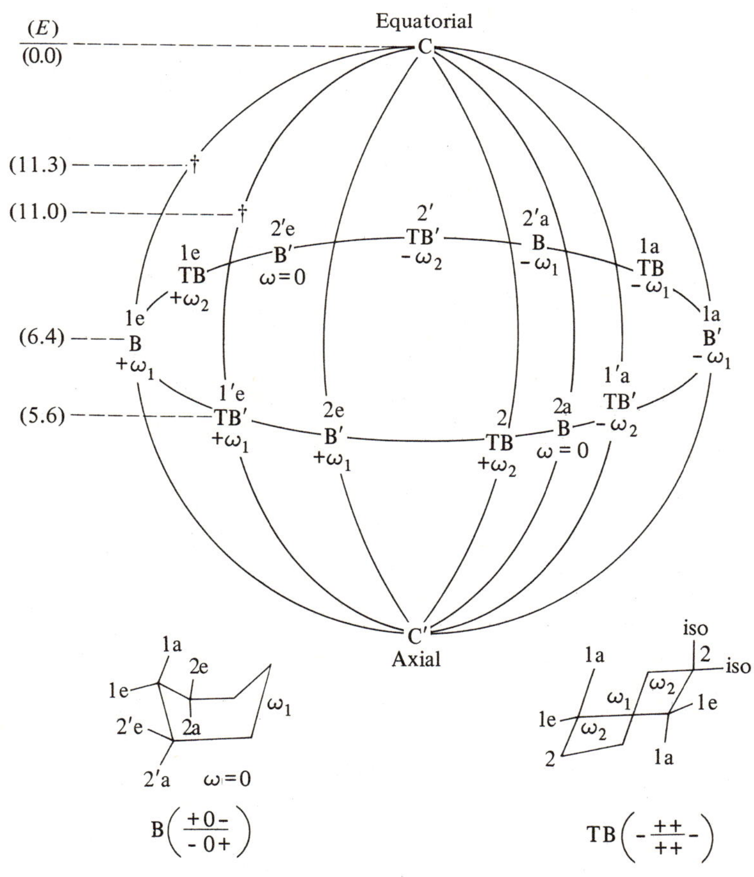


*FIGURE 12.10*
*Map of cyclohexane interconversions (ring energies in parentheses).*

[7] J. B. Hendrickson, *J. Amer. Chem. Soc.*, *89*, 7047 (1967).

The longitudinal energy barrier (‡) corresponds to the transition state for the interconversion of cyclohexane.

CYCLOHEPTANE

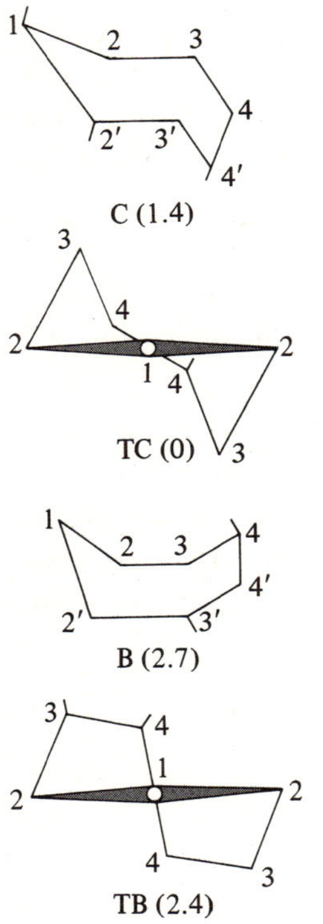


*FIGURE 12.11*
*Conformations of cycloheptane and their relative energies.*

Cycloheptane can exist in four favorable conformations, which have been called chair (C), twist-chair (TC), boat (B), and twist-boat (TB). The conformation of lowest energy appears to be the twist-chair; however, the calculated conformational energies of the remaining three forms are within a range of 3.0 kcal/mole (Figure 12.11).

The four conformations of cycloheptane form two pairs of pseudorotating families. The twist-chair (minimum) and chair (maximum) form one pair, and the twist-boat (minimum) and boat (maximum) the other. The two pseudorotating families in turn are interconverted via a symmetrical transition, maintaining an axis of symmetry, from the favored twist-chair form to the twist-boat form with a calculated activation energy of 8.1 kcal/mole. In analogy with cyclohexane, the TC and TB forms comprise energy minima and the C and B forms are saddle points on the overall energy profile.

There is experimental support[8] for these calculations. The $^{19}F$ nmr spectrum of 1,1-difluorocycloheptane remains unchanged down to $-180°C$, indicating either that pseudorotation and inversion are rapid processes or that the two fluorines occupy equivalent positions. Two geminal substituents related via the symmetry axis in the favored twist-chair form would be equivalent. The fact that the $^{19}F$ nmr of 1,1,3,3-tetrafluorocycloheptane also remains unchanged with temperature indicates that pseudorotation (or inversion) is very facile since only two of the four fluorines can be on the equivalent axis positions at any one time. On the other hand the dramatic change (Figure 12.12) in the $^{19}F$ nmr of 1,1-difluoro-4,4-dimethylcycloheptane, where the bulkier methyl groups force the fluorines into nonequivalent positions, shows that rotation (or inversion) can be prevented if the barrier is made sufficiently large by bulky substituents.

CYCLOOCTANE[9]

Cyclooctane, because of the existence of numerous forms of comparable energy, is probably the conformationally most complex medium-ring cycloalkane. Within an energy range of about 2.0 kcal/mole there may be as many as six different conformations. Their geometries and calculated relative energies are summarized in Figure 12.13, together with three forms of higher energy. Due to the closeness of the relative conformations of cyclooctane, Hendrickson's calculations indicate that it probably exists as a mixture of conformations slightly dominated by the boat-chair form. The familiar boat, chair, and twist-chair forms are all high-energy forms because of torsional eclipsing around several bonds and therefore can probably be ignored.

The interconversions among the various forms of cyclooctane are quite complex. There are four pseudorotating pairs: chair/twist-chair, twist-chair-chair/chair-chair,

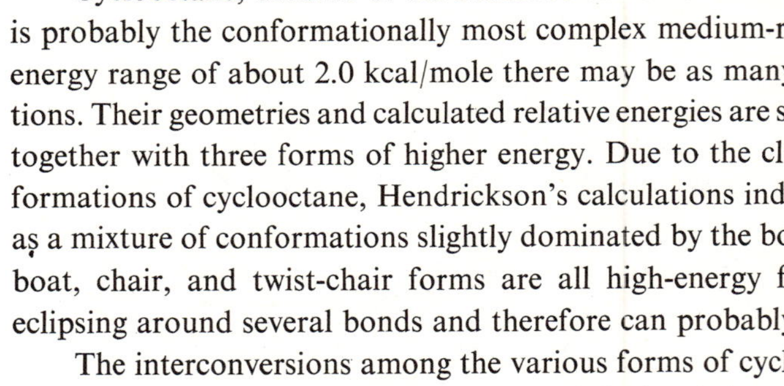


*FIGURE 12.12*
*Changes in experimental and calculated $^{19}F$ spectra of 1,1-difluoro-4,4-dimethylcycloheptane as a function of inversion rate.*

[8] J. D. Roberts, *Chem. Brit.*, 529 (1966).
[9] J. E. Anderson, E. S. Glazer, D. L. Griffith, R. Knorr, and J. D. Roberts, *J. Amer. Chem. Soc.*, *91*, 1386 (1969).

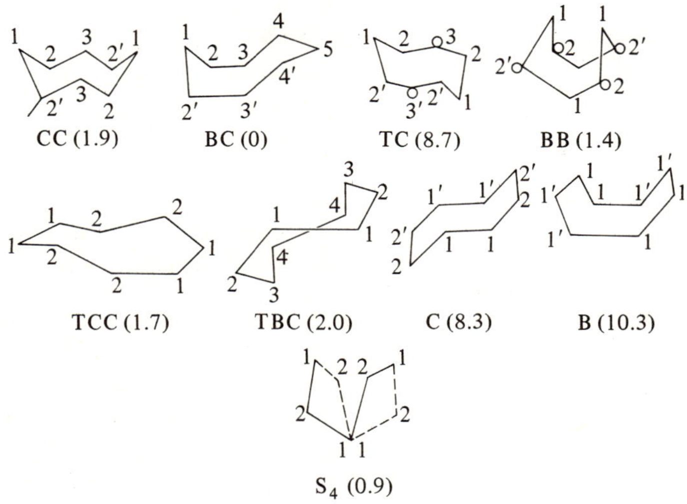


*FIGURE 12.13*
*Conformations of cyclooctane and their relative energies.*

boat-chair/twist-boat-chair, and boat-boat/boat, where the chair, twist-chair-chair, boat-chair, and boat-boat are minima of the respective pseudorotation pairs. Furthermore, in the twist-chair-chair/chair-chair itinerary a complete cycle can interconvert only the axial or the equatorial positions with no crossover possible between axial and equatorial substituents via pseudorotation alone. An analogous situation exists in the boat chair/twist boat-chair itinerary. Interconversions between pseudorotating families in turn take place via higher energy transition states that comprise true maxima on the overall energy surface analogous to the cyclohexane and cycloheptane situation.

Existing x-ray data on substituted cyclooctanes confirm the boat-chair form as the favored conformation. The nmr results of Anet[10] on various deuteriocyclooctanes and methyl and *t*-butylcyclooctane are also best explained on the basis of a conformational mixture with the boat-chair predominating. The nmr results further indicate that pseudorotation is a viable process down to −145°C.

CYCLONONANE AND CYCLODECANE

For cyclononane, calculations indicate that there are five conformations within an energy range of about 10 kcal/mole (Figure 12.14). There are other conformations possible but they possess higher energies and are not expected to be important. Interconversion among the various conformations is complicated; the major forms of pseudorotation possible among the various cyclononanes involve three pairs.

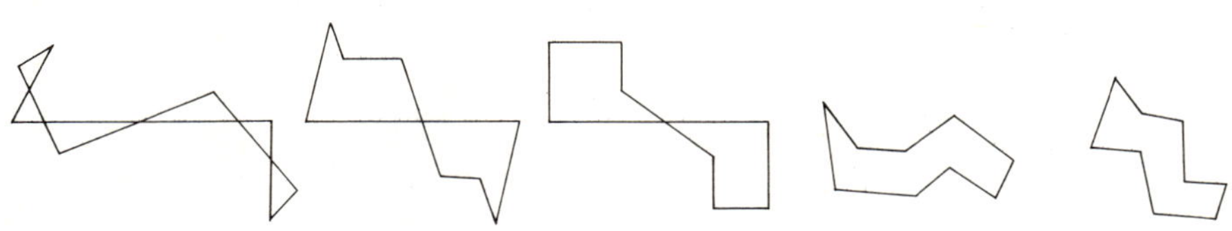

*FIGURE 12.14*
*Possible conformations of cyclononane.*

Cyclodecane has three possible conformations within an energy range of about 10 kcal/mole (Figure 12.15). Although there are numerous other possible conformations, calculations indicate that they all possess high conformational energies, arising from either severe H—H nonbonded interactions or unfavorable torsional interactions, and can therefore be ruled out (see also Problem 7).

[10] F. A. L. Anet *et al.*, *J. Amer. Chem. Soc.*, *88*, 2586 (1966).

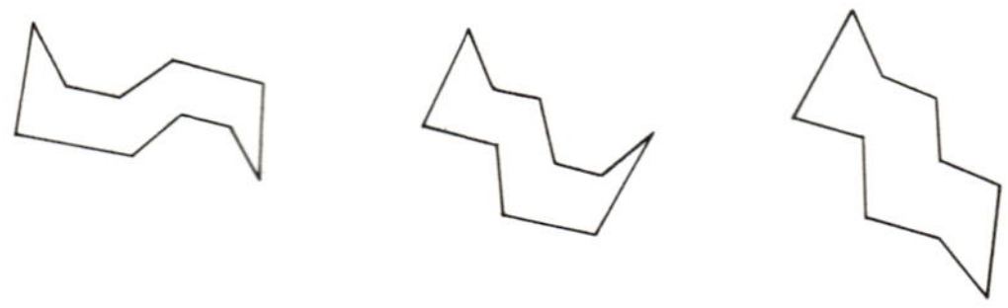

*FIGURE 12.15*
*Possible conformations of cyclodecane.*

## 12.3 *PROBLEMS*

SAMPLE PROBLEM

The best thermodynamic parameters for the cyclohexane ring inversion were obtained with $C_6D_{11}H$, decoupling the deuterium nuclei:

$$\Delta G^* = 10.22 \text{ kcal/mole} \qquad \Delta H^* = 10.8 \text{ kcal/mole} \qquad \Delta S^* = +2.8 \text{ e.u.}$$

The value for the enthalpy of activation is in very good accord with the calculated energy difference between the chair and the half-chair forms of cyclohexane, but how should one account for the observed entropy of activation?

ANSWER

Ignoring contributions to $\Delta S^*$ from differences in vibrational and rotational partition functions between the reactant and the activated complex, six different routes are available to the reactant for ring inversion:

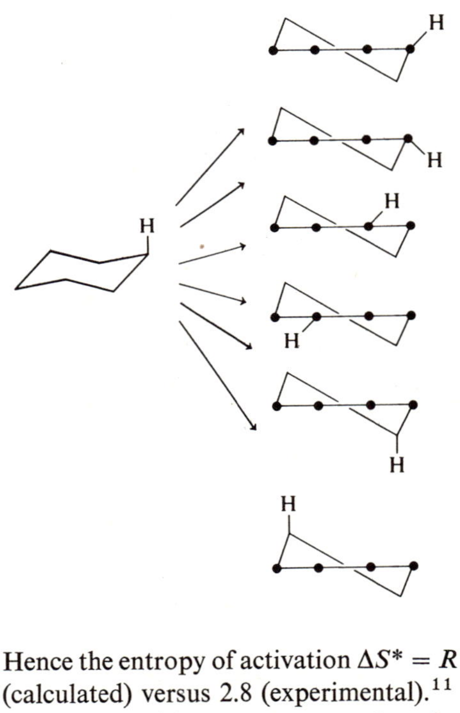


(three pairs of enantiomers)

Hence the entropy of activation $\Delta S^* = R \ln 6$, so $R(2.3 \log 6) = 4.6 \times 0.78 = +3.6$ e.u. (calculated) versus 2.8 (experimental).[11] Another *equivalent* way to get the number 6 is to consider the symmetry numbers for the reactant and for the activated complex: $n = (w\ddagger/w) \times (\sigma/\sigma\ddagger)$, where $w$ is statistical weight and $\sigma$ is a symmetry number.

reactant 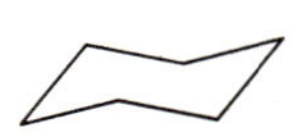 $\sigma = 6;\ w = 1$

$n = 6$

act. complex 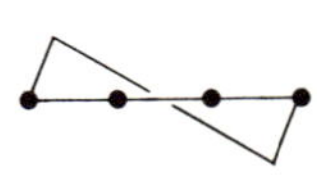 $\sigma\ddagger = 2;\ w\ddagger = 2$

---

[11] R. J. Abraham and D. B. McDonald, *Chem. Comm.*, 188 (1966).

1. Assuming that there are three possible rotational states around each internal C—C single bond (anti with $\phi = 0°$ and two gauche states with $\phi = 120°$) how many conformations are possible for *n*-pentane, *n*-hexane, *n*-heptane, and *n*-dodecane?
2. The free energy of interconversion between an equatorial and axial methyl, ethyl or isopropyl cyclohexane are *all* around $2.0 \pm 0.2$ kcal/mole whereas that for the *t*-butyl group is around 12.0 kcal/mole. Explain these values.
3. The barrier to ring inversion of (1) is 0.9 kcal/mole lower than for the isomeric (2). How do you account for this difference?

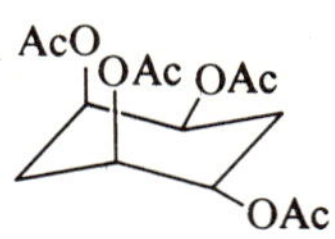


1

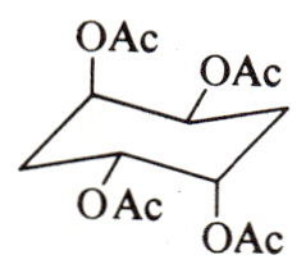


2

4. In the solvolysis of some cyclooctyl tosylates the following relative rates were obtained:

| Tosylate | Relative rate |
|---|---|
| Cyclooctyl | 1 |
| *trans*-5-*t*-Butylcyclooctyl | 0.4 |
| *cis*-5-*t*-Butylcyclooctyl | 22.0 |

The *trans* isomer gives 88 percent 5-*t*-butylcyclooctene and 12 percent 1-*t*-butyl-cyclooctene as product, whereas the *cis* isomer gives 99 percent 1-*t*-butylcyclo-octene. Account for the relative rates and products.
5. Both *cis*- and *trans*-cyclooctene have been prepared. One of the isomers shows an extraordinarily large dipole moment (0.8 D) for an olefin—which one, and why? One of the isomers has been prepared in optically active form—which one, and why can it sustain optical activity?
6. Identify *each* of the different positions of substitution for each of the possible conformations of cycloheptane, cyclooctane, and cyclononane.
7. In the cyclohexane series, it is well known that a *t*-butyl substituent can be used as a conformation-holding group. Since there is now a body of evidence suggesting that the favored conformations for cyclooctane and for cyclodecane are, respectively, the "boat-chair" and the "armchair" ($C_{2h}$), respectively, indicate what conformation-holding groups could be devised in these systems.

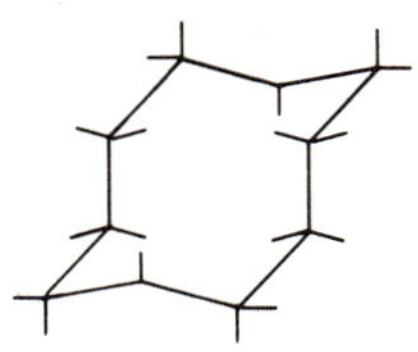

cyclooctane

cyclodecane

8. Figure out the favored conformation of *cis,cis*-cyclodeca-1,6-diene. Hint: Use the conformation of cyclodecane from Problem 7 as a starting point.
9. A polycyclic saturated hydrocarbon $C_9H_{12}$ is characterized by a mass spectrum with major peaks at $m/e = 120$(M), 105, 91, 79, and 69. The pmr spectrum has only

two broad singlets at $\delta = 0.60$ (6H) and 2.27 (6H), with widths of 3.5 and 4.0Hz, respectively. The ir spectrum has peaks at 3032 and 3019 $cm^{-1}$, 2938, 2895, and 2861 $cm^{-1}$, and at 1043, 958, 890, and 755 $cm^{-1}$, *and is otherwise blank*. What structure can you formulate on the basis of this data?

## 12.4 *SUGGESTED READINGS*

J. D. Roberts, "Studies of Conformational Equilibria and Equilibration by Nuclear Magnetic Resonance Spectroscopy," *Chem. Brit.*, 529 (1966).

J. E. Williams, P. J. Stang, and P. von R. Schleyer, "Quantitative Conformational Analysis: Calculation Methods," *Ann. Rev. Phys. Chem.*, *19*, 531 (1968).

J. B. Hendrickson, "Molecular Geometry. V. Evaluation of Functions and Conformations of Medium Rings," *J. Amer. Chem. Soc.*, *89*, 7043, 7047 (1967).

A. J. Jones, D. M. Grant, and W. M. Litchman, "Carbon-13 Magnetic Resonance. XV. Non-alternant Hydrocarbons," *J. Amer. Chem. Soc.*, *92*, 2386 (1970).

G. Borgen and J. Dale, "The Preferred Conformation of 1,1,4,4-tetramethylcyclononane and some Derivatives: Low Temperature Nuclear Magnetic Resonance Study," *Chem. Comm.*, 1105 (1970).

P. Dowd *et al.*, "On the Existence of Polar Conformations of Cycloheptane, Cyclooctane, and Cyclodecane," *J. Amer. Chem. Soc.*, *92*, 6325 (1970).

# CHAPTER 13 UNSATURATED HYDROCARBONS

Let us again start by defining the structural elements to be recognized:

1. The number of carbon-carbon double bonds.
2. Whether they are mono-, di-, tri-, or tetrasubstituted.
3. If the number of substituents is two or more, whether they are part of a ring system or not; and whether they are endocyclic or exocyclic.
4. The configuration of an individual double bond (cis-trans-isomerism) and the relative configurations if several double bonds are present (s-cis/s-trans isomerism).
5. The ring size.

## 13.1 *Units of Unsaturation and Number of Olefinic Protons*

Fortunately, there are good and, in general, unambiguous means of recognizing the presence of double bonds and the total number of olefinic protons and carbon atoms in a molecule. The molecular weight of a hydrocarbon is given by the parent peak in the mass spectrum and the corresponding elemental formula indicates the total number of double bonds and/or rings.

If a double bond is incorporated in a cyclic molecule, especially cyclohexenes, it will usually influence the mass spectrum by a special mode of fragmentation corresponding to the retro Diels–Alder reaction:

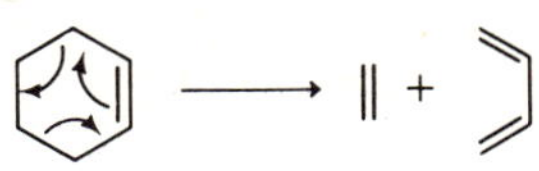

For example, the carotenoid pigments, very important naturally occurring compounds present in plants and animals, are hydrocarbons with a $C_{40}$ skeleton. It is a priori a difficult task to elucidate the exact structure of such molecules with a rather high molecular weight and in many cases containing only carbon and hydrogen. It is difficult, for instance, to distinguish, particularly by chemical means, between the two carotenes shown below:

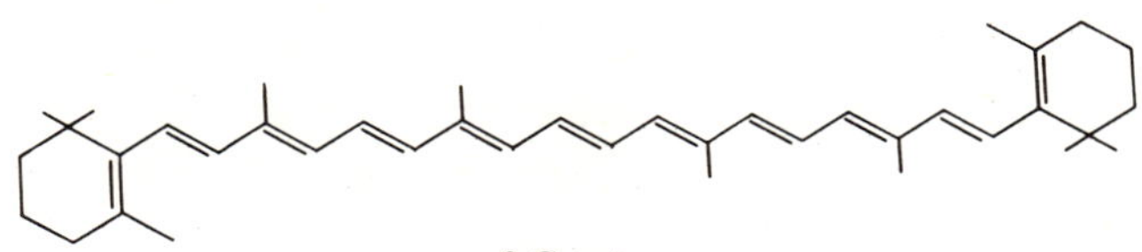

β-Carotene

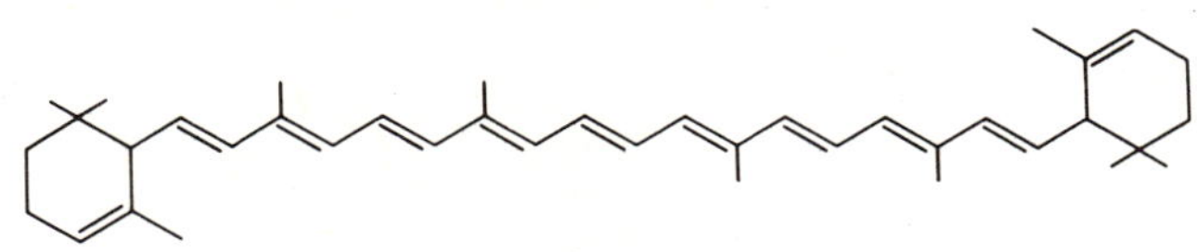

α-Carotene

α-Carotene displays in its mass spectrum a prominent peak at M-56 due to the retro-Diels–Alder fragmentation:

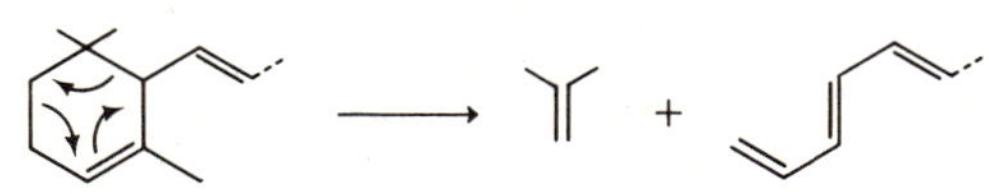

whereas the spectrum of β-carotene does not have such a feature. The next step is to count, in the pmr spectrum, the number of protons on trigonal carbon atoms in the range of 5–7 ppm: the integrated spectrum thus reveals 14 olefinic hydrogens in β-carotene versus 16 in the isomeric α-carotene. A valuable complement comes from the carbon-13 spectrum, in which olefinic and aromatic carbons absorb 40–90 ppm upfield from the $CS_2$ reference, whereas saturated carbons are found at much higher fields: 150–200 ppm. Another structural observation is derived from the fact that, in pmr spectra, allylic positions are characterized by the following chemical shifts:

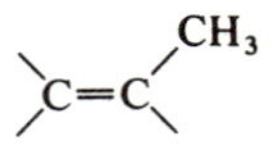
$\delta \sim 2$

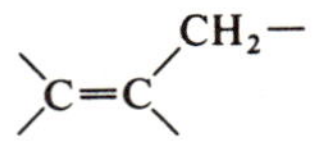
$\delta \sim 2.5$–$3.0$

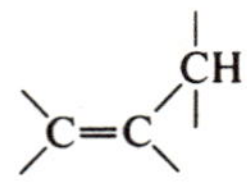
$\delta \sim 3.0$–$4.0$

The type and total number of allylic protons, as derived from the integration, will indicate the substitution of the double bonds (for example, 22 in β-carotene versus 24 in α-carotene).

### 13.2 *Substitution of the Double Bond*

In special cases strong arguments can be made concerning double-bond substitution from the appearance of the mass spectrum. In general, however, this problem is most easily solved by resorting to another spectroscopy. Proton magnetic resonance spectroscopy is extremely characteristic for vinyl groups $H_2C{=}CHR$ and while both chemical shifts and coupling constants depend upon the nature of R, the following base values are useful:

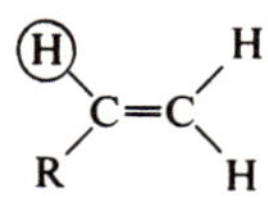
$\delta \sim 6$

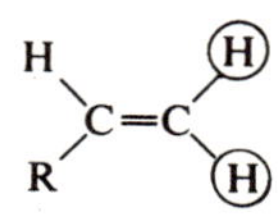
$\delta \sim 5$
terminal methylene

$^3J_{trans} \sim 18$ Hz; $^3J_{cis} \sim 11$ Hz; $^2J_{gem} \sim 2$ Hz

A close examination of the olefinic region, therefore, usually suffices to determine the configurations of disubstituted double bonds, particularly when both isomers are available, or to recognize whether a given double bond is endocyclic or exocyclic.

### 13.3 *Ring Size of Cyclic Olefins*

There are a number of properties that can be used to ascertain ring size. The ir spectrum of a cyclic olefin,

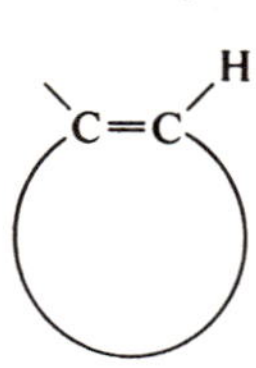
or
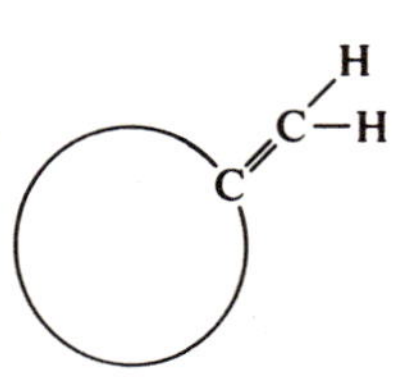

Endocyclic double bond    Exocyclic double bond

will show a $\nu_{C=C}$ stretch (strong, provided that the double bond is unsymmetrically substituted), a $\nu_{C-H}$ stretch, and a $\gamma_{C-H}$ out-of-plane "wagging" deformation band. The following values in particular are diagnostic of olefinic systems:

$\nu_{C=C}$ $\sim 1650\ cm^{-1}$

$\nu_{C-H}$ $> 3000\ cm^{-1}$

$\gamma_{C-H}$ $675$–$1000\ cm^{-1}$

A number of important values for these frequencies are gathered in Table 13.1.

TABLE 13.1

*Infrared Frequencies for Cyclic Olefins*

| $CH_2{=}C{=}CH_2$ (allene) |  |  |  | 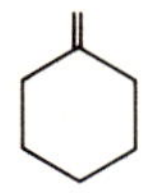 | ($cm^{-1}$) |
|---|---|---|---|---|---|
| 2993 | | | 3073 | 3072 | $\nu_{C-H}$ |
| 1980, 1070[a] | 1781 | 1680 | 1657 | 1648 | $\nu_{C=C}$ |
| | | | 879 | 892 | $\gamma_{C-H}$ |

| H—C≡C—H |  |  |  |  | |
|---|---|---|---|---|---|
| 3374 | 3076 | 3060 | 3045 | 3017 | $\nu_{C-H}$ |
| 1974 | 1641 | 1566 | 1611 | 1646 | $\nu_{C=C}$ |
| | | | 695 | 718 | $\gamma_{C-H}$ |

[a] This splitting is due to the coupling between the two C=C's.

The data in this table manifest the increased "stiffness" of double bonds incorporated into already strained rings. The stretching frequency $\nu$ is of course proportional to the square root of the force constant $k$ between two atoms with masses $m_1$ and $m_2$. For the different carbon-carbon bonds in general, the stretching frequencies are

| | |
|---|---|
| C≡C | ~2200 $cm^{-1}$ |
| C=C | ~1650 $cm^{-1}$ |
| C—C | ~1100 $cm^{-1}$ |

If one of the carbon atoms of the double bond is part of a strained ring, the force constant for the C=C bond is increased, since the stretching of the bond is restricted by the ring, hence increasing $\nu_{C=C}$. A similar argument applies when both trigonal carbons are part of a small ring. For exocyclic methylenes, one can again relate $\nu_{C=C}$ to the size of the endocyclic C–C–C angle $\theta$, subtended by the double bond. The exact relationship[1] is given by equation (1),

$$\nu_{C=C} = 1773 - 1.1\theta \tag{1}$$

where $\nu_{C=C}$ is expressed in reciprocal centimeters and $\theta$ in degrees of angle.

The ring size is also expressed in a functional dependence of the $^1J_{C-H}$ coupling constant on the interatomic angle (Table 13.2)[2] very similar to that obtained for the cycloalkanes (see Section 3.2.4).

Ring size also has a marked effect upon the vicinal olefinic coupling as indicated by the following examples[3]:

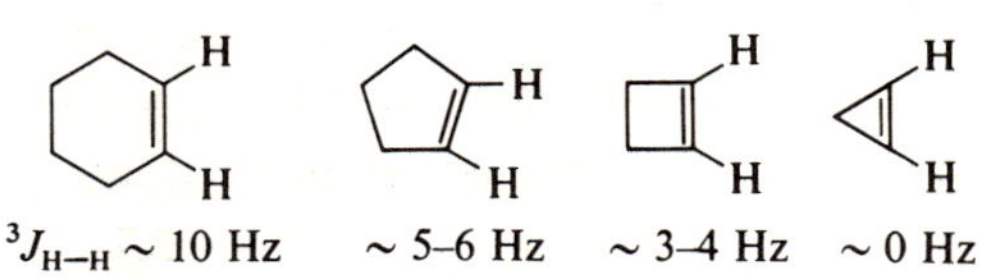


$^3J_{H-H}$ ~ 10 Hz  ~ 5–6 Hz  ~ 3–4 Hz  ~ 0 Hz

TABLE 13.2

*Correlation of $^1J_{^{13}C-H}$ with Endocyclic Angle $\theta$ for Cycloalkenes*

| Ring Size $n$ | $^1J_{^{13}C-H}$ (Hz) | $\theta$ (deg) |
|---|---|---|
| 3 | 220 | 65 |
| 4 | 170 | 94 |
| 5 | 160 | 111.7 |
| 6 | 157 | 122.5 |
| 8[a] | 155 | 126.8 |

[a] Cyclooctatetraene.

[1] S. Bank and W. D. Closson, *Tetrahedron, 24,* 381 (1968).
[2] P. Laszlo and P. von R. Schleyer, *J. Amer. Chem. Soc., 86,* 1171 (1964).
[3] P. Laszlo and P. von R. Schleyer, *J. Amer. Chem. Soc., 85,* 2017 (1963).

## 13.4 *Conjugated Double Bonds*

Conjugated unsaturation has been partly discussed already in the chapter on simple application of MO theory (Chapter 6). Such molecules with extended delocalization are characterized in the ir by $\nu_{C=C}$ bands at 1600 and 1650 cm$^{-1}$. The lower frequency band is usually more intense, and may hide or overlap the higher frequency band. As already stressed, extended conjugation shifts the uv absorption to higher wavelengths; Woodward has formulated empirical rules for these shifts (Table 13.3).

*TABLE 13.3*

*Woodward's Rules for Polyenes*

Base values:

| s-trans 214 nm | s-cis 253 nm |
|---|---|
| A double bond extension | 30 nm |
| Exocyclic nature of any double bond | 5 nm |
| Each alkyl substituent or ring residue | 5 nm |
| Other substituents: | |
| OCOR′ | 0 nm |
| OR | 6 nm |
| SR | 30 nm |
| Cl, Br | 5 nm |
| $NR_2$ | 60 nm |

An example of application to steroidal dienes would be as follows:

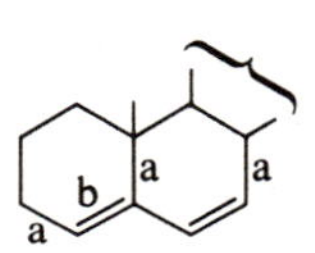


| | |
|---|---|
| base | 214 |
| exo (b) | 5 |
| subst (a) | 15 (3 × 5) |
| calc. | 234 nm (exp: 235 nm, ε = 19,000) |

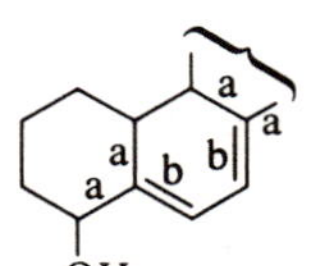


| | |
|---|---|
| base | 253 |
| subst (a) | 20 (4 × 5) |
| exo (b) | 10 (2 × 5) |
| calc. | 283 nm (exp: 282 nm) |

This is a means for detecting the s-*cis*/s-*trans* character of a diene, supplementing interproton coupling constants. Notice that in Table 13.3 the difference between the s-*cis* and the s-*trans* base values is 39 nm, which is precisely the increment corresponding to a homoannular diene element in the Woodward rules for α,β-unsaturated ketones (see Chapter 15). Of course, strain does modify these increments:

| Hydrocarbon | $\lambda_{max}$ (nm) |
|---|---|
|  | 253 |
|  | 238.5 |
| 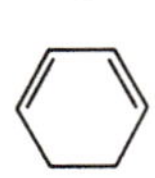 | 256 |
| 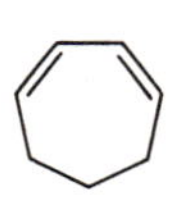 | 248 |

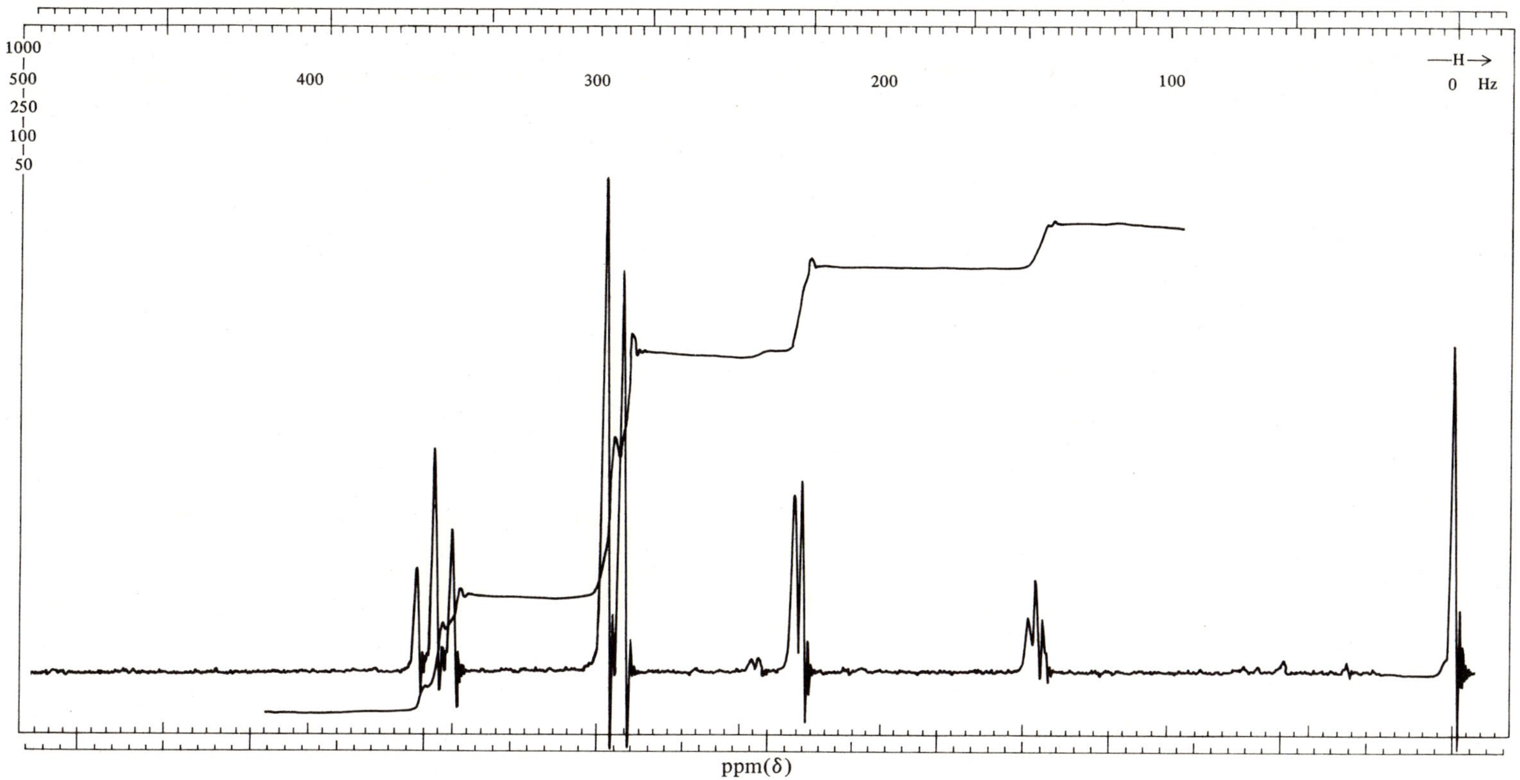


*FIGURE 13.1*
*The 60-MHz nmr spectrum of a mixture of allenyl and propargyl bromide.*

## 13.5 *ILLUSTRATIVE EXAMPLES*

When propargyl bromide, is heated with $HC{\equiv}CCH_2Br$, CuBr it isomerizes to allenyl bromide, $H_2C{=}C{=}CHBr$, and one obtains a mixture of the two compounds. Figure 13.1 shows the nmr spectrum of this mixture. Note the position of the various absorptions; the two low-field resonances are due to the allenyl bromide, the two high-field resonances to the propargyl bromide. Note in particular the position of the acetylenic hydrogen at 2.43 ppm, very much upfield from the normal olefinic region of 4.5–6.5 ppm, because of the diamagnetic shielding from the carbon-carbon triple bond. Also, the long-range coupling is larger in the allenyl isomer, $J \sim 6.1$ Hz, than in the propargyl isomer, $J \sim 2.1$ Hz, a general phenomenon in allenyl versus acetylenic compounds. Finally the integration indicates that the mixture consists of about 74 percent allenyl bromide and 26 percent propargyl bromide.

The 60-MHz nmr spectrum of norbornadiene is shown in Figure 13.2. The chemical shift assignments are as follows: the olefinic protons at 6.6 ppm, the bridgehead protons at 3.55 ppm and the methylene protons at 2.0 ppm. Note also the splitting pattern due to long-range coupling.

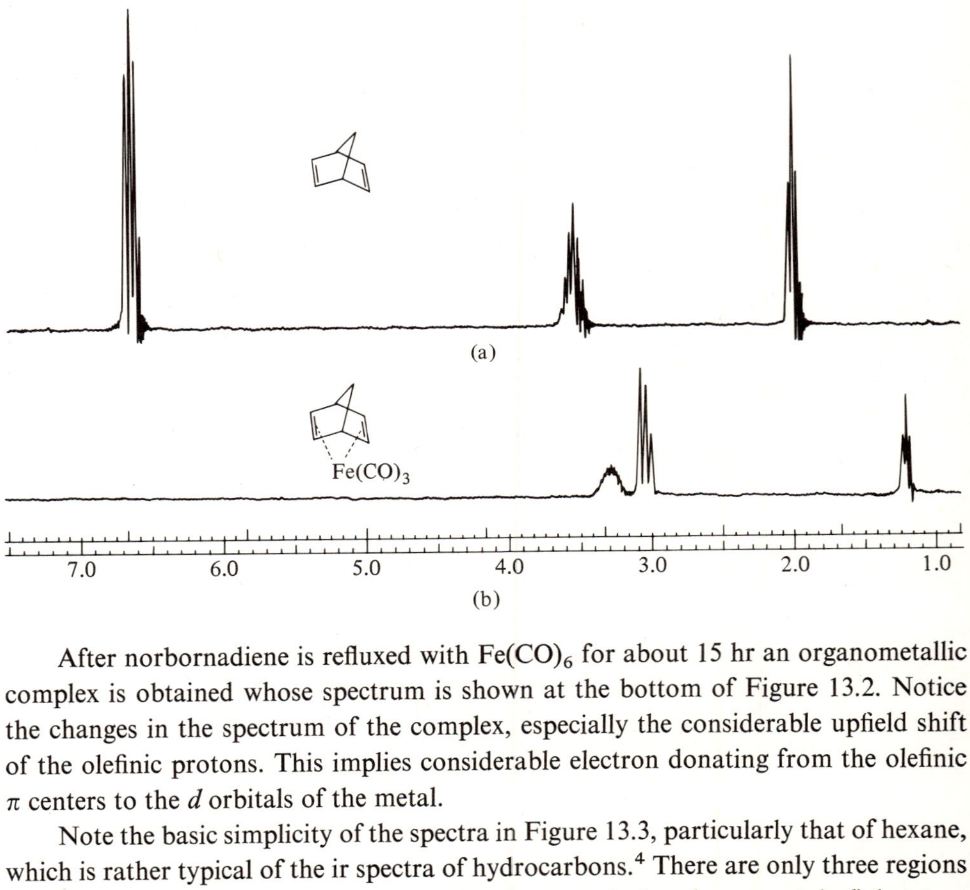


*FIGURE 13.2*
*The 60-MHz nmr spectrum of norbornadiene (a) and its* $Fe(CO)_3$ *complex (b). Intensity scales are different.*

After norbornadiene is refluxed with $Fe(CO)_6$ for about 15 hr an organometallic complex is obtained whose spectrum is shown at the bottom of Figure 13.2. Notice the changes in the spectrum of the complex, especially the considerable upfield shift of the olefinic protons. This implies considerable electron donating from the olefinic $\pi$ centers to the $d$ orbitals of the metal.

Note the basic simplicity of the spectra in Figure 13.3, particularly that of hexane, which is rather typical of the ir spectra of hydrocarbons.[4] There are only three regions of significance; the C—H stretching region (symmetrical and asymmetrical) between 2800–3000 $cm^{-1}$, the $CH_2$ bending region around 1470 $cm^{-1}$ and the band characteristic of a $CH_3$ group at 1380 $cm^{-1}$. These are usually the only bands of significance in saturated hydrocarbons. In the spectrum of 1-hexene three additional regions of importance appear which are characteristic of olefins. The =C—H stretching absorption just above 3000 $cm^{-1}$ (3100 $cm^{-1}$ for 1-hexene) the C=C bond stretching around

[4] Unless otherwise indicated, all ir spectra were taken in 10–20 percent $CCl_4$ solution in a 0.05-mm cell on a Beckman IR-20 spectrometer.

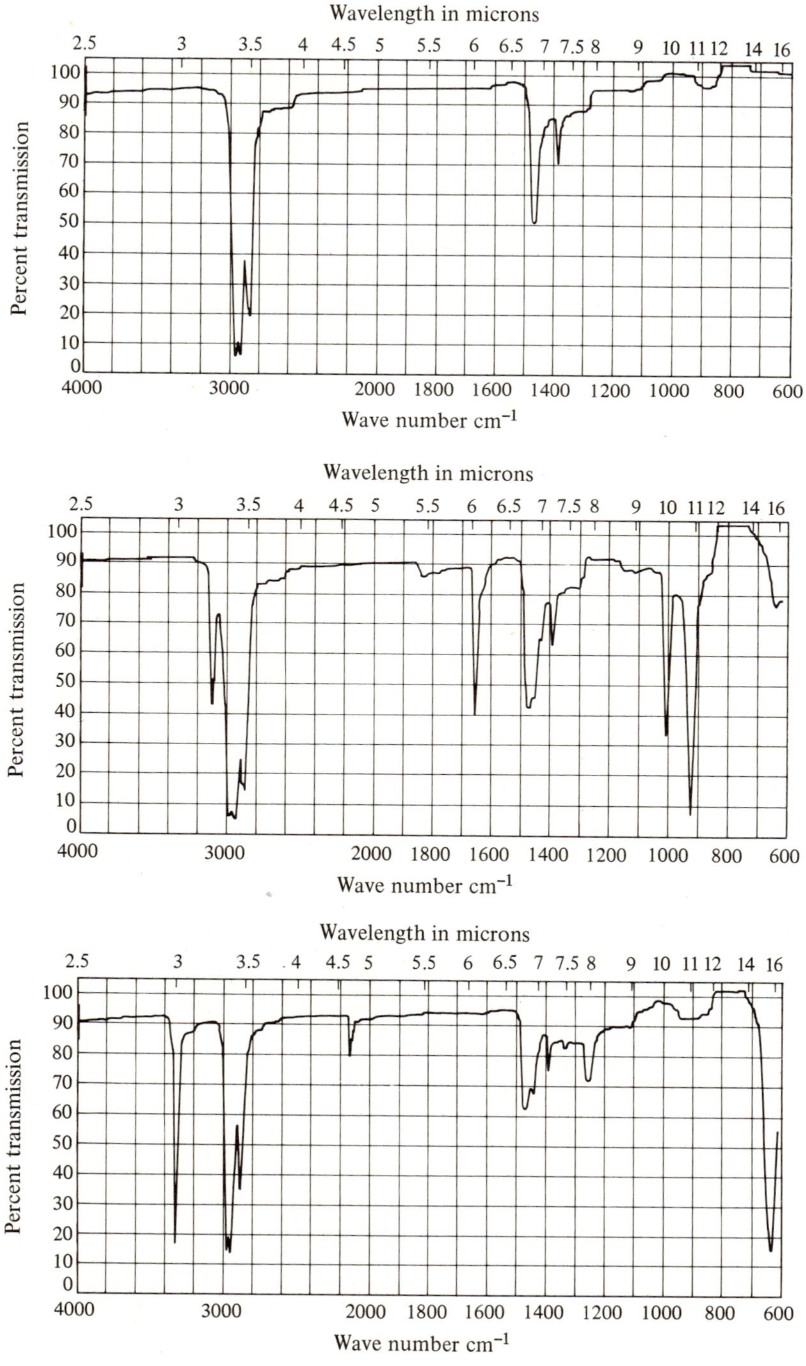


*FIGURE 13.3*

*Infrared spectrum of n-hexane (top), 1-hexene (middle), and 1-hexyne (bottom).*

1650 $cm^{-1}$, and the strong olefinic C—H bending modes around 920 $cm^{-1}$ and 1000 $cm^{-1}$. The small absorption around 1840 $cm^{-1}$ is an *overtone* of the C—H bending mode around 920 $cm^{-1}$. In 1-hexyne the unsaturated C—H stretching is even higher at 3330 $cm^{-1}$ and rather strong. The C≡C triple bond stretching is at 2140 $cm^{-1}$. The band at around 630 $cm^{-1}$ is the unsaturated C—H bending mode and its first overtone appears at about 1260 $cm^{-1}$. However, if instead of 1-hexyne the molecule had been the symmetrical 3-hexyne there would be no absorption above 3000 $cm^{-1}$ nor any around 600–700 $cm^{-1}$, because there would be no acetylenic hydrogen. Furthermore there would be no absorption between 2100–2250 $cm^{-1}$ since, because of symmetry, the C≡C absorption is ir-inactive.

## 13.6 *PROBLEMS*

SAMPLE PROBLEM

The C═C stretching frequency is at a minimum for cyclobutene in the series cyclopropene, cyclobutene, cyclopentene, cyclohexene (Table 13.1). Yet cyclohexene is assuredly less strained than cyclobutene. How would you account for this observation?

ANSWER

Clearly another factor than strain must be involved. For cyclopropene, a very strained molecule, it is misleading to consider the band at 1641 $cm^{-1}$ as *the* C═C stretch; it is really a vibration of the whole ring. This is the key to the explanation for these paradoxical values. There is coupling between the vibration of the C═C and of the other bonds in these rings. In cyclobutene, the C═C bond is almost perpendicular to the two adjacent C—C single bonds and the coupling is negligible. In fact, the degree of coupling follows a $\cos^2 \theta$ relationship, with $\theta$ the angle between the coupled vibrating bonds. This accounts qualitatively for cyclopropene and cyclohexene both having higher C═C stretches than cyclobutene and $\theta$'s differing by $\pm 30°$.

1. Using Woodward's rule for dienes calculate $\lambda_{max}$ for the following steroids:

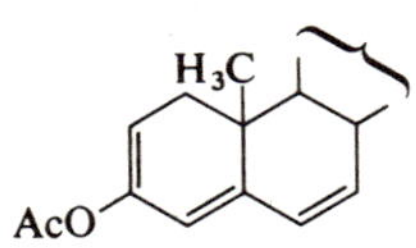
1

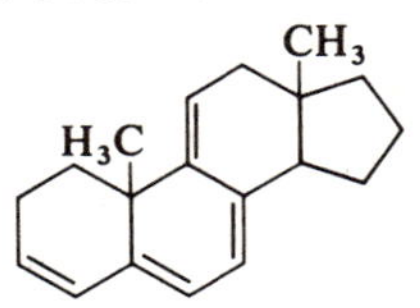
2

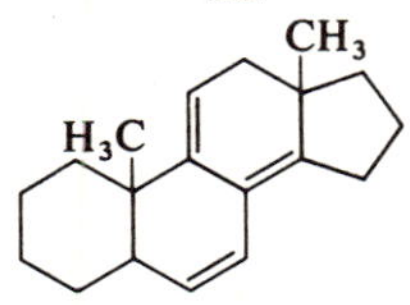
3

2. The carbon-carbon stretching frequency is observed at 1566 $cm^{-1}$ in cyclobutene, at 1641 $cm^{-1}$ in 1-methylcyclobutene, and at 1685 $cm^{-1}$ in 1,2-dimethylcyclobutene. Likewise, the corresponding cyclohexene values are 1646, 1670, and 1685 $cm^{-1}$ Why?
3. 2,2,5,5-tetramethyl-1,3-cyclohexanedione is transformed (ethanedithiol + $BF_3$-etherate) into the corresponding bisethylenethioketal (1). Raney nickel hydrogenolysis of (1) produces hydrocarbons (2) (70 percent), (3) (20 percent), and (4) (10 percent). The spectral data follow:

(1) 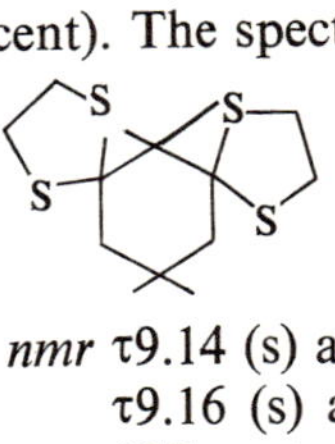

(2) *nmr* τ9.14 (s) and 8.76 (s), relative areas 3:2, in $CCl_4$ at 27°C.
τ9.16 (s) and 9.12 (s), plus a broad complex multiplet at 8.78 and 8.72 ($CS_2$, at −78°C).

| | | | | | | | | | | | |
|---|---|---|---|---|---|---|---|---|---|---|---|
| *mass m/e* | 41 | 43 | 55 | 56 | 57 | 69 | 70 | 71 | 83 | 125 | 140 (M) |
| (percent base peak) | 77 | 42 | 55 | 17 | 12 | 100 | 38 | 43 | 26 | 93 | 2 |

(3) *nmr* τ9.05 + 8.57 + 4.84, relative areas 6:2:1.

| | | | | | | | | | | | | |
|---|---|---|---|---|---|---|---|---|---|---|---|---|
| *mass m/e* | 48 | 50 | 67 | 69 | 77 | 79 | 81 | 82 | 91 | 95 | 123 | 138 (M) |
| (percent base peak) | 13 | 15 | 17 | 12 | 10 | 8 | 36 | 28 | 10 | 14 | 100 | 12 |

(4) *ir* = 764 $cm^{-1}$, very strong (approx. 50 $cm^{-1}$ "too high").
*uv* = intense end absorption.
*nmr* τ4.7 (s) + 9.0 (s), relative areas 1 : 3.

| | | | | | | | | | | | | |
|---|---|---|---|---|---|---|---|---|---|---|---|---|
| *mass m/e* | 51 | 53 | 65 | 77 | 79 | 91 | 105 | 106 | 119 | 121 | 122 | 136 (M) |
| (percent base peak) | 8 | 6 | 5 | 12 | 8 | 16 | 33 | 10 | 5 | 100 | 10 | 4.5 |

Write the structures of (2), (3), and (4), and comment on any features of interest in the data above.

4. *cis-cis-cis*-1,4,7-Cyclononatriene displays a temperature-dependent pmr spectrum. At 37°C the spectrum consists of a multiplet at $\delta = 5.3$ (area 6) and a broad band at $\delta = 2.9$ (area 6). While the position and general appearance of the low-field band do not change with temperature, the high-field band at lower temperature splits into two equally intense multiplets at $\delta = 2.19$ and $\delta = 3.68$, symmetrical about an axis at 2.92 ppm. At higher temperatures, the broad room-temperature peak sharpens into a triplet centered at $\delta = 2.9$ ppm. Explain these facts.
5. The pmr spectrum of 1,6-ethylene- [10]-annulene is analyzed to give the following values for the ring vicinal coupling constants:

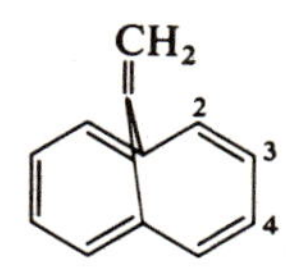


What can you conclude regarding the electronic structure of this molecule?

$J_{23} = 8.85$ Hz

$J_{34} = 9.43$ Hz

6. Suggest an explanation for the fact that the *cis* compound shown below dissolves in strong acids to give a deep yellow solution $\lambda_{max}$ (60 percent $H_2SO_4$) = 386, 427 nm, whereas the *trans* fixed ixomer shows no strong absorption beyond 300 nm in 60 percent sulfuric acid.

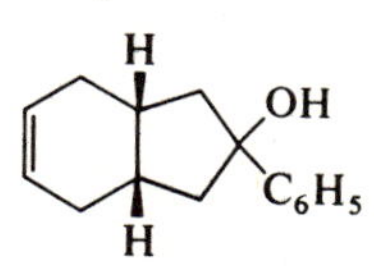


7. Determine the molar ratio of naphthalene, tetralin, and decalin.

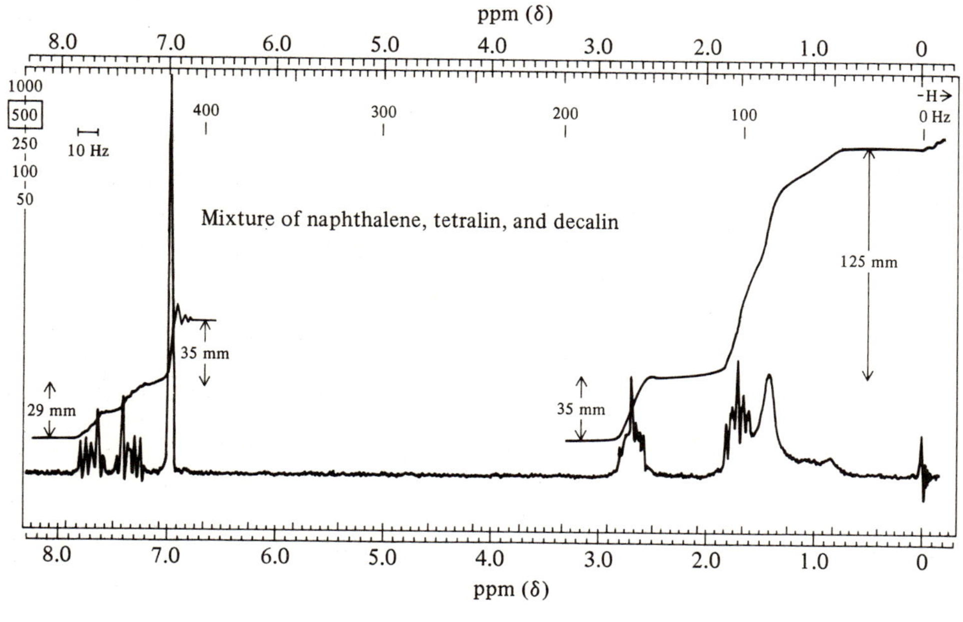

8. 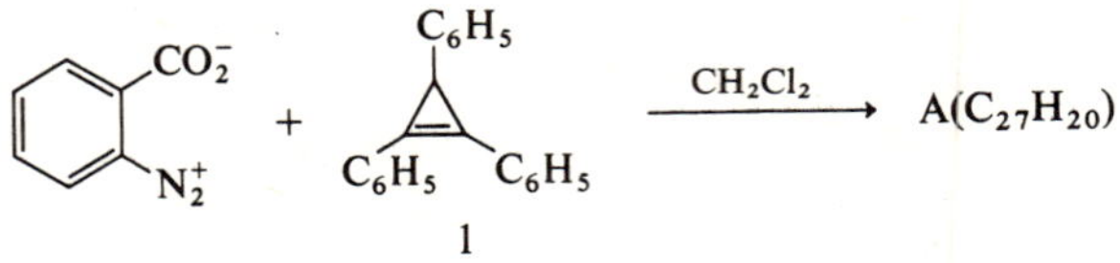


1

For A

uv $\lambda_{max}$(EtOH) 229, 303, 316, 333 nm
log $\varepsilon$ 4.52, 4.32, 4.36, 4.32

nmr only 7.3 and 7.7 ppm (from TMS)

ir max 5.5 $\mu$

For comparison (1)

uv $\lambda_{max}$(EtOH) 228, 305 sh, 314, 331 nm
log $\varepsilon$ 4.44, 4.37, 4.45, 4.33

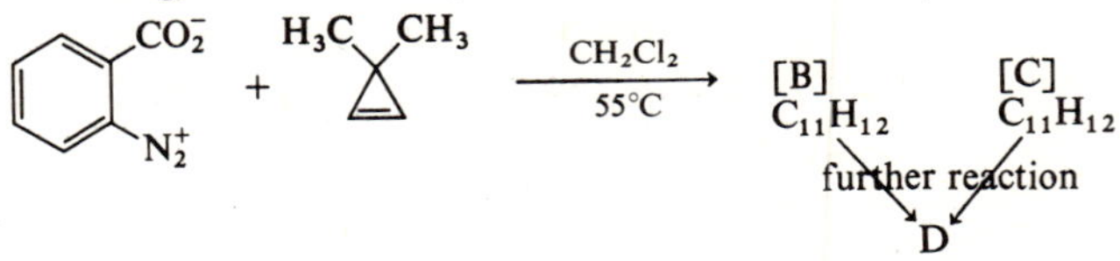


For D

Mass spec. $M^+$ at 220 *m/e*

nmr 6.9 ppm (m, 8H)
3.63 ppm (s, 2H)
0.90 ppm (s, 6H)

uv $\lambda_{max}$(EtOH) 274, 281 nm
log $\varepsilon$ 3.25, 3.37

Deduce the structure of A and D. Can you think of possible mechanisms? B and C are reactive intermediates.

9. Treatment of naphthalene, $C_{10}H_{10}$, with sodium in liquid ammonia yields a hydrocarbon $C_{10}H_{12}$. Addition of dichlorocarbene, generated from chloroform with potassium *t*-butoxide, to this hydrocarbon yields a compound $C_{11}H_{12}Cl_2$, which is then dechlorinated by sodium in liquid ammonia to another hydrocarbon $C_{11}H_{12}$. Treatment with 2 moles of bromine then leads to a tetrabromide $C_{11}H_{12}Br_4$. Dehydrobromination by ethanolic potassium hydroxide gives a hydrocarbon $C_{11}H_{10}$. The pmr spectrum of this $C_{11}H_{10}$ hydrocarbon consists of a symmetrical multiplet for eight hydrogens centered at $\delta = 7.1$ and a singlet for two hydrogens at $\delta = -0.5$. What is the structure of this $C_{11}H_{10}$ hydrocarbon?

## 13.7 *SUGGESTED READINGS*

J. Sicner, M. Svoboda, J. Závada, R. B. Turner, and P. Goebel, "Stereochemical Studies. XXXVI. An Approach to Conformational Analysis of Medium Ring Compounds. Unsaturated Ten-Membered Ring Derivatives," *Tetrahedron*, *22*, 659 (1966).

W. Luttke and A. de Meijere, "Conformations of *trans*-1,2-Dicyclopropylethylene and Vinylcyclopropane," *Angew. Chem. Int. Ed.*, *5*, 512 (1966).

B. C. L. Weedon, "Carotenoids: Recent Advances," *Chem. Brit.* 424 (1967).

E. W. Garbisch, Jr., and M. G. Griffith, "The Conformation of 1,4-Cyclohexadiene from Stereoisomeric Allylic-Allylic Proton Couplings," *J. Amer. Chem. Soc.*, *90*, 3590 (1968).

G. D. Newsoroff and S. Sternhell, "A Re-Evaluation of the Steric Dependence of Allylic Coupling Constants," *Tetrahedron Lett.*, 6117 (1968).

G. J. Abruscato and T. T. Tidwell, "Tri-*t*-butylethylene," *J. Amer. Chem. Soc.*, *92*, 4125 (1970).

D. A. Lightner and F. S. Steinberg, "Charge Localization in the Retro-Diels-Alder Fragmentation of Tetralins," *Org. Mass Spectry.*, *3*, 1095 (1970).

# CHAPTER 14 EFFECT OF SUBSTITUENTS

The main elements composing organic molecules are carbon, hydrogen, oxygen, and nitrogen. Due to their importance, individual chapters have been reserved for oxygen- and for nitrogen-containing systems. There are common organic molecules, however, such as natural products, which contain other important elements: halogens, phosphorus, silicon, sulfur, and so on. Our aim in this book cannot be all embracing; the format of the Encyclopaedia Britannica would be needed in order to do justice to the material currently available. Instead our ambition has been to be comprehensive, not in our coverage of the facts, but rather in the description of the methods available to relate the facts to one another or to a theory. This chapter will therefore be limited to a few important applications which will serve to indicate the kind of concepts utilized.

Qualitatively, substituting another element for hydrogen in a molecule has in general the following implications:

1. The mass and composition of the molecule are changed, a fact immediately reflected by the mass spectrum, both in the molecular and in fragment peaks.
2. The distribution of electrical charges is modified, since the substituent will attract (or repel) electrons in a different way than hydrogen. This redistribution of electron densities will be seen most easily in changes of chemical shifts for the hydrogen and the carbon nuclei.
3. If the substituent possesses a magnetic (and possibly a quadrupolar) moment, it will itself be affected by the rest of the molecule to which it is attached in terms of chemical shifts (think of $^{19}F$), coupling constants ($^{7}Li$, to take another example, has a spin number of $\frac{3}{2}$), or quadrupolar frequencies (which can be used to estimate for example the polarity of $^{35}Cl—C$ bonds).
4. The geometry of *other* parts of the molecule can be affected by substitution. Conformational changes are observed due to the steric and electronic requirements of a substituent.

From this description and from the characteristics of the various spectroscopies, it follows that the energy states between which transitions are observed will be raised or lowered in energy depending upon the nature of the substitution. The lifting of orbital degeneracy in benzene by substituents as studied by esr and by photoelectron spectroscopy has already been discussed (Chapter 6). When these energy changes are small enough and localized in one part of the molecule they can be treated conveniently as if they were additive. We have already presented a semiempirical additivity scheme which has been immensely successful—the Woodward rules for relating the uv absorption maxima of conjugated molecules to their constituent parts. Let us deal now with other additivity relationships which have been found to be useful, that is, valid for a very large number of cases.

## 14.1 *The Effect of Substitution upon Chemical Shifts*

From the intuitive reasoning above, it is quite clear that additivity of chemical shifts is likely to be observed only if the effect of substitution is restricted to the electrical redistribution of charges to the exclusion of large geometrical distortions. Therefore, such an additivity is to be looked for in rigid molecular systems which will be prevented from undergoing large changes of geometry. The corollary is that such additivity relationships will be an important structural tool only in rigid molecules. There are many such relationships; they will be illustrated here first with an example for a saturated hydrocarbon framework.

In the proton magnetic resonance spectra of steroids, which can belong to four great families,

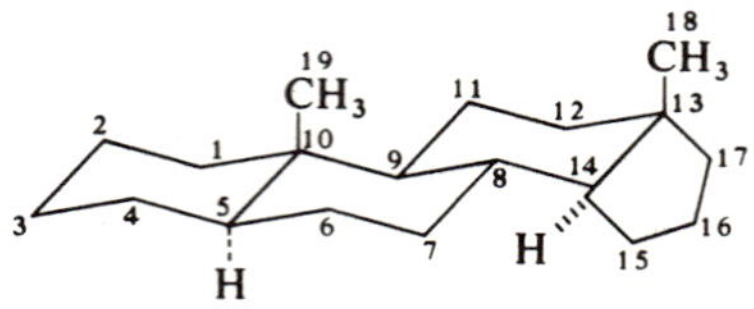


5α, 14α-Androstane

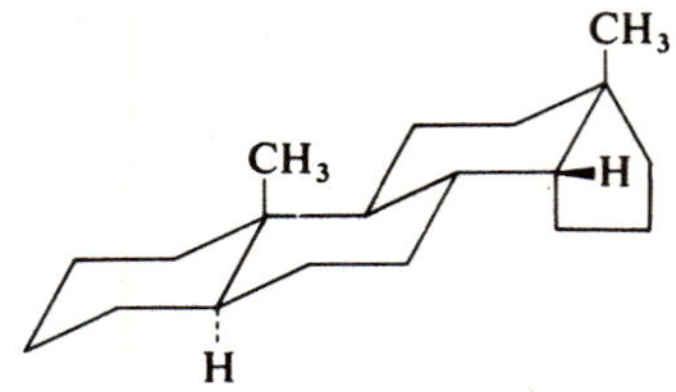


5α,14β-Androstane

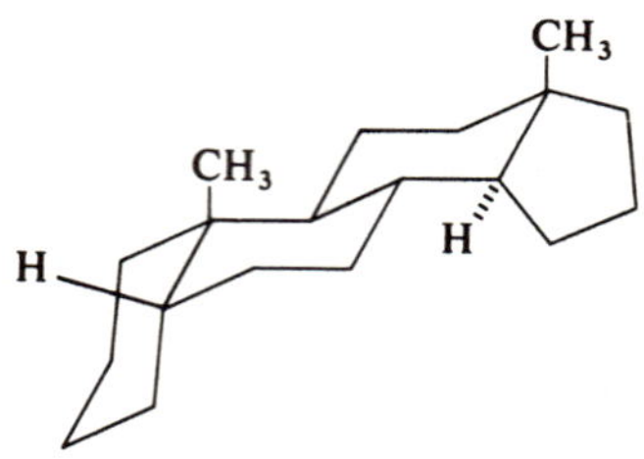


5β,14α-Androstane

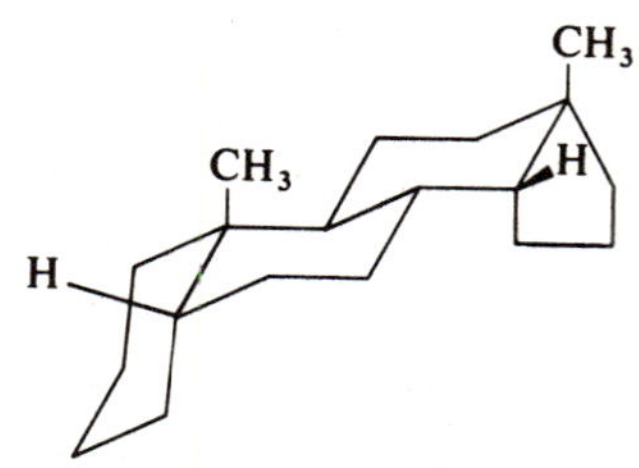


5β,14β-Androstane

the singlet resonances for the 18 and 19 methyl groups can be distinguished easily from the more complex peaks of the other protons in the molecule and their position can be determined very accurately. The steroid framework is fairly rigid and these systems have molecular dimensions such that a substituent in ring A, at position 2 for instance, may be very distant from a substituent in ring D, for example at $C_{15}$. Zürcher[1] has indeed been able to show that increments can be attached to substituents in the various positions on either the α or the β face of the steroid. Using these increments, the chemical shift of either the 18 or the 19 methyl group can be calculated with an accuracy of ±0.02–0.03 ppm, a remarkable feat considering that to date hundreds of molecules have been so correlated.

A further very useful generalization due to Zürcher is the notion of equivalent positions: To a good approximation, certain positions are viewed as equivalent by an observer placed on either methyl 18 or methyl 19. This pseudosymmetry is best visualized by considering the planar formula for a steroid:

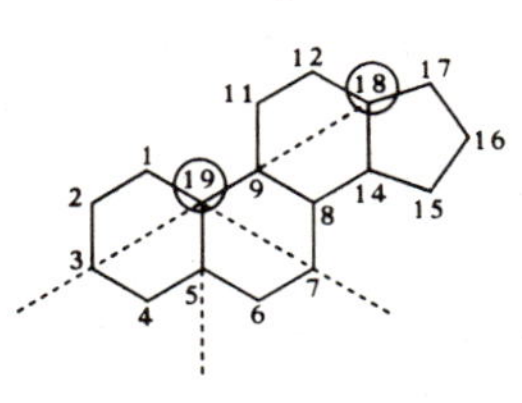


thus positions 2 and 4, or 3 and 7, are equivalent positions with respect to the 19 angular methyl group; likewise, positions 8 and 11 are equivalent when seen from the 18 methyl group. The effects of substituents in equivalent positions are compared for a few cases in Table 14.1.

Another example is from the field of aromatic systems: The equalization of bond lengths enforced by the electron delocalization in these molecules is a powerful safeguard against the occurrence of large structural distortions. There is a large body of literature covering the incremental shifts associated with the effect of substituents in various positions of benzenoid hydrocarbons on the chemical shifts of the ring

[1] R. F. Zürcher, *Helv. Chim. Acta.*, *44*, 1380 (1961); *46*, 2054 (1963).

*TABLE 14.1*

*Incremental Shifts for the 19-Methyl due to Substituents in Equivalent Positions in the (5α)-Steroids*

| Substituent (ring A) | $\Delta\delta^a$ | Substituent (ring B) | $\Delta\delta^a$ |
|---|---|---|---|
| 3-Keto | 14.0 | 7-keto | 17.0 |
| 4-Keto | −1.5 | 6-keto | −2.5 |
| $\Delta^4$-3-Keto | 24.5 | $\Delta^5$-7-keto | 24.0 |
| $\Delta^2$ | 0.5 | $\Delta^7$ | 0.5 |
| 3-β-Hydroxy | 2.5 | 7-β-hydroxy | 1.0 |

[a] In hertz at 60 MHz, to be added to a base value of 46.5 Hz for the 19-methyl group in (5α)-androstane.

protons. Some of the references are given at the end of this chapter. Since the interest of a compilation of data resides in its use, no further comment is necessary here. We shall limit ourselves to indicating the great generality of such an approach by quoting the additivity observed for the carbon-13 chemical shifts in five-membered nitrogen heterocycles. Weigert and Roberts[2] have shown that the following four-parameter equation is sufficient to reproduce the experimentally observed $^{13}$C chemical shifts (in parts per million, referred to the resonance of the cyclopentadienide anion):

$$\delta_C = n_\alpha C_\alpha + n_\beta C_\beta + n_{\alpha\beta} C_{\alpha\beta} + n_{\beta\beta} C_{\beta\beta}$$

where $n$ indicates how many times a particular nitrogen structural feature occurs in the molecule relative to the carbon of interest. The four parameters used are $C_\alpha = -15$ ppm, $C_\beta = -4$ ppm, $C_{\alpha\beta} = -10$ ppm, and $C_{\beta\beta} = +7$ ppm. Table 14.2 compares the experimental and the calculated values.

*TABLE 14.2*

*Correlation of $^{13}$C Chemical Shifts in the Five-Membered Nitrogen Heterocycles*

| Compound | Carbon | $n_\alpha$ | $n_\beta$ | $n_{\alpha\beta}$ | $n_{\beta\beta}$ | δ (calc.) | δ (obs.) | Structure |
|---|---|---|---|---|---|---|---|---|
| Pyrrole | 2 | 1 | 0 | 0 | 0 | 15 | 15 |  |
| | 3 | 0 | 1 | 0 | 0 | 4 | 4 | |
| Pyrazole | 3 | 1 | 1 | 1 | 0 | 29 | 29 | 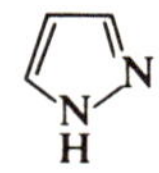 |
| | 4 | 0 | 2 | 0 | 1 | 1 | 1 | |
| Imidazole | 2 | 2 | 0 | 0 | 0 | 30 | 32 |  |
| | 4 | 1 | 1 | 0 | 0 | 19 | 18 | |
| 1,2,3-Triazole | | 1 | 2 | 1 | 1 | 26 | 27 | 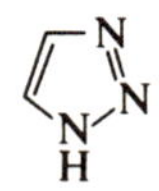 |
| 1,2,4-Triazole | | 2 | 1 | 1 | 0 | 44 | 44 |  |
| Tetrazole[a] | | 2 | 2 | 2 | 1 | 51 | 41 | 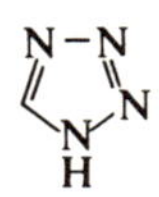 |

[a] This empirical relationship starts to break down in tetrazole, because of the accumulation of nitrogen heteroatoms.

## 14.2 *Substituent Effects on One-Bond Nuclear Magnetic Resonance Coupling Constants*

Numerous authors have reported that one-bond couplings such as $^{13}$C—H, $^{29}$Si—H, $^{13}$C—$^{19}$F or $^{29}$Si—$^{19}$F for molecules of the type A—B(XYZ), in which A and B have spins of ½ and X, Y, Z are substituents, are to a good approximation additive

[2] F. Weigert and J. D. Roberts, *J. Amer. Chem. Soc.*, *90*, 3543 (1968).

in terms of the increments characteristic of the X, Y, Z residues. An example is afforded by the Malinowski rules for the $^1J_{C-H}$ coupling constants in substituted methanes.[3] For trisubstituted methanes CHXYZ $^1J_{C-H} = \xi_x + \xi_y + \xi_z$. Xi values for various substituents are given in Table 14.3, together with some examples of the calculated and experimental chemical shifts. With very electronegative substituents, such as fluorine, this empirical correlation breaks down, meaning that it is no longer possible to attribute a definite *s* character affinity to any substituent. This correlation has been generalized to other types of molecules such as aldehydes.

*TABLE 14.3*

*Malinowski Parameters and Their Application to Substituted Methanes*

| X | $\xi_X$ | | Calc. | Exp. |
|---|---|---|---|---|
| H | 41.7 | $CH_2ClCN$ | 163 | 161 |
| F | 65.6 | $CH_2I_2$ | 177 | 173 |
| Cl | 68.6 | $CH_2(C_6H_5)_2$ | 127 | 127 |
| Br | 68.6 | $C_6H_5CH_2F$ | 150 | 151 |
| I | 67.6 | | | |
| OH | 59.6 | $CH_2F_2$ | 173 | 185 |
| CN | 52.6 | $CHF_3$ | 197 | 238 |
| $C_6H_5$ | 42.6 | $CHFCl_2$ | 203 | 220 |
| | | $CHF_2Cl$ | 200 | 231 |

What is the precise origin of these increments, and how are they determined? The answers to this question have tended to be either in terms of the electronegativities of the substituents X, Y, Z, as they modify the nuclear charge on carbon, or in terms of the rehybridization attendant upon substitution of the central carbon atom, which changes the percent *s* character in the carbon-hydrogen bond under study. It is now taken for granted that the former is the main component, since otherwise unrealistic hybridizations would be calculated for compounds such as fluoroform ($CHF_3$ would appear by this criterion to be close to *sp* hybridized). As a semiempirical device these $^1J_{C-H}$ coupling constants can be used as a measurement of electronegativity somewhat more reliably than chemical shifts since they are not influenced by the magnetic anisotropies of the substituents. Recently, MO methods have become available for the calculation of nuclear spin-spin coupling constants and have been applied to one-bond couplings. One of the most interesting studies, that of Jensen and Murrell[4] derives the increments $\alpha_X, \alpha_Y, \alpha_Z$ (similar to Malinowski's) from a functional form including quadratic terms. Of course such a procedure gives better results than the truncation to the first term used previously by Malinowski and other investigators. For the M—H coupling in compounds of the type MHXYZ (M = $^{29}$Si, $^{13}$C) $^1J_{M-H} = J(MH_4) + A(\alpha_X + \alpha_Y + \alpha_Z) + B(\alpha_X^2 + \alpha_Y^2 + \alpha_Z^2) + C(\alpha_X\alpha_Y + \alpha_Y\alpha_Z + \alpha_Z\alpha_X)$ and the parameters *A*, *B*, and *C*, as well as the $\alpha$'s are optimized to give the best fit to the experimental values (Table 14.4).

*TABLE 14.4*

*The Jensen Parameters for Substituted Silanes and Methanes*

| Substituent | $\alpha_X$ (silane) | $\alpha_X$ (methane) (eV) |
|---|---|---|
| F | −2.67 | −3.57 |
| Cl | −2.50 | −2.83 |
| Br | −2.53 | −2.83 |
| I | −2.04 | −2.93 |
| $OCH_3$ | −1.49 | −2.13 |

[3] J. W. Emsley, J. Feeney, and L. H. Sutcliffe, *High Resolution Nuclear Magnetic Resonance Spectroscopy*, Pergamon Press, Oxford, 1966, vol. 2, pp. 1017–1020.
[4] R. Ditchfield, M. A. Jensen, and J. N. Murrell, *J. Chem. Soc.* (A), 1674 (1967).

The best values for $A$, $B$, and $C$ are

| | | | |
|---|---|---|---|
| $A = 18.26$ Hz/eV | $B = 1.78$ Hz/eV$^2$ | $C = -2.98$ Hz/eV$^2$ | (silanes) |
| $A = -8.28$ Hz/eV | $B = -0.21$ Hz/eV$^2$ | $C = 0.78$ Hz/eV$^2$ | (methanes) |

With such a formulation, even the coupling constants involving highly electronegative substituents are accounted for. The parameter $\alpha_X$ characteristic of each substituent has the precise meaning of the Coulomb integral or effective electronegativity for that substituent X, and a vast amount of data is correlated within 2 percent of the experimental determinations. The success of such an approach can again be attributed to the relative constancy of the local environment: the spin densities at carbon and hydrogen for a C—H bond change little; only the mutual polarizability of the $2s$ orbital on carbon and of the $1s$ orbital on hydrogen varies greatly, hence the possibility of such an empirical relationship as devised by Jensen.[5]

## 14.3 *The Nitrogen-14 Probe of Substituent Effects: Example of the Azabenzenes*

As indicated in Chapter 2, nqr frequencies for a quadrupolar nucleus (with a quadrupolar moment $Q$) are a very sensitive function of the chemical environment in a molecule. They are proportional to $e^2qQ$, in which $q$ is the electrical field gradient at the quadrupolar nucleus. Because of this dependence, nqr can accurately mirror very small changes in the electronic distribution within the molecule.

Let us consider the example of nitrogen in heterocycles of the pyridine type. To render calculations simpler, the following assumptions will be made here: The molecular geometry is that of a regular hexagon; the heterocyclic nitrogen is $sp^2$ hybridized, two of the $sp^2$ hybrids being used to form $\sigma$ bonds with the neighboring atoms, and the remaining $sp^2$ hybrid being occupied by two unshared electrons (the nitrogen lone pair). The remaining $p$ orbital enters the $\pi$-electronic system delocalized over the other ring atoms. In a molecule such as pyridine, the field gradient $q$ at nitrogen is due to the presence of two unshared electrons in the lone pair while only about one electron is contained in each of the $\sigma$ orbitals. Hence, the field gradient $q$ which is represented by a tensor has its principal axis (largest value) $q_{zz}$ directed towards the nitrogen lone pair. Let us define a referential triad accordingly:

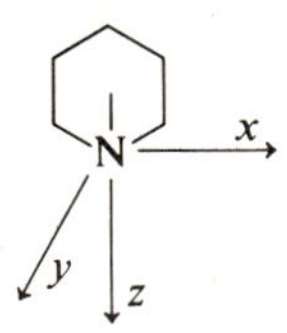


The wave functions of interest in this coordinate system will be

$$\psi_1 = \frac{\sqrt{3}}{3}\phi_{1s} + \frac{\sqrt{2}}{2}\phi_{p_x} - \frac{\sqrt{6}}{6}\phi_{p_z} \qquad \sigma \text{ bond } (b)$$

$$\psi_2 = \frac{\sqrt{3}}{3}\phi_{1s} - \frac{\sqrt{2}}{2}\phi_{p_x} - \frac{\sqrt{6}}{6}\phi_{p_z} \qquad \sigma \text{ bond } (b)$$

$$\psi_3 = \frac{\sqrt{3}}{3}\phi_{1s} + \frac{\sqrt{6}}{6}\phi_{p_z} \qquad \text{lone pair } (2)$$

$$\psi_\pi = \phi_{p_y} \qquad \pi \text{ bond } (a)$$

[5] M. A. Jensen, *J. Organometal. Chem.*, *11*, 423 (1968).

Let us denote the numbers of electrons effectively present in each of these orbitals as 2 for the lone pair, $a$ for the $\pi$ bond, and $b$ for each of the $\sigma$ bonds. These numbers refer to the electrons in the vicinity of nitrogen. They are so-called Mulliken occupation numbers; that is, for each bonding orbital the electrons are partitioned arbitrarily between each of the bonded atoms on which they are then fictitiously localized as "net charges." The quadrupolar coupling constant can then be calculated as[6]

$$e^2q_zQ = e^2q_pQ(\tfrac{1}{6}b - \tfrac{1}{2}a + \tfrac{4}{3})$$

where $e^2q_pQ$ refers to the charge distribution of a single $2p$ electron in a neutral nitrogen atom along the $2p$ symmetry axis and its value is $e^2q_pQ = 10\text{–}12$ MHz. Another experimentally accessible value is the asymmetry parameter, for which the expression is[7]

$$\eta = \frac{q(a-b)}{3a+b-8}$$

From the experimental determination of $e^2q_zQ$ and $\eta$, it is therefore possible to derive the occupation numbers $a$ and $b$, that is, the molecular data of interest to the organic chemist. Their values are listed in Table 14.5 for a few selected cases. The first conclusion to be drawn from this table is the considerable polarization of the electronic charge onto the heteroatom. The excess charge on nitrogen is consistently of the order of 0.9 electron. Notice to what extent these variations conform to our chemical intuition. The $\pi$ electrons are indeed much more polarizable than the $\sigma$ electrons, the total spread in Table 14.5 is 0.090 for the $\pi$ electrons versus 0.033 for the $\sigma$ electrons. Methyl groups in the meta positions do not affect the nitrogen electron densities, but methyl groups in the ortho and para positions increase them strongly (by inductive as well as by resonance electron releasing). Likewise, meta chlorines little affect the $\pi$ and $\sigma$ occupation numbers; ortho chlorines are electron withdrawing inductively ($b$ is decreased) and electron releasing by resonance ($a$ is increased):

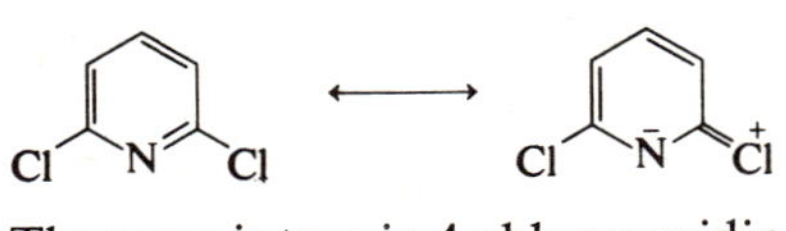

The same is true in 4-chloropyridine, but the inductive electron withdrawal is decreased by the distance. The resonance effect can be formulated as:

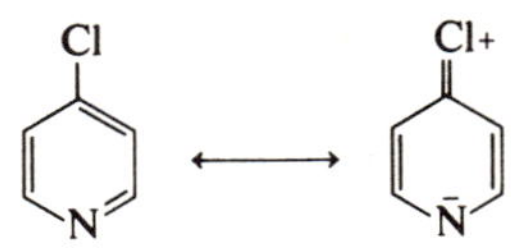

This example of the chloropyridines is a useful check on the entire method, since the

TABLE 14.5

*$\pi$ and $\sigma$ Occupation Numbers of the Orbitals at Nitrogen*

| Compound | $a(\pi)$ | $b(\sigma)$ | $qxs^a$ |
|---|---|---|---|
| Pyridine | 1.200 | 1.333 | 0.866 |
| 4-Methylpyridine | 1.245 | 1.349 | 0.943 |
| 3,5-Dimethylpyridine | 1.192 | 1.331 | 0.854 |
| 2,4,6-Trimethylpyridine | 1.273 | 1.349 | 0.971 |
| 4-Chloropyridine | 1.209 | 1.339 | 0.887 |
| 3,5-Dichloropyridine | 1.197 | 1.323 | 0.843 |
| 2,6-Dichloropyridine | 1.282 | 1.316 | 0.914 |

[a] Excess charge on nitrogen = $(a - 1.000) + (2b - 2.000)$.

[6] E. Schempp and P. J. Bray, *J. Chem. Phys.*, *46*, 1186 (1967).
[7] *Ibid.*

postulated changes in the polarity of the C—Cl bond ortho, meta, or para to the heterocyclic nitrogen could be verified by $^{35}Cl$ nuclear quadrupole resonance. This nqr method strongly resembles the fluorophenyl label method of Taft. Both are precise and objective tools with which to gauge the effects of chemical substitution. As a concluding remark to this chapter, one could say that whereas 10 or 20 years ago chemists contented themselves with characterizing the *presence* of a given substituent (by its group frequency, for instance), not only are we now in a position to *locate* a substituent in the molecule (mainly by arguments based upon the mass spectrum and the nmr spectrum), but we can also have some idea of the electron distribution, and how it varies in the presence or absence of a given substituent. Progress in this area has allowed us to probe deeper and deeper into the intricacies of molecular structure.

## 14.4 PROBLEMS

1. The $\sigma$ and $\pi$ charge densities $l$ and $m$ can be determined by nqr for the nitrogen atom of the cyano group in a variety of compounds:

| Compound | $l$ | $m$ |
|---|---|---|
| Acetonitrile | 1.261 | 1.001 |
| Benzonitrile | 1.305 | 1.033 |
| 4-Methoxybenzonitrile | 1.305 | 1.000 |
| 2-Cyanopyridine | 2.309 | 1.022 |
| 3-Cyanopyridine | 1.305 | 1.032 |
| 4-Cyanopyridine | 1.288 | 1.004 |

Can you rationalize these observations?

2. The occupation numbers of the orbitals at nitrogen are found to be 1.195 ($\pi$) and 1.320 ($\sigma$) in pyridine, by a method differing slightly from that in Table 14.5. For 2-chloro-, 2-cyano-, 2-amino-, and 2-methylpyridine the pairs of values found are

| | |
|---|---|
| 1.187 | 1.299 |
| 1.219 | 1.322 |
| 1.227 | 1.308 |
| 1.334 | 1.343 |

Can you attribute these to the appropriate substituent?

3. The occupation numbers of the orbitals at nitrogen are 1.166 ($\pi$) and 1.321 ($\sigma$) in 4-cyanopyridine. The pyridine values are 1.200 and 1.333, respectively. Explain.

## 14.5 SUGGESTED READINGS


U. E. Matter, C. Pascual, E. Pretsch, A. Pruss, W. Simon, and S. Sternhell, "Estimation of the Chemical Shifts of Olefinic Protons Using Additive Increments. II. The Compilation of Additive Increments for 43 Functional Groups," *Tetrahedron*, *25*, 691 (1969).

D. F. Ewing and K. A. W. Parry, "Studies of Geometrical Isomerism in Some Chloro- and Dichloro-Olefins by Nuclear Magnetic Resonance," *J. Chem. Soc.* (*B*), 970 (1970).

A. W. Douglas, "$^{13}C$—H Coupling Constants. II. Periodic Table Variation of $J_{^{13}C-H}$ and Carbon Hybridization Effects on Its Value," *J. Chem. Phys.*, *45*, 3465 (1966).

R. M. Hammaker, "$J_{^{13}C-H}$ and Electronegativity," *J. Chem. Phys.*, *43*, 1843 (1965).

C. P. Yue, "An Empirical Periodic Correlation of $J_{^{13}C-H}$ for the Methyl Group," *Can. J. Chem.*, *46*, 2675 (1968).

R. Ditchfield, M. A. Jensen, and J. N. Murrell, "Effect of Substituents on Methane $^{13}C$—H and $^{29}Si$—H Coupling Constants," *J. Chem. Soc.*, (A), 1674 (1967).

M. A. Jensen, "The Prediction of $^{29}Si$—H and $^{13}C$—H Coupling Constants in Substituted Silanes and Methanes," *J. Organometal. Chem.*, *11*, 423 (1968).

R. Ikeda, S. Onda, D. Nakamura, and M. Kubo, "Nuclear Quadrupole Resonance of Nitrogen-14 in Some Pyridine Derivatives," *J. Phys. Chem.*, *72*, 2501 (1968).

L. Guibé and E. A. C. Lucken, "The Electron Distribution at the Nitrogen Atom in Substituted Pyridines and Related Compounds, as Measured by their $^{14}$N Quadrupole Coupling Constants," *Mol. Phys.*, *14*, 79 (1968).

T. L. Brown, W. G. McDugle, Jr., and L. G. Kent, "Vibrational and Pure Nuclear Quadrupole Spectra of Hexahalometalates," *J. Amer. Chem. Soc.*, *92*, 3645 (1970).

G. W. Smith, "High Resolution Proton Magnetic Resonance Studies of Para-disubstituted Benzenes," *J. Mol. Spectry.*, *12*, 146 (1964).

J. G. Traynham and G. A. Knesel, "Correlation of Polar Effects of Substituents with Nuclear Magnetic Resonance Spectra of Phenols," *J. Org. Chem.*, *31*, 3350 (1966).

K. Hayamizu and O. Yamamoto, "The Analysis of the Proton Magnetic Resonance Spectra of Monosubstituted Benzenes, Alkyl Benzenes and an Empirical Interpretation of the Ring Proton Chemical Shifts," *J. Mol. Spectry.*, *29*, 183 (1969).

C. H. Yoder, R. H. Tuck, and R. E. Hess, "Nuclear Magnetic Resonance Studies of the Bonding in Aromatic Systems. A Correlation of Hammett $\sigma$ Constants with Methyl $^{13}$C—H Coupling Constants and Chemical Shifts," *J. Amer. Chem. Soc.*, *91*, 539 (1969).

# CHAPTER 15 OXYGEN-CONTAINING MOLECULES

Oxygen-containing molecules can be divided into two broad classes: compounds containing $\sigma$-bonded oxygens such as in alcohols and ethers, and $\pi$-bonded oxygens such as in ketones and aldehydes. Compounds such as carboxylic acids, anhydrides, esters, and lactones can be looked upon as composites of these functional groups. Within these broad categories it is usually desirable to specify further the exact position and configuration of the heteroatom, the ring size in the case of cyclic compounds, and the nature and influence of neighboring substituents.

In this chapter as in previous sections, we shall be concerned not so much with the tabulation of characteristic absorption frequencies as with the general rules that have been developed for structure elucidation of oxygen-containing molecules, and the methods involved in the formulation of such rules.

## 15.1 σ-Bonded Oxygen

### 15.1.1 Alcohols and Phenols

The unique feature of hydroxylic compounds is the labile hydrogen capable of undergoing exchange and participation in hydrogen bonding. This property naturally affects physical measurements and shows up in the nmr, mass spectrum, and infrared and provides an easy handle for identification. Let us look at these in turn.

In the proton magnetic resonance spectra, hydroxyl protons normally show up as broad featureless bands, which are markedly temperature and concentration dependent, a case of almost too much information in one message. The reason is, of course, the interchange of the hydroxylic protons between various molecules in the medium, at an intermediate rate of exchange under ordinary conditions. The temperature-dependent spectrum of methanol (Figure 15.1) illustrates this behavior (note that coupling between the methyl and the hydroxylic protons vanishes when exchange is rapid). This exchange is facilitated by traces of acidic or basic impurities present in the solvent. However, if one resorts to rigorous purification of the solvent or to use of strong hydrogen-bond acceptors, such as dimethylsulfoxide, pyridine, or hexamethylphosphoramide as solvent it becomes possible to standardize conditions and obtain reproducible spectra and meaningful parameters for the hydroxylic protons.

Of course this exchange can be turned to advantage and serve as a test for the presence of hydroxylic protons (or other labile protons) by shaking the sample with a few drops of $D_2O$ and observing the disappearance of the signal due to the —O—H proton. Addition of a few drops of acid, on the other hand, results in a marked sharpening and downfield shift of the proton resonance signal due to rapid acid-catalyzed exchange and to the low-field resonance of acidic protons. The exact chemical shift of the hydroxylic proton is dependent on the strength of the hydrogen bonding, the nature of R or Ar, and the acidity of the O—H bond. A summary of chemical shifts for hydroxylic protons is given in Table 15.1. Note in particular that phenols, which are considerably more acidic than alcohols, display chemical shifts at much lower fields.

*TABLE 15.1*

*Chemical Shifts of Hydroxylic Protons in "Inert" Solvents*

| Compound | Chemical shift (ppm) |
|---|---|
| Alcohols RO—H | 2–3.5 |
| Phenols ArO—H | 5–10 |
| Carboxylic acids RCOO—H | 10–15 |

*FIGURE 15.1*
*Nuclear magnetic resonance spectrum of methanol at $-74°$ and $+8°C$.*

In the mass spectrometry of alcohols, prior exchange of the active hydroxylic protons by introducing $D_2O$ in the inlet together with the sample will shift the parent peak by 1 mass unit. Furthermore, the spectra are usually characterized by loss of water (M-18). These eliminations of water usually proceed by an intramolecular hydrogen abstraction via a cyclic mechanism:

The "formation" of five- and six-membered rings appears to be most favorable. Virtually no 1,2 elimination of $H_2O$ is observed, as indicated by the retention of the isotopic label at the $\alpha$ and $\alpha'$ positions of secondary alcohols:

$$\mathrm{CH_3CD_2{-}\underset{\substack{|\\ OH}}{C}HCD_2C_{10}H_{21}} \longrightarrow \mathrm{H_2O + CH_3CD_2{-}\overset{CD_2-(CH_2)_n}{\overset{|\qquad\quad |}{CH\; - \;CH}}{-}(CH_2)_{8\text{-}n}{-}CH_3}$$

In the infrared the hydroxyl group is characterized by the O—H stretching vibrations which normally occur between 3650–3590 $cm^{-1}$ for the free O—H band and are shifted to lower wavelength by hydrogen bonding. Absorption due to hydrogen bonding usually occurs between 3200 and 3550 $cm^{-1}$, depending on the type and strength of the particular hydrogen bond. Let us then assume that the presence of a hydroxylic function has been established and see how we can use this information to obtain other structural information about the molecule under consideration.

In the nmr, useful structural information can be obtained from the vicinal coupling constants, a number of which are listed in Table 15.2. The regularity inherent in the values of the coupling constants listed in Table 15.2 is that *trans* coupling constants corresponding to a dihedral angle of 180° are large, and of the order of 12 Hz, *cis* coupling constants (0°) are also large, but *gauche* couplings (60°) are much smaller. The trend is schematized in Figure 15.2.

*TABLE 15.2*

*Vicinal Coupling Constants $^3J_{HCOH}$ in Hydroxylic Molecules*

| Molecule | (Reference) | $^3J_{cis}$ (0°) | $^3J$ (dihedral angle) | $^3J_{trans}$ (180°) |
|---|---|---|---|---|
| 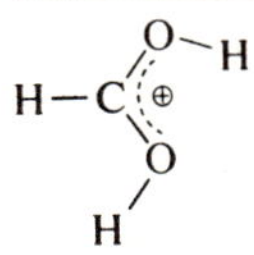 | *a* | 3.2 | — | 13.9 |
| 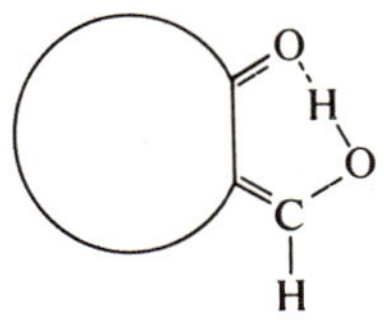 | *b* | — | — | ~11.5 |
| 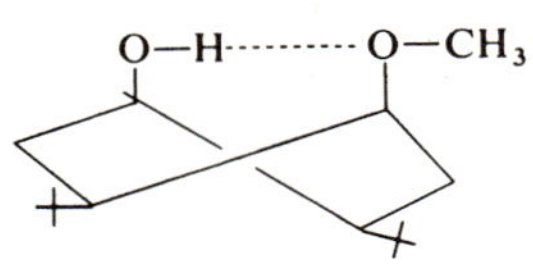 | *c* | — | 11.4 (167 ± 2) | — |
| 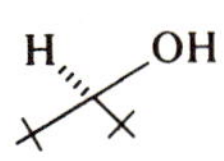 | *d* | — | 1.95 (60) | 12.3 |
| 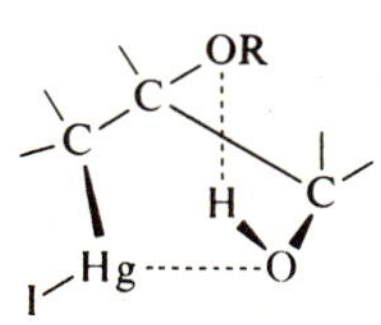 | *e* | — | 2.8 (60) | 11.0 |

[a] H. Hogeveen *et al., Chem. Comm.*, 899 (1966).
[b] E. W. Garbisch, Jr., *J. Amer. Chem. Soc., 85*, 1696 (1963).
[c] R. D. Stolow and A. A. Gallo, *Tetrahedron Lett.*, 3331 (1968).
[d] R. M. Hammaker *et al., J. Phys. Chem., 72*, 4346 (1968).
[e] E. F Kiefer *et al., J. Amer. Chem. Soc., 90*, 6246 (1968).

A functional dependence of the type $^3J_{HCOH} = A \cos^2 \phi_{HCOH}$ is a good zero-order approximation to this dihedral-angle dependence of the vicinal coupling constants. Such a functional dependence was to be expected (see Chapter 3). This functional dependence of the coupling constant on the dihedral angle can serve as a valuable tool in deducing molecular geometry, provided that the observed $^3J_{HCOH}$ of a given dihedral angle, $\phi_{HCOH}$, is a true intramolecular quantity and is unaffected by hydrogen bonding to the solvent or by other medium effects. That this is indeed the case can be verified experimentally. The $^3J_{HCOH}$ coupling constant is the same for cyclohexanol in a strongly hydrogen-bonding solvent such as dimethylsulfoxide and in an inert solvent such as carefully purified and dried carbon tetrachloride.

In the ir the already mentioned frequency shift $\Delta\nu_{OH}$ due to hydrogen bonding can be used as a sensitive structural probe. In fact, hydrogen bonding can be either

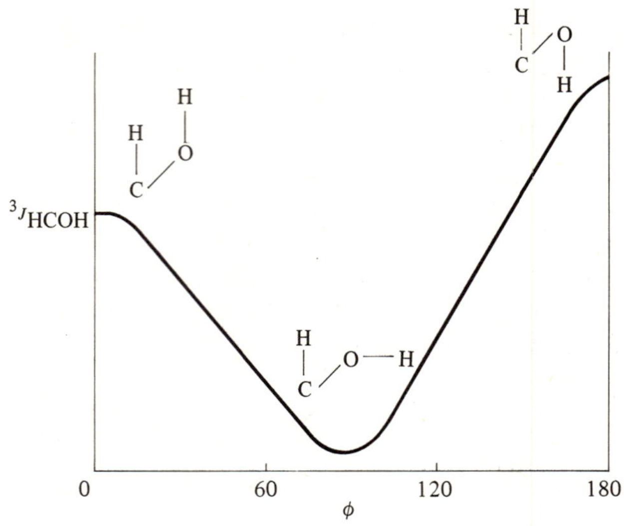


*FIGURE 15.2*
*Vicinal coupling constants as a function of dihedral angle.*

intermolecular or intramolecular. Intermolecular hydrogen bonding refers to association between two or more molecules and is therefore concentration dependent. Intramolecular hydrogen bonding refers to bonding association with a suitable acceptor site within the same molecule; it is therefore concentration independent. The latter type of hydrogen bonding is most useful in structure determination, as indicated by a few examples below. For *cis*-cyclohexane-1,3-diol, $\Delta\nu$ is halved by the introduction of an ethane residue bridging positions 4 and 6:

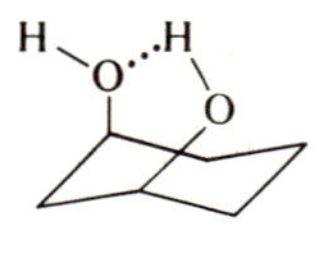

$\Delta\nu = 76\ \text{cm}^{-1}$

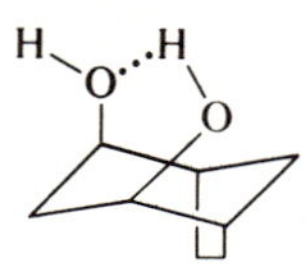

$\Delta\nu = 39\ \text{cm}^{-1}$

Since carbon atoms 4 and 6 are brought together by the bridging, carbon atoms 1 and 3 have a tendency to compensate for this deformation by spreading apart, and thus increasing the distance between the two alcohol functions, as evidenced by the decrease of the hydrogen bond shift.[1] 1,2-Diols in the bicyclic series also have values of $\Delta\nu$ strongly dependent upon the proximity of the two alcoholic functions[2] (Table 15.3). It is clear that the strength of the hydrogen bond and therefore the shift in

*TABLE 15.3*
*$\Delta\nu$ as a Function of Structure*

| Compound | Isomer | Dihedral angle (deg) | $\Delta\nu$ (cm$^{-1}$) |
|---|---|---|---|
| 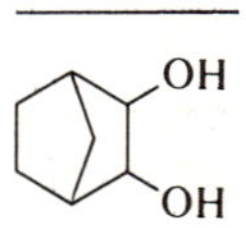 | *cis* | 0 | 102 |
| | *trans* | ~120 | 0 |
| 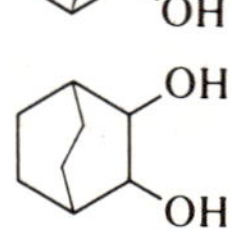 | *cis* | ~10 | 82 |
| | *trans* | ~120 | 0 |
| 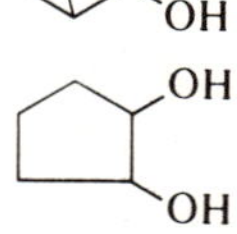 | *cis* | $0 < \phi < 60$ | 61 |
| | *trans* | ~120 | 0 |
| 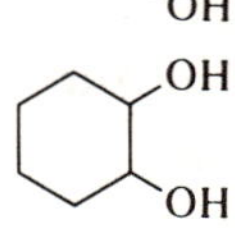 | *cis* | <60 | 38 |
| | *trans* | >60 | 33 |

[1] J. G. Durocher and H. Favre, *Can. J. Chem.*, *42*, 260 (1964).
[2] L. P. Kuhn, P. von R. Schleyer, W. F. Baitinger, and L. Eberson, *J. Amer. Chem. Soc.*, *86*, 650 (1964).

frequency $\Delta\nu$ is once again a function of the dihedral angle and can thereby serve as a useful empirical tool for assigning conformations. In the acyclic 1,2-diols a different kind of structural deformation accounts for the observed frequency shifts (Table 15.4). As the alkyl groups R become bulkier and bulkier, they tend to repel each other, thus bringing the oxygen functions into closer proximity and increasing the strength of the hydrogen bond.

*TABLE 15.4*

*$\Delta\nu$ as a Function of Structure in Acyclic 1,2-Diols*

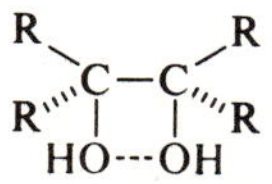

| R | Me | Et | *i*-Pr | *t*-Bu |
|---|---|---|---|---|
| $\Delta\nu$ | 46 | 62 | 96 | 170 |

*15.1.2 Ethers*

LINEAR ETHERS AND CYCLIC ETHERS OF RING SIZE GREATER THAN THREE

In dealing with ethers, allowance must be made for the special case of epoxides: They owe their unique character to the presence of the three-membered ring.

Characterization of ethers is best accomplished with the spectral bands summarized in Table 15.5. The downfield nmr shift is due predominantly to the "inductive" effect of the electronegative oxygen atom, and can be dissected into anisotropic, van der Waals, and electric-field terms. The mass spectra of ethers are dominated by fragmentations triggered by the presence of the oxygen atom which can accomodate either a positive charge or an unpaired electron, according to whether the $\alpha$ or the $\beta$ bond breaks:

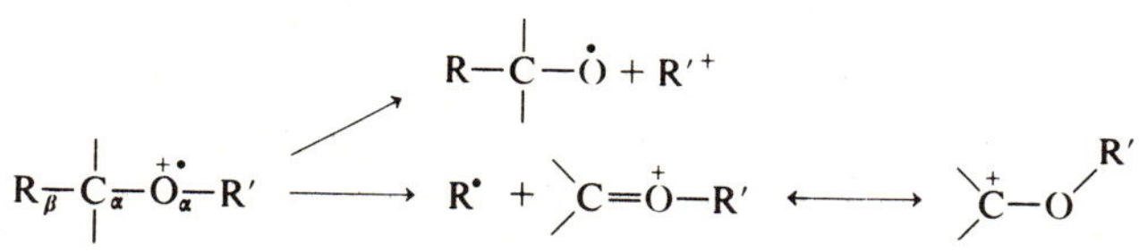

Obviously, the relative importance of the two modes depends upon the stabilities of the $R^+$ and $>C=\overset{+}{O}-R'$ cations which are formed. Also, like alcohols, ethers can lose small neutral alcohol (ROH) fragments; however, such extrusion reactions are somewhat less important than for alcohols.

*TABLE 15.5*

*Characteristic Absorption Frequencies of Ethers*

| | Residue | Absorption | Comments |
|---|---|---|---|
| pmr | $-OCH_3$ | $\delta \sim 3.3$ | singlet (3H) |
| | $-C(=X)OCH_3$ | $\delta \sim 3.7–3.8$ | singlet (3H) |
| ir | $ROCH_3$ | 2820 $cm^{-1}$ | methyl sym. stretch |
| | C—O—C | 1150–1070 $cm^{-1}$ | asym. stretch |
| | $ArOCH_3$ | 2850 $cm^{-1}$ | methyl sym. stretch |
| | =C—O—C | 1274–1200 $cm^{-1}$ | asym. stretch |

For cyclic ethers, the ring size can be determined through a variety of means, depending upon the problem at hand. Among the direct means are the observation of the asymmetric stretch, $C \rightarrow O \rightarrow C$, in the ir spectrum at 1070–1150 $cm^{-1}$ for ethers five-membered or larger, whereas four-membered ethers, such as trimethylene oxide, absorb around 1000 $cm^{-1}$. These ring sizes can be confirmed through the examination of $^{13}C$—H coupling constants for the CH groups $\alpha$ to the oxygen heteroatom (Table 15.6).

*TABLE 15.6*

*$^{13}C$—H Coupling Constants in Cyclic Ethers*

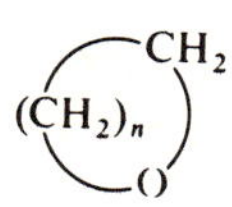

| *n* | $^1J_{C-H}$ (Hz) |
|---|---|
| 3 | 150 |
| 4 | 145 |
| 5 | 140 |

**EPOXIDES**

Three-membered ring ethers (epoxides) are readily characterized by proton chemical shifts around 2.3–2.5 ppm for the hydrogens in the oxirane ring, a compromise between the high-field position of cyclopropane protons (~0.2 ppm) and the low-field position of α hydrogens in ethers (~3.7 ppm). The vicinal interproton coupling constants also reflect both the oxygen electronegativity and the special geometry of the ring, with dihedral angles of 0° (*cis*) and ~130° (*trans*):

$$^3J_{\text{cis}} = 4 \text{ Hz}$$
$$^3J_{\text{trans}} = 2 \text{ Hz}$$

The carbon-hydrogen coupling constant is also very typical of the ring size with $^1J_{\text{C-H}} = 187\text{–}185$ Hz.

The mass spectra of epoxides are dominated not only by loss of oxygen to give the corresponding olefin but also by ring opening to the corresponding cations and by β fragmentation. An example is[3]

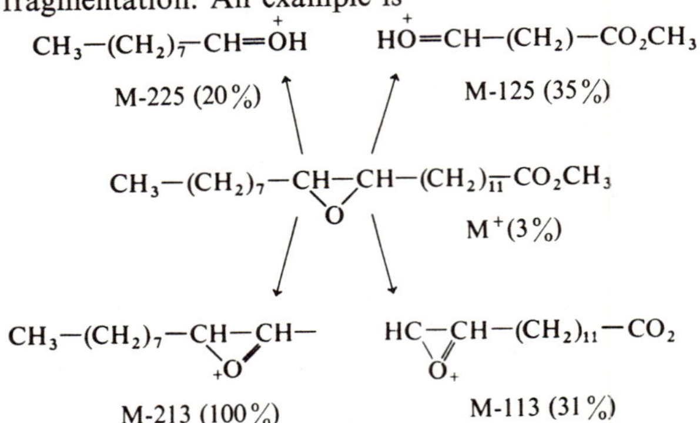


This represents a good method for the localization of double bonds within long aliphatic chains through simple conversion of the olefin to the epoxide, which is then submitted to mass spectrometry.

In general, cleavage of a bond β to a heteroatom,

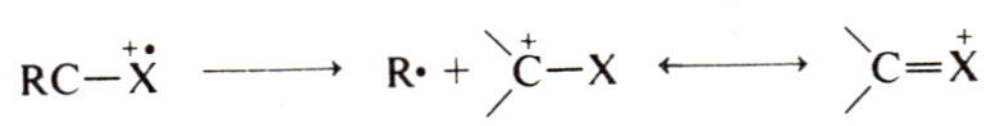

represents a very efficient fragmentation pathway for mass-spectral reactions. Additional important applications of such β fragmentations will be presented in subsequent sections.

## 15.2 *π-Bonded Oxygen*

π-Bonded oxygen refers to the carbonyl group

Molecules containing this group only are ketones and aldehydes. The distinction between aldehydes and ketones is best achieved in practice by the observation in the pmr spectra of aldehydic protons around 10 ppm. It is virtually the only group to absorb in that region, and this resonance is also little dependent upon substitution. The presence of an aldehyde can be confirmed by observation in the ir spectrum of a pair of rather strong bands at 2720 and 2820 cm$^{-1}$ due to stretching vibrations. Ketones, on the other hand, give rise to pmr signals between 2 and 2.5 ppm for protons on tetrahedral carbons bound to trigonal centers:

[3] R. T. Aplin and L. Coles, *Chem. Comm.*, 858 (1967).

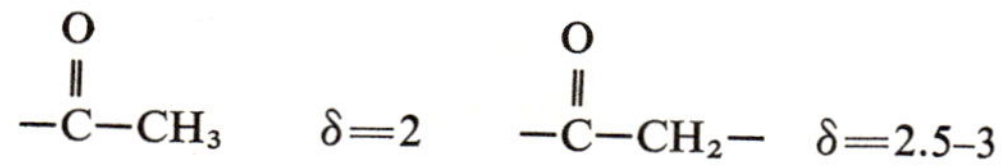


Methyl ketones, in particular, are easily recognized by the presence of a singlet at ~2 ppm downfield from TMS. In general, the various spectroscopies are influenced directly by the nature of the environment of the carbonyl group. Optical rotatory dispersion and circular dichroism have already been described (Chapter 7) and various other spectral methods will now be considered.

### 15.2.1 Ring Size of Ketones

The ring size of cyclic ketones is best obtained from the empirically observed linear relationship between the ir carbonyl stretching frequency $\nu_{C=O}$ and the magnitude of the endocyclic angle $\theta$ at the carbonyl carbon[4]:

$$\nu_{C=O}(\text{cm}^{-1}) = 1947 - 2.2\,\theta$$

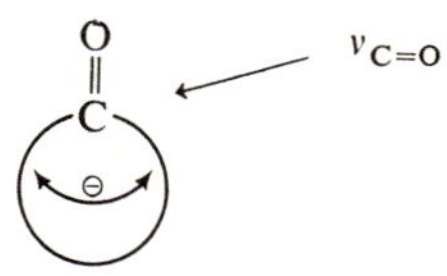


Typical values of $\nu_{C=O}$ and $\theta$ are gathered in Table 15.7 for a few important systems.

TABLE 15.7

*Correlation of Endocyclic Angles with Carbonyl Stretching Frequencies*

| Compound | $\nu_{C=O}$ (cm$^{-1}$) | $\theta$ (deg) |
|---|---|---|
| Cyclopropanone | 1813 | 60 |
| Cyclobutanone | 1791 | 90 |
| Cyclopentanone | 1740 | 102 |
| Cyclohexanone | 1716 | 117 |
| Cycloheptanone | 1705 | 122 |
| Cyclooctanone | 1703 | 123 |
| Cyclodecanone | 1704 | 123 |
| Norbornanone (2-bicyclo [2.2.1] heptanone) | 1751 | 101 |
| 2-Bicyclo [2.2.2] octanone | 1731 | 110 |
| 2-Bicyclo [3.3.0] octanone | 1738 | 107 |

### 15.2.2 Substituents on the Carbonyl Group

There is an elegant and general method for determining the molecular weights of the R and R′ residues in a ketone RCOR′. The ketone is reacted with ethylene glycol, in the presence of a trace of acid, to give the corresponding ethylene ketal:

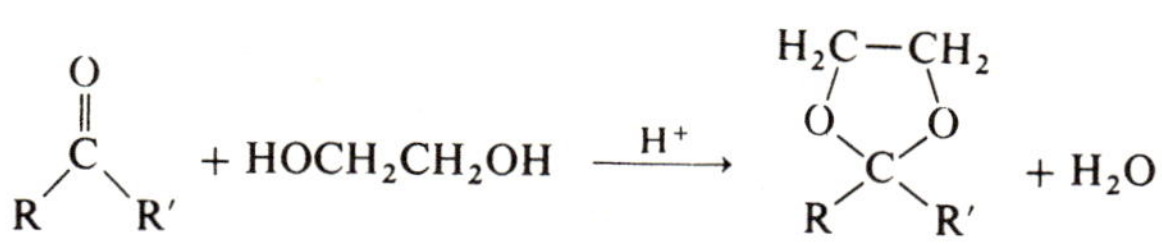


The mass spectral fragmentations[5] of ethylene ketals are dominated to a remarkable extent by cleavage of the R or the R′ fragment:

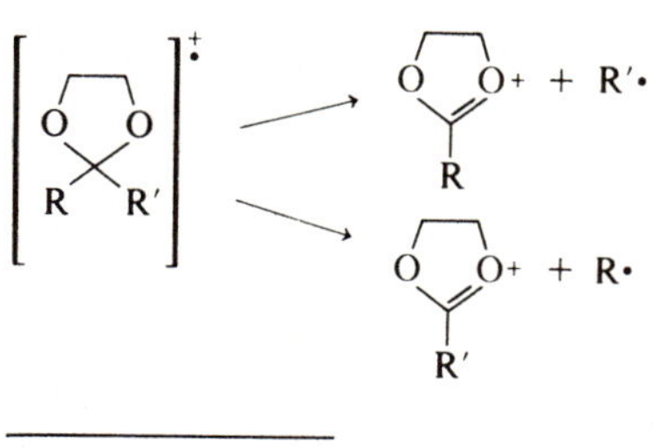


[4] S. Bank, W. D. Closson, and L. T. Hodgins, *Tetrahedron, 24,* 381 (1968).

[5] H. Audier, M. Fétizon, J. C. Gramain, T. Schalbar, and B. Waegell, *Bull. Soc. Chim. Fr.,* 1880 (1961); Z. Pelah, D. H. Williams, H. Budzikiewicz, and C. Djerassi, *J. Amer. Chem. Soc., 86,* 3722 (1966).

This is just another case of breaking a bond $\beta$ to a heteroatom and placing the charge on the heteroatom, as described above for ethers. For example, the mass spectrum of the ethylene ketal of 3-heptanone (M.W. = 158) displays major peaks at $m/e = 129$, 101, and 57. The first two correspond to the loss of the ethyl and butyl fragments and provide easy identification for the location of the carbonyl within the chain. The fragmentation mechanism for cyclic ketones is exemplified by 4-methylcyclohexanone, whose ethylene ketal has a very prominent mass spectral peak at $m/e = 99$, corresponding to fragmentation of the $\alpha$ bond followed by hydrogen abstraction:

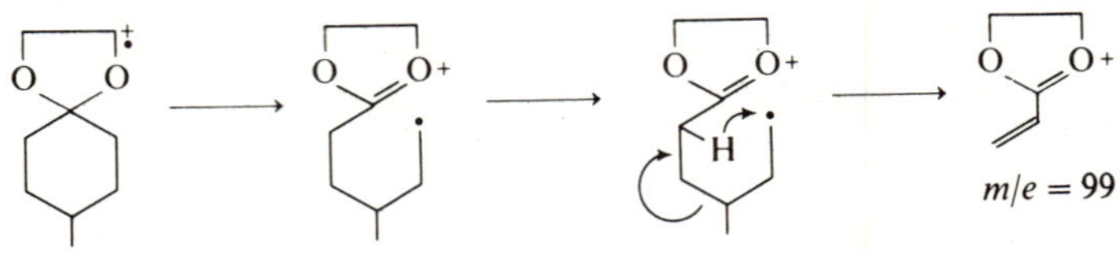


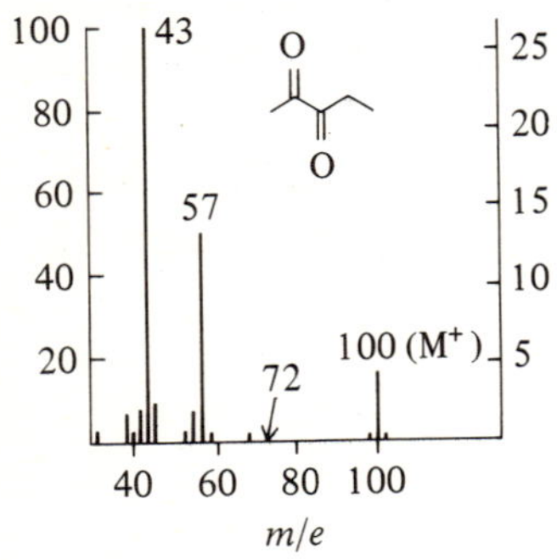


*FIGURE 15.3*
*Mass spectrum of 2,3-pentane-dione.*

The ethylene ketal of 4-methylcyclohexanone perdeuterated at positions 2 and 6 has this peak displaced to $m/e = 100$, verifying the indicated mode of fragmentation. It should be noted that ketones themselves possess an important fragmentation mode $\alpha$ to the carbonyl, leading to the resonance-stabilized acylium ion (Figure 15.3):

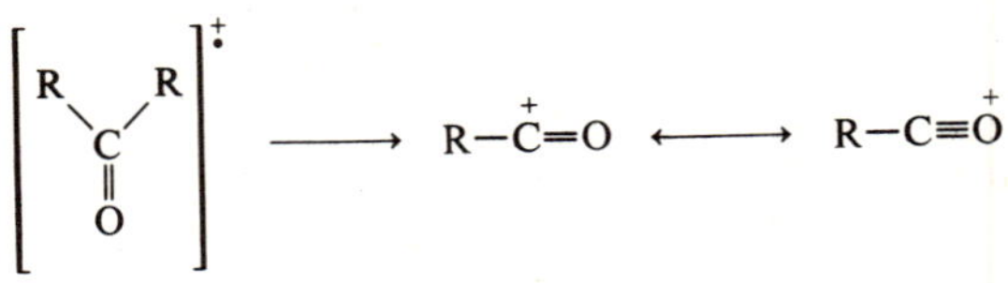

However, transformation of ketones into ethylene ketals leads to much clearer mass spectra, characterized by only a few dominant peaks, hence the interest in this method (Figure 15.4).

The electronic transitions for the carbonyl group also provide a detailed picture of how the corresponding energy states are affected by the chemical environment of the carbonyl. The reader is referred again to the hybridization scheme in which $\pi$ and $\pi^*$ orbitals are generated from the interaction of a $p_z$ atomic orbital on carbon and a $p_z$ atomic orbital on oxygen (see Figure 7.1).

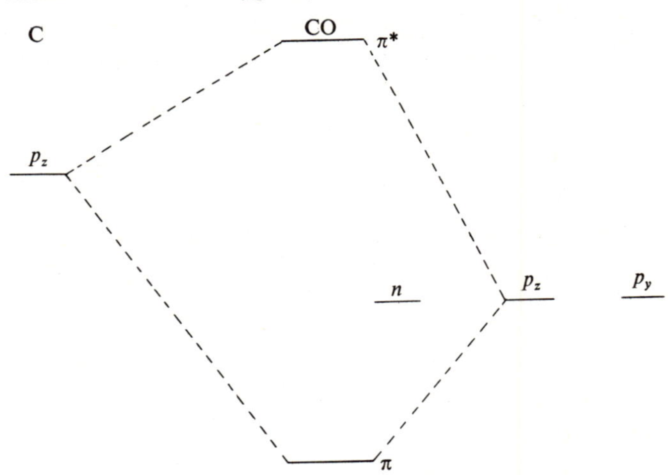


Upon the contraction of the RCR′ angle, as in a small ring, the energy of the $p_z$ orbital on carbon is increased, but that of the $p_z$ orbital on oxygen or of the nonbonding orbitals on oxygen are left practically unchanged. We should therefore expect that the $\pi$ energy state will be higher than in an unstrained molecule. However, the $\pi^*$ energy state will be raised to an even greater extent. One would predict that the transition

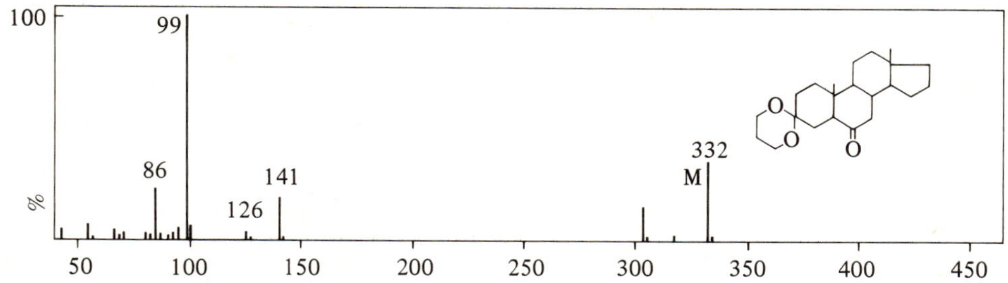


FIGURE 15.4
Mass spectrum of an ethylene ketal from a steroidal diketone.

energy $h\nu$ for the $n \rightarrow \pi^*$ transition would be greater in small ring ketones, while $\lambda_{max}$, of course, will vary in the opposite direction. This is indeed what is observed experimentally (Table 15.8). The complementary prediction is that a deviation of the RCR′ angle from its $\sim 120°$ trigonal value to larger values as in large ring ketones, will also lead to greater values of $h\nu(n \rightarrow \pi^*)$ (see Table 15.8). The carbonyl $^{13}C$ chemical shifts follow a very similar trend,[6] since they are governed by the magnitude of the energy separation between the ground and excited states. The point of this comparison is to underline the similar factors which influence two different spectroscopies, in this case $^{13}C$ chemical shifts and ultraviolet transitions.

TABLE 15.8

Carbonyl $^{13}C$ Chemical Shifts and $n \rightarrow \pi^*$ Maximum Wavelengths of Some Cyclanones

| Compound | $^{13}C{=}O$ shift[a] | $\lambda_{max}$[b] |
|---|---|---|
| Cyclobutanone | −79.0 | 281.5 |
| Cyclopentanone | −89.3 | 300.1 |
| Cyclohexanone | −81.0 | 291.4 |
| Cycloheptanone | −83.7 | 291.8 |
| Cyclooctanone | −86.7 | 290.5 |
| Cyclononanone | −86.6 | 292.6 |
| Cyclodecanone | −83.8 | 288.3 |
| Cyclopentadecanone | −81.0 | 285.9 |

[a] In parts per million from the benzene reference.
[b] In isooctane solution.

The electronic energy states for the carbonyl group are also markedly affected by α substitution. The effect of an adjacent carbon-carbon double bond, as in α,β-unsaturated ketones, is particularly strong. In these compounds the resonance $C{=}C{-}C{=}O \leftrightarrow C^{+}{-}C{=}C{-}O^{-}$ is responsible for raising the absorption maximum to ~210 nm, in a readily accessible spectral range. Further conjugation or substitution will further shift the absorption maximum to higher wavelengths, according to the empirical rules formulated by Woodward[7] (Table 15.9). These so-called Woodward's rules have been particularly useful in the steroid field, since uv and ir spectroscopy were virtually the only physical methods available for the development of this important branch of chemistry in the 1940–1950 decade.

For instance, one can apply the Woodward rules to compounds such as

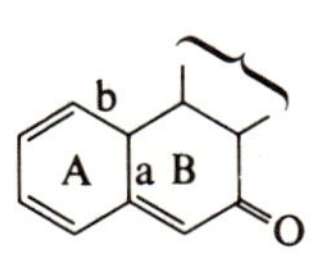


and add to the base value of 215 nm increments of 60 nm (two additional conjugated double bonds), 12 nm and 18 nm [for bonds indicated as (a) and (b) on the structure),

[6] G. B. Savitsky, K. Namikawa, and G. Zweitel, *J. Phys. Chem.*, *69*, 3105 (1965).
[7] R. B. Woodward, *J. Amer. Chem. Soc.*, *63*, 1123 (1941); *64*, 76 (1942).

*TABLE 15.9*

***Woodward's Rules for the Ultraviolet Absorption of α,β-Unsaturated Ketones***

δ γ β α
Solvent-ethanol C=C–C=C–C=O

Basic values

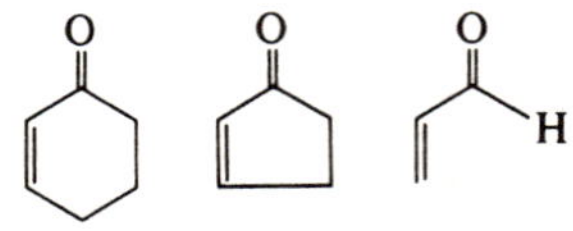


| | 215 | 202 | 207 nm |
|---|---|---|---|

| Increments | | nm |
|---|---|---|
| Conjugated C=C | | 30 |
| R, Ring residue, OR, OAc | α | 10 |
| | β | 12 |
| | γ and δ | 18 |
| Exocyclic nature of double bond | | 5 |
| Homoannular diene component | | 39 |
| Change to a hydrocarbon solvent | | 11 |

5 nm for the exocyclic nature of the α,β carbon-carbon double bond (it is exocyclic to ring A), and 39 nm for the homoannular diene moiety in ring A. These add to a calculated value of 349 nm, to be compared to the experimental value of 348 nm (ε = 11,000). Another example, similar in appearance, is

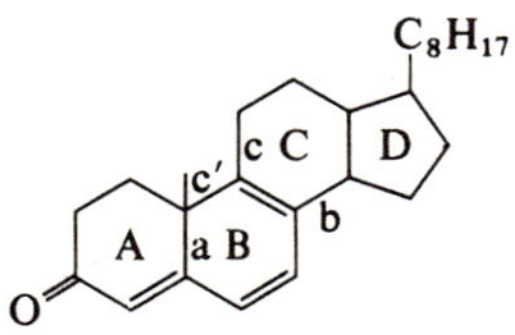


The experimentally observed absorption for this compound is $\lambda_{max}$ = 388 nm, with contributions from the various structural features listed below:

| | |
|---|---|
| 215 nm | (basis cyclohexenone) |
| 60 nm | (two conjugated double bonds) |
| 12 nm | bond (a) |
| 18 nm | bond (b) |
| 36 nm | bonds (c) and (c′) |
| 5 nm | exocyclic nature of the α,β C=C (to ring B) |
| 39 nm | homoannular diene (within ring B) |
| 385 nm | |

That α,β-unsaturated ketones are resonance hybrids of the type

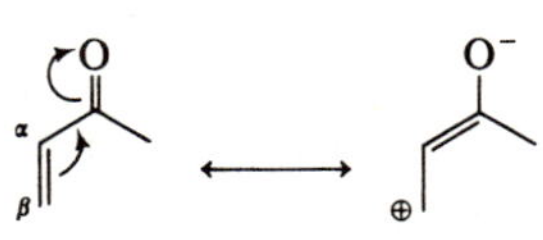


is also very important in determining the corresponding nmr chemical shifts. For protons, $H_\beta$ is usually found 1.0–1.5 ppm downfield from $H_\alpha$ (which has the normal olefinic position), due to the positive charge on the β carbon. This phenomenon is reflected as well by the $^{13}C$ spectra, as indicated by the comparison shown below:

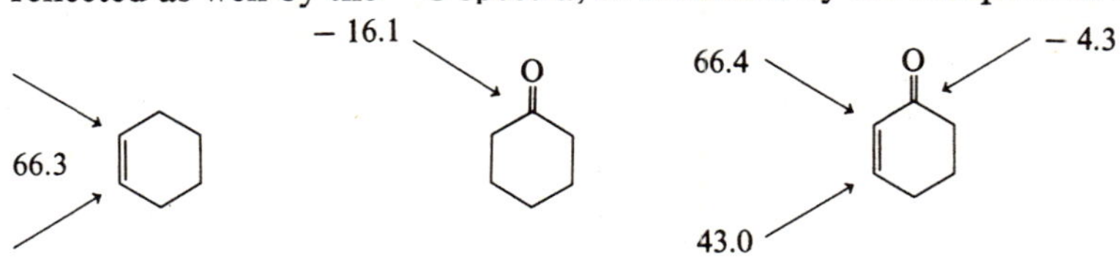

(The values shown are $^{13}C$ chemical shifts in parts per million relative to $CS_2$). The $\beta$ carbon is deshielded by the presence on it of a positive charge, whereas the carbonyl carbon is shielded by the presence of an adjacent negatively charged oxygen (but it is also influenced by the change in the $\pi$–$\pi^*$ excitation energy, as discussed with reference to Table 15.8).

The presence of $\alpha,\beta$ unsaturation is also a prime influence on the magnitude of the carbonyl stretching frequency in the ir spectrum. It measures in a direct fashion the strength of the carbon-oxygen bond. For instance, the bonding in carbon monoxide can be described by the resonance hybrid:

$$:C{=}\ddot{\underset{\cdot\cdot}{O}} \longleftrightarrow :\overset{-}{C}{\equiv}\overset{+}{O}:$$

The corresponding carbonyl stretching frequency is observed at 2143 $cm^{-1}$. Upon bonding to a charge-deficient species it increases markedly, indicating the augmented importance of the triply bonded resonance form:

$$BH_3 + CO \longrightarrow H_3B^{-}{-}C{\equiv}O^{+} \qquad \nu_{C=O} = 2165\ cm^{-1}$$
$$CH_3^{+} + CO \longrightarrow H_3C{-}C{\equiv}O^{+} \qquad \nu_{C=O} = 2299\ cm^{-1}$$

On the other hand the C—O bond is much weaker in metal carbonyls such as iron pentacarbonyl (2025 and 2004 $cm^{-1}$), which points to the importance of back donation from the *d* orbitals of the metal into the antibonding orbital of the C—O group:

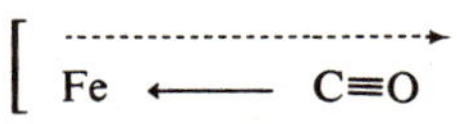

Likewise, the carbonyl stretching frequency, which is 1715 $cm^{-1}$ in acetone, decreases to 1675 $cm^{-1}$ in $\alpha,\beta$-unsaturated ketones and to 1690 $cm^{-1}$ in acetophenone. In cyclopropylmethyl ketone, it is 1695 $cm^{-1}$, indicating that the cyclopropane ring has an electronic character intermediate between that of a normal alicyclic residue and that of a double bond.

An adjacent heteroatom with a lone electron pair would be expected to show a similar effect due to resonance of the type:

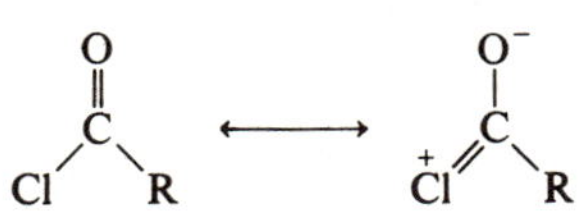

However, it is often observed in such cases that the carbonyl frequency is increased rather than lowered, a consequence of the inductive effect of the substituent prevailing over its mesomeric effect:

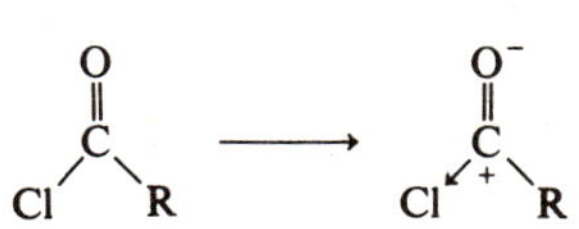

Acyl chlorides, for instance, show carbonyl absorption around 1795 $cm^{-1}$ and acyl bromides around 1810 $cm^{-1}$.

*5.2.3 Interactions Due to ie Dipolar Character of the 'arbonyl Group*

Strong ir stretching vibrations are observed for bonds with a large dipole moment. One should therefore expect in the typical case of the carbonyl stretch that introduction of an adjacent dipolar group would modify the frequency of the absorption if the adjacent group is coplanar with the carbonyl group, but not if it is perpendicular to the carbonyl. Conversely, the $n \rightarrow \pi^*$ uv absorption should be unchanged when a halogen substituent is in the nodal plane of the carbonyl, but markedly influenced when it is parallel to the *p* orbitals on carbon and on oxygen (Table 15.10).

*TABLE 15.10*

*Infrared and Ultraviolet Shifts for α-haloketones*

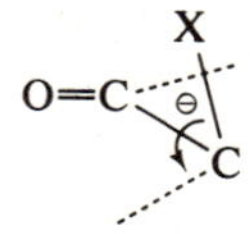


| X | Δν (cm⁻¹) θ = 0° | Δν (cm⁻¹) θ = 90° | Δλ (nm) θ = 0° | Δλ (nm) θ = 90° |
|---|---|---|---|---|
| Cl | 25 | 0 | 0 | +30 |
| Br | 20 | 0 | 0 | +25 |
| I | 10 | 0 | 0 | +20 |

In the pmr spectrum of α-haloketones the axial proton is found downfield and the equatorial proton upfield,[8] contrary to the general observation in cyclic hydrocarbons.

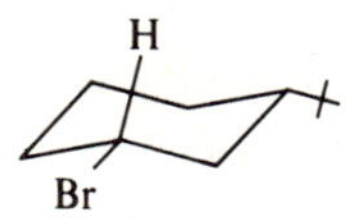


$\delta = 3.8$

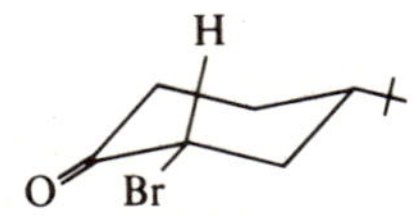


$\delta = 4.7$ (deshielding)

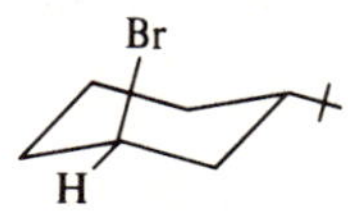


$\delta = 4.6$

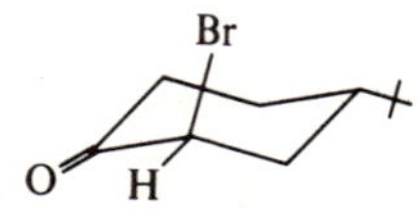


$\delta = 4.3$ (shielding)

Consideration of the local electrostatic field at the —CHBr proton (see Chapter 4) explains these results: the molecular dipole moment has its projection along the C—H bond pushing electron density towards the hydrogen nucleus in the equatorial H isomer, and withdrawing electron density from it in the axial epimer.

*15.2.4 Dipole-Induced Dipole Interactions with Aromatic Solvents*

When the pmr spectrum of a ketone, or a polar molecule in general, is obtained in an aromatic solvent such as benzene, and compared to that in a more "inert" solvent such as cyclohexane or carbon tetrachloride, substantial shifts of the various resonances can be observed. These ASIS (aromatic solvent-induced shifts) span the range of 0–1.5 ppm. They are greater by an order of magnitude than the usual solvent shifts seen in pmr because of the large magnetic anisotropy of the aromatic rings. Their origin is still somewhat mysterious; in order for these shifts to occur a specific interaction must take place between the solute and the solvent molecules since nuclei (protons) differently situated with respect to the carbonyl are affected to a different extent. For methyl groups adjacent to the carbonyl, the change from a deuteriochloroform to a benzene solution will result in an *upfield* shift (0.2–0.3 ppm) of an axial methyl group, while an equatorial methyl group is shifted *downfield* by 0.05–0.10 ppm. Perhaps the easiest way to visualize the interaction is to think of the mobile $\pi$ electrons in the solvent molecule being polarized by the dipolar solute which has thus induced into the solvent molecule a smaller dipole moment antiparallel to itself. One could align these two dipoles either in a single common plane or on two parallel planes. One does

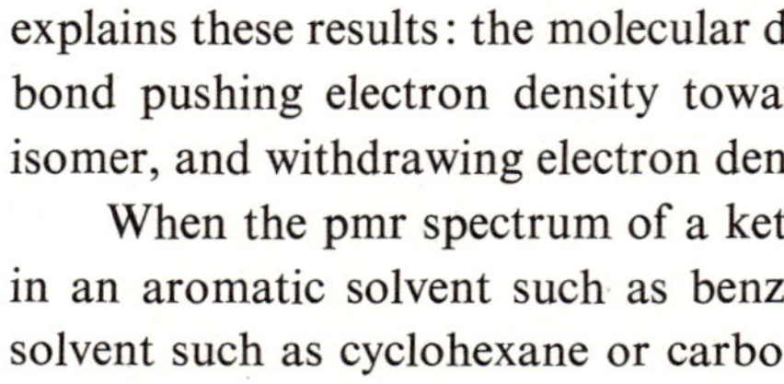


not, however, know the microscopic details of the interaction, such as its duration (assuredly very short, since these complexes are short lived and collapse many times

[8] K. M. Wellman and F. G. Bordwell, *Tetrahedron Lett.*, 1703 (1963).

during a pmr transition), its stoichiometry (how many solvent molecules are involved with a given solute molecule), and its geometry. Despite this ignorance, such solvent shifts are easily measured, are very often additive when several polar groups are present, are relatively little affected by steric hindrance of the dipole by bulky groups, and are often very characteristic of the location of a proton in the molecule with respect to the polar group. One can therefore *use* these ASIS as a useful "black box" to elucidate details of stereochemistry. To give an example, the ASIS (carbon tetrachloride to benzene, in hertz at 60 MHz) for the methyl groups α to the carbonyl in a number of α-methyl cyclohexanones are given below:

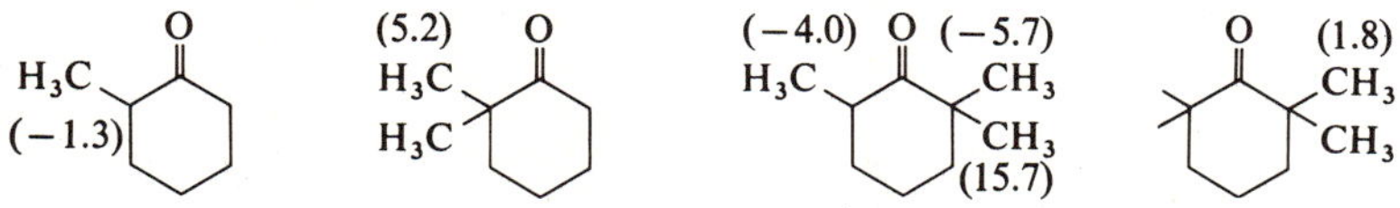


The analysis is straightforward: the 2,2,6-trimethylcyclohexanone has the 6-methyl predominantly equatorial, in order to avoid the syn-axial repulsion with the axial 2-methyl; therefore, one can take −4.8 Hz (the average of −4.0 and −5.7) as the ASIS for an equatorial methyl α to the carbonyl, and +15.7 Hz for its axial counterpart. The 2,2-*gem*-dimethyl molecule would have a calculated ASIS of $\frac{1}{2}$ (15.7 − 4.8) = 5.4 Hz since its methyl groups are interconverted very rapidly between the axial and the equatorial position in good agreement with the experimental values of 5.2 Hz. For α-methylcyclohexanone, this treatment indicates that it exists mainly (85 percent) as the equatorial conformer, whereas 2,2,6,6-tetramethylcyclohexanone adopts a flattened conformation to relieve the syn axial interaction of the 2 and 6 methyl groups, since the observed value of 1.8 Hz is much smaller than the value of 5.4 Hz expected if the additivity of ASIS really holds. This flattening is supported further by the magnitudes of $\nu_{C=O}$ and $\lambda_{max}$ ($n \rightarrow \pi^*$) for the tetramethyl case as compared to the others.

## 5.3 *Iolecules with σ-Bonded nd π-Bonded Oxygen*

This class of compounds with σ-bonded and π-bonded oxygen includes carboxylic acids, anhydrides, esters, and lactones. These compounds usually show the properties characteristic of both functional groups although in a somewhat modified form. The ir carbonyl absorption of esters, for instance, is in the range of 1750–1735 cm$^{-1}$ and for anhydrides in the range of 1850–1800 cm$^{-1}$, considerably higher than the ketone carbonyl range of 1725–1705 cm$^{-1}$. This is probably due to the inductive polarizing effect of the σ-bonded electronegative oxygen relative to the alkyl group of a ketone. A similar effect is observed in cyclic esters so that the characteristic frequencies for lactones,

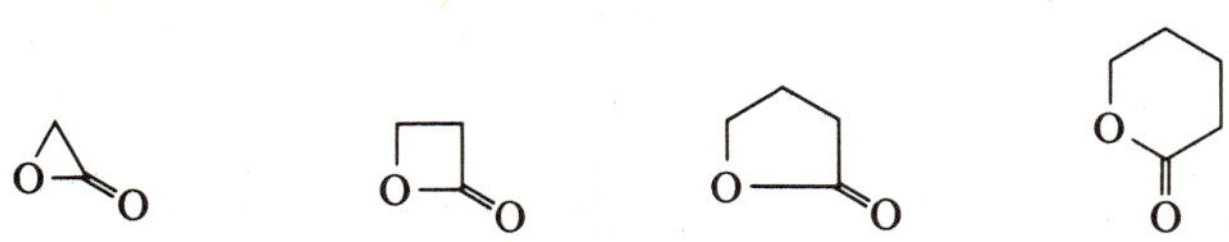

> 1900 cm$^{-1}$ 1841 cm$^{-1}$ 1770 cm$^{-1}$ 1738 cm$^{-1}$

are due to a combination of ring size (see Table 15.7) and the effect of an electronegative substituent resulting in a shift to higher frequency relative to the corresponding cyclic ketones. Carboxylic acids, on the other hand, show carbonyl stretching vibrations at lower frequencies, 1725–1700 cm$^{-1}$ than esters and anhydrides, a phenomenon usually attributed to the dimeric nature of carboxylic acids in solution due to hydrogen bonding.

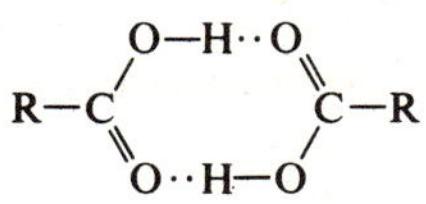

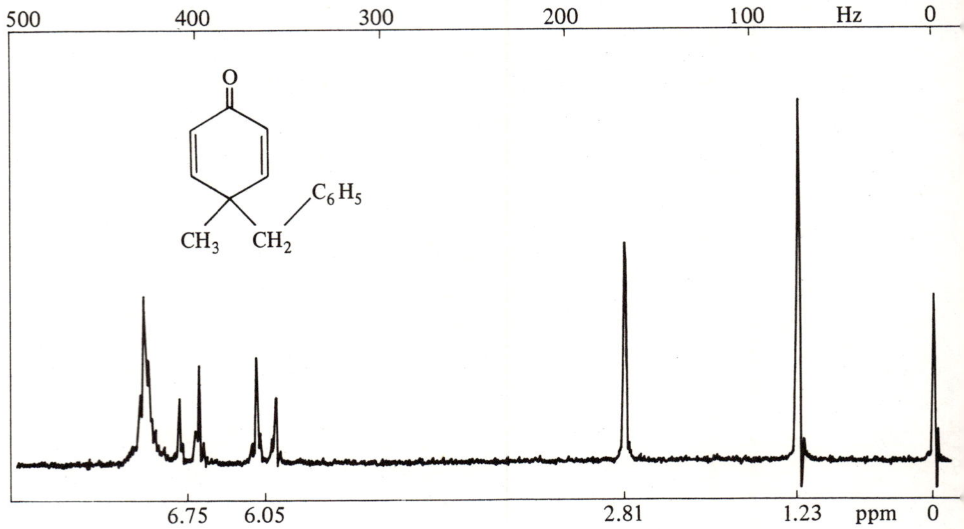


*FIGURE 15.5*
*The 60-MHz nmr spectrum (in $CCl_4$) of the indicated cyclohexadienone.*

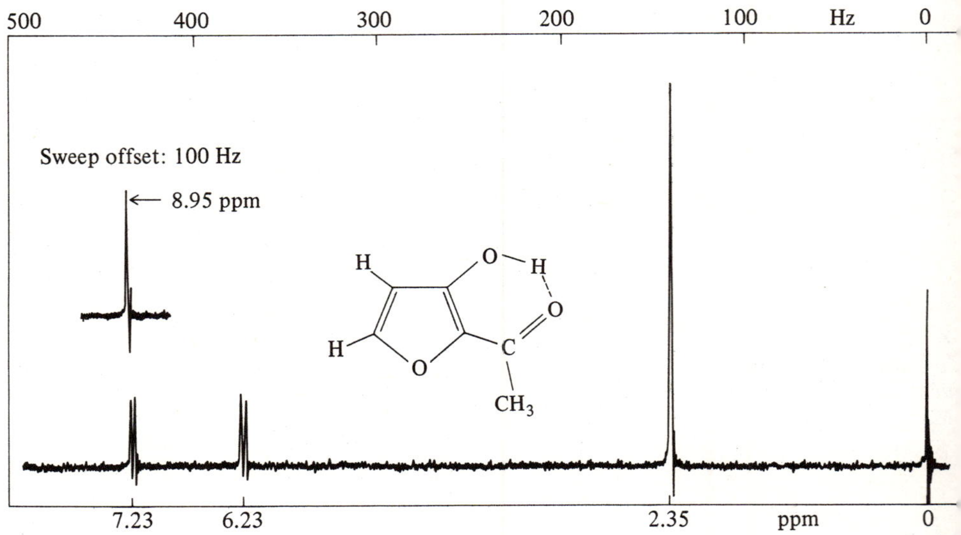


*FIGURE 15.6*
*The 60-MHz nmr spectrum (in $CCl_4$) of the indicated hydroxyfuran.*

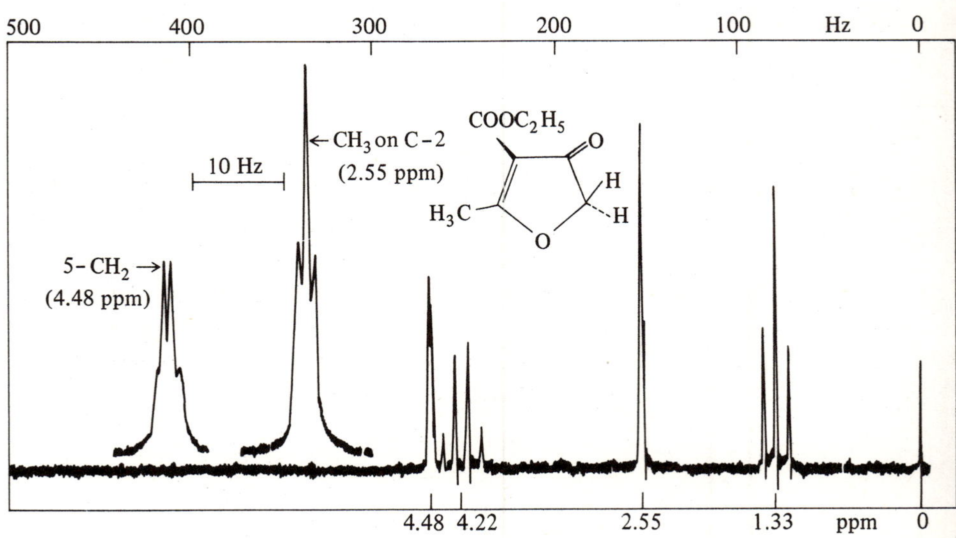


*FIGURE 15.7*
*The 60-MHz nmr spectrum (in $CCl_4$) of the indicated $C_8H_{10}O_4$ compound.*

Such hydrogen bonding, of course, also shifts the O—H stretching to lower frequencies, usually in the range of 3200–2500 $cm^{-1}$, well below that of the free O—H stretch (see Section 15.1.1 on alcohols). Conjugated unsaturation in this class of compounds lowers these characteristic carboxylic absorptions in analogy with ketones.

In the nmr spectrum, carboxylic protons are considerably further downfield (because of their acidity) than the OH protons of alcohols, the usual values being in the range of 10–15 ppm. Protons on esters are also considerably further downfield than in the corresponding ketones; the methoxy protons of methyl esters, for instance, are in the range of 4.0 ppm downfield whereas methyl ketones are only 2.0 ppm from TMS. Table 15.11 gives a variety of chemical shifts for methyl esters. As seen from the table the position of the methyl proton resonance is not affected markedly by substitution and is of good diagnostic value for methyl esters.

*ABLE 15.11*

*H and $^{13}C$ Chemical Shifts of e Methyl Group in $RCO_2Me$*

| R | $^1H$ (ppm) | $^{13}C$ (ppm) (relative) |
|---|---|---|
| H | 3.69 | 0.0 |
| Me | 3.61 | 0.75 |
| Ph | 3.85 | 1.45 |
| $PhCH_2$ | 3.61 | 1.27 |
| $CH_2Cl$ | 3.90 | 2.38 |
| $CCl_3$ | 4.00 | 3.70 |
| *o*—HO—$C_6H_4$ | 3.91 | 1.41 |

SOURCE: W. McFarlane, *J. Chem. Soc.* (B) 28 (1969), p. 28.

## 5.4 *LUSTRATIVE XAMPLES*

In Figure 15.5 the 1.23-ppm signal is due to the $CH_3$ group and the 2.81-ppm signal to the benzylic methylene. The doublet at 6.05 ppm is due to the protons $\alpha$ to the carbonyl while the doublet at 6.75 ppm is due to the protons $\beta$ to the carbonyl. Note the difference in chemical shift between the $\alpha$- and $\beta$-olefinic protons which is typical of $\alpha,\beta$-unsaturated carbonyls. The low-field multiplet is due to the aromatic protons.

In Figure 15.6 the singlet at 2.35 ppm is due to the acetyl methyl group, its low-field position indicating conjugation of the acetyl group. The doublet at 6.23 ppm is due to the hydrogen $\alpha$ to the ring oxygen and the doublet at 7.23 ppm is due to the $\beta$ hydrogen. The difference in chemical shifts between the $\alpha$ and $\beta$ proton is typical of furans as is the coupling. Note the low-field, 8.95-ppm, position of the O—H proton typical of intramolecular hydrogen bonding. In what respect would the spectrum be different if this compound existed in the $\beta$-diketo form?

In Figure 15.7 the triplet at 1.33 ppm is due to the ethoxy methyl group and the quartet at 4.22 ppm to the ethoxy methylene hydrogens, a very typical pattern and position for ethoxy groups. The signal at 2.55 ppm is due to the ring methyl group and the one at 4.48 to the ring methylene group. The inserts are reproductions of these latter two resonances under expanded scale and high-resolution conditions. Note the long-range coupling of the ring methyl group with the ring methylene protons. Does the spectrum rule out the alternate enolic form (a furan) of this compound?

The most significant feature of the spectrum of an alcohol or phenol is the O—H stretching region (Figure 15.8). If it is taken of the pure alcohol without solvent it always appears as a strong broad band between 3600–3200 $cm^{-1}$ due to intermolecular hydrogen bonding. If the spectrum is taken in solution it will usually show a free O—H band just above 3600 $cm^{-1}$ as well as the band due to hydrogen bonding; in very dilute solution only free O—H will be observed. There is also a band around 1050–1200 $cm^{-1}$ due to C—O stretching which is characteristic of alcohols or phenols.

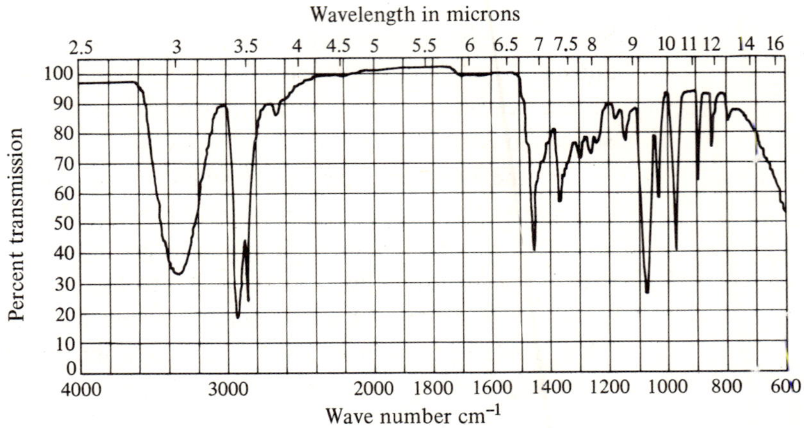


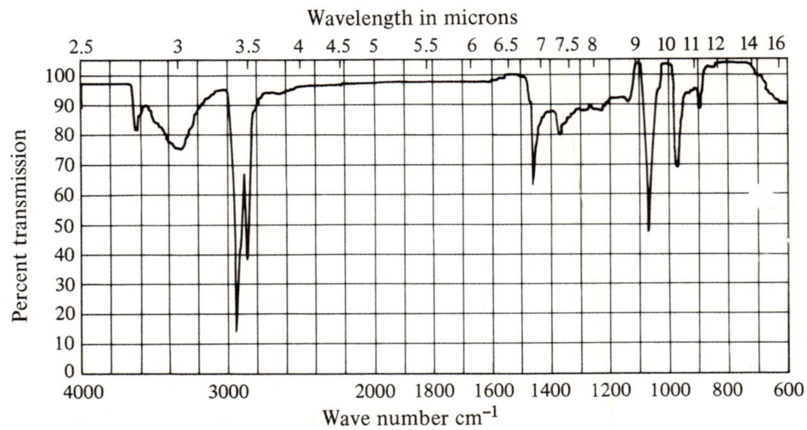


*FIGURE 15.8*
*The ir spectrum of pure cyclohexanol (top) and in* $CCl_4$ *solution (bottom).*

The strongest absorption for a ketone occurs in the carbonyl stretching regio (Figure 15.9 *opposite*). For cyclohexanone this band is at 1715 cm$^{-1}$ while for cyclc hexenone it is at 1675, because conjugation lowers the transition energy. Notice th cyclohexenone also has absorption at 3060 cm$^{-1}$, indicating unsaturated C—H stretc ing as well as a band at 1630 cm$^{-1}$ due to C=C, both of which are absent in cyclc hexanone. The small high-field bands around 3400 cm$^{-1}$ in each spectrum are n due to impurities or moisture in the sample but are the first overtones of the corre ponding carbonyl absorptions.

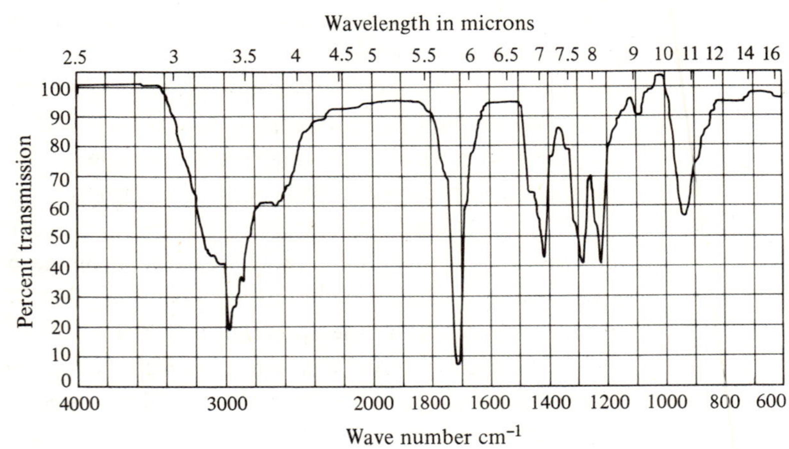


*FIGURE 15.10*
*The ir spectrum of n-butanoic acid.*

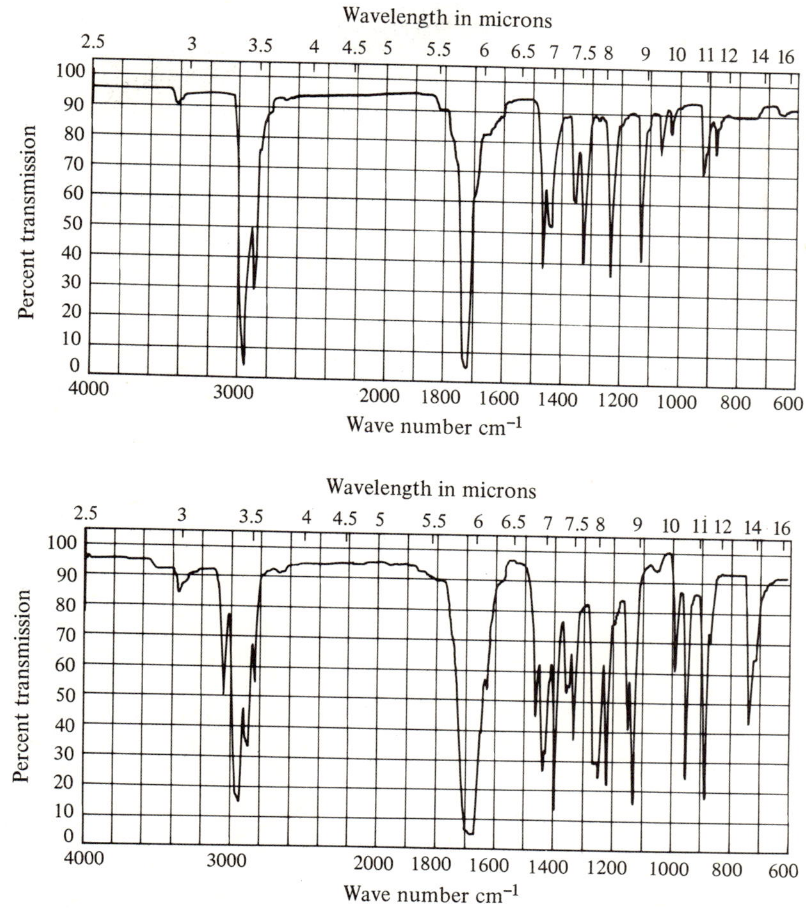


*FIGURE 15.9*
*The ir spectrum of cyclohexanone (top) and cyclohexenone (bottom).*

In Figure 15.10 (*opposite*) the strong broad band at 3200–2600 $cm^{-1}$ overlapping the C—H stretching region is very characteristic of hydrogen bonded carboxylic O—H. In conjunction with a strong carbonyl band around 1710–1760 $cm^{-1}$ it unmistakably points to the presence of a carboxylic acid.

In Figure 15.11, note the strong C=O stretching absorption at 1740 $cm^{-1}$, in conjunction with the strong C—O—C absorption at 1235 $cm^{-1}$ it is very indicative of the presence of an ester functionality.

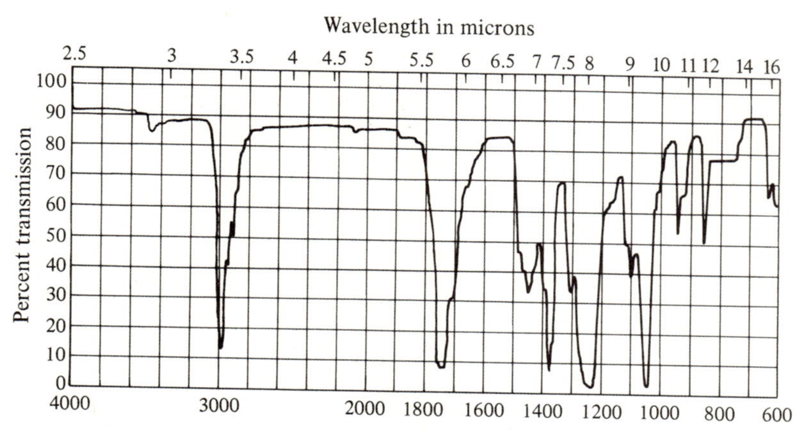


*FIGURE 15.11*
*The ir spectrum of ethyl acetate.*

15.5
*PROBLEMS*
SAMPLE PROBLEM

The widths at half-height for the CHOH methine proton resonances for a number of alcohols are

| | |
|---|---|
| *cis*-2-*t*-butylcyclohexanol (1) | 7.7 Hz |
| *trans*-2-*t*-butylcyclohexanol (2) | 22.3 Hz |
| *trans*-2-*t*-butyl-4-ketocyclohexanol | 15 Hz |

Explain these results.

ANSWER[9]

$^3J_{HCCH}$ coupling constants follow a very similar dihedral-angle dependence as $^3J_{HCOH}$ coupling constants, with $^3J_{aa}$ = 10–15 Hz, $^3J_{ae}$ = 3.5 Hz, and $^3J_{ee}$ = 1–3 Hz.

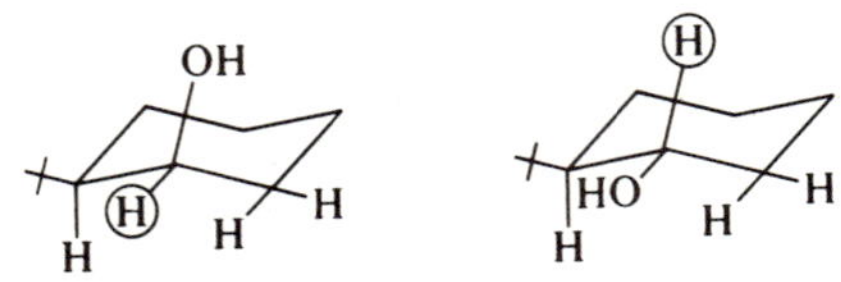


1 2

The cyclohexanone must be less than 60 percent in the chair form (2), with the remaining 40 percent in the boat form (3) due to the favorable hydrogen bonding conformation:

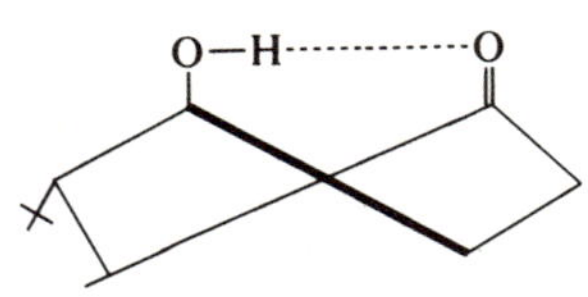


3

1. Two structures have been proposed for α-cyperone, a ketone of natural origin:

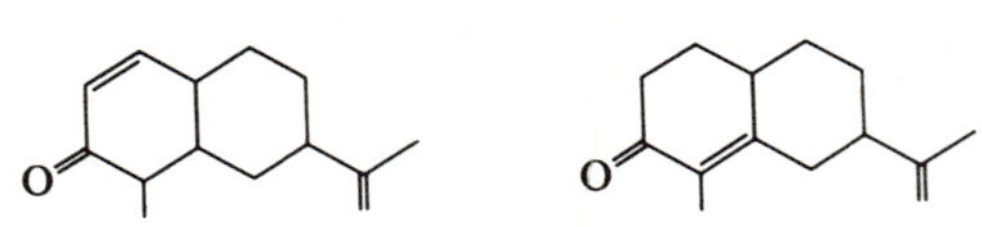


1 2

Do you favor (1) or (2), knowing that the uv spectrum of this compound has a maximum at 252 nm (ε = 19,000)?
2. Bromination of the steroid derivative

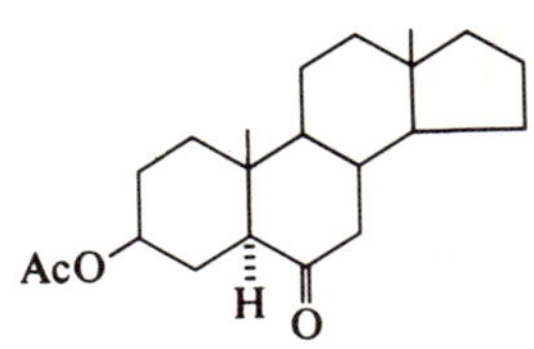


$\lambda_{max}$ = 280 nm (ε = 40)

leads to only two monobromo compounds with $\lambda_{max}$ of 308 nm (ε = 125) and 310 nm (ε = 160). What are possible structures for these compounds?

[9] R. D. Stolow, T. Groom, and M. Gerace, *J. Amer. Chem. Soc.*, *90*, 3290 (1968).

3. Using Woodward's rules calculate $\lambda_{max}$ for the following compounds:

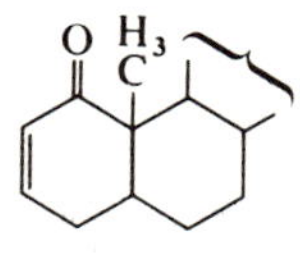


1

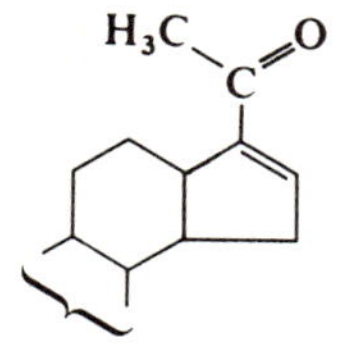


2

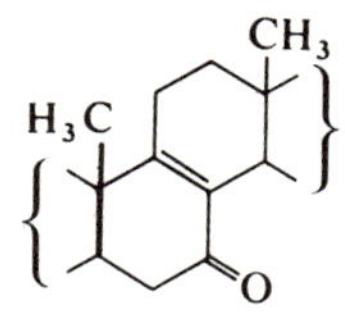


3

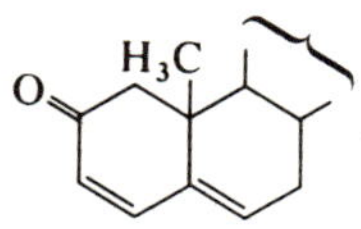


4

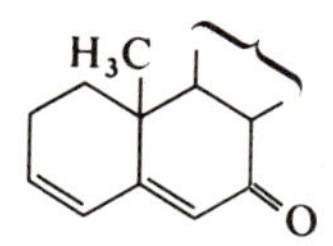


5

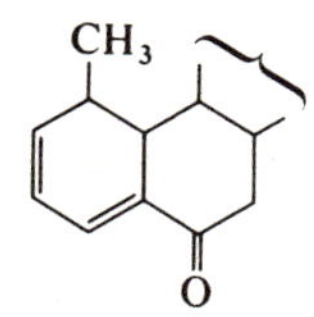


6

4. Explain the differences between the chemical shifts of the phenolic proton in

| | |
|---|---|
| 1-hydroxy-2-acetylnaphthalene | $\delta = 13.98$ |
| 1-acetyl-2-hydroxynaphthalene | $\delta = 13.42$ |
| 2-acetyl-3-hydroxynaphthalene | $\delta = 11.52$ |

5. When phenylacetone is placed in an $SO_2{-}FSO_3H{-}SbF_5$ medium (9:7:2 mole ratio) at low temperatures, two proton signals are seen in an approximate 1:1 intensity ratio at $\delta = 13.47$ (triplet; $J = 1.1$ Hz) and at $\delta = 14.60$ (quartet; $J < 1.5$ Hz). How do you interpret these observations? (Note: Only the absorption for the hydrogen on the protonated ketone is given; the aromatic and aliphatic hydrogens are ignored.)
6. Irradiation of a cyclohexane solution of an eight-membered ring compound (1) with light of wavelengths above 300 nm leads to 80 percent conversion of (1) to the isomeric (2). Spectral characteristics are as follows.

| | Compound 1 | Compound 2 |
|---|---|---|
| uv | $\lambda_{max}$ 223 nm ($\varepsilon = 7{,}600$) | absent |
| | $\lambda_{max}$ 321 nm ($\varepsilon = 70$) | 283 nm ($\varepsilon \sim 60$) |
| ir | 1675 $cm^{-1}$ | 1727, 1634, 1161, 988 $cm^{-1}$ |
| nmr | 1H $\sim$ 3.1$\tau$ d $J = 12.8$ Hz | 1H 3.5$\tau$ d $J = 18$ Hz |
| | 1H $\sim$ 4.2$\tau$ d.t $J = 12.8$ Hz | 1H 4.3$\tau$ d.t $J = 18$ Hz |
| | 4H $\sim$ 8$\tau$ | 4H $\sim$ 8$\tau$ |
| | 6 remaining protons absorb between 8 and 9$\tau$ | 6 remaining protons absorb between 8 and 9$\tau$ |

What can you say about molecules 1 and 2? (d.t. = doublet of triplet.)
7. One of the products of the reaction between 2,2-dimethylcyclopropanone and 1,1-dimethoxyethylene is a molecule $C_{18}O_6H_{32}$. Its pmr spectrum displays the following peaks:

$\delta = 0.87$ 4H AB quartet $J = 5.5$ Hz
1.06 6H singlet
1.23 6H singlet
2.39 4H AB quartet $J = 16$ Hz
3.17 6H singlet
3.22 6H singlet

What is the structure and stereochemistry of this molecule?

8. The mass spectrum of flavone,

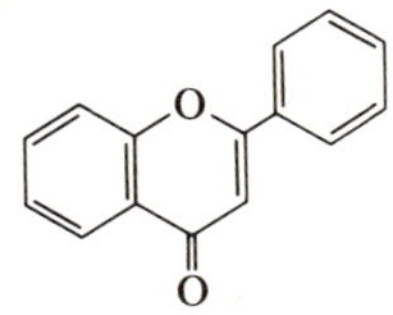

has important peaks at $m/e = 222$, 194, 120, 105, 102, 97, and 92. Write plausible mechanisms for generation of the corresponding fragments.

9. A fermentation process yields a mixture of ethyl stearate and palmitate, $C_{20}H_{40}O_2$ and $C_{18}H_{36}O_2$. The dominant peak in the mass spectrum occurs at $m/e = 88$, and can be attributed to the $C_4H_8O_2$ fragment. Account for its formation.

10. The methoxycyclohexanol

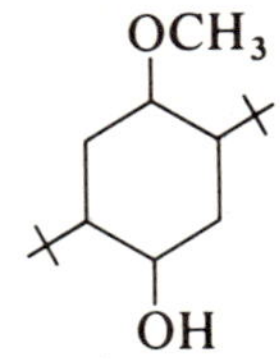

with two *cis-t*-butyl groups and oxygen functions shows two O—H bands in the ir, a very weak one at 3620 $cm^{-1}$ with only about 2 percent of the intensity of the lower frequency band at 3463 $cm^{-1}$. Interpret this observation.

11. *Trans*-2-phenyl-4-hydroxy-1,3-dioxane displays a free O—H peak at 3633 $cm^{-1}$ and a bound O—H peak at 3602 $cm^{-1}$. The *cis* isomer shows only a bound O—H peak. Explain.

12. A trimethylsilyl derivative $C_{19}H_{36}O_3Si_2$ could be isolated from the grasshopper *Romalea microptera*. Upon hydrolysis a $C_{13}H_{20}O_3$ diol was obtained with the following properties. Deduce its structure.

| | |
|---|---|
| ir | 3450, 1942, 1667 $cm^{-1}$ |
| uv | 232 nm ($\varepsilon = 12{,}500$) |
| nmr | ($CDCl_3$) 1.16 + 1.36 + 1.40 three $CH_3$ singlets |
| | 2.15 (3H singlet) |
| | 5.80 ppm (1H singlet) |
| | 4.28 ppm (1H $\nu_{1/2} \sim 32$ Hz) |
| | 1.40–1.55 ppm, two protons exchangeable with $D_2O$ |

13. Pyrolysis at 100° of methylcyclopropenone yields a compound $C_8H_8O_2$:

| | |
|---|---|
| nmr | 6.60 ppm 1H $J_{C-H} = 174$ Hz |
| | 7.25 ppm 1H $J_{C-H} = 232$ Hz |
| | + two methyl resonances ~2 ppm |
| ir | 3080, 3028, 2950, 1745 (strong), 1768 (medium), |
| | 1620 (medium-weak), 790 $cm^{-1}$ |
| uv | $\lambda_{max} = 241$ nm |
| mass | $m/e = 136$ (M) 108 96 68 |

Deduce a structure from the above data.

14. A compound of molecular formula $C_{13}H_{20}O_2$ is responsible for the fragrant odor of tea. The uv spectrum has a $\lambda_{max}$ (EtOH) 234 nm ($\varepsilon = 16{,}000$). The ir spectrum does not show any hydroxyl bands but displays absorption at 1660 and 1100 $cm^{-1}$. The pmr spectrum shows a singlet for three hydrogens at $\delta = 0.92$, another singlet for three hydrogens at $\delta = 1.10$, a doublet, $J = 6.1$ Hz for three hydrogens at $\delta = 1.28$, another doublet, $J = 1.0$ Hz for three hydrogens at $\delta = 1.97$. Between $\delta = 1.5$ and 2.5 broad bands for six protons are observed. At $\delta = 4.1$ there is a

multiplet for one hydrogen and at $\delta = 5.73$ a sharp quartet, $J = 1.0$ Hz, integrated for one hydrogen. Irradiation of the 4.1 multiplet decoupled the $\delta = 1.28$ doublet. The mass spectrum has the parent peak at 208 and shows a very strong peak at 152 mass units. Give a structural formulation consistent with all of these spectral observations.

15. 2,3,5-Tri-*t*-butylfuran is isomerized upon irradiation with light.

| | | | | |
|---|---|---|---|---|
| $M^+$ | 236 | | | |
| $\nu_{max}$ ($CCl_4$) | 3060 + 1719 $cm^{-1}$ | | | |
| $\lambda_{max}$ (EtOH) | 206.5 ($\varepsilon$ 6540) | | | |
| | 291 nm ($\varepsilon$ 143) | | | |
| $\tau$($CDCl_3$) | 9.39 | d | $J = 5.5$ | (1H) |
| | 8.61 | d.d | $J = 5.5$, $J' = 1.9$ | (1H) |
| | 8.88 | s | | (9H) |
| | 8.82 | s | | (9H) |
| | 9.00 | s | | (6H) |
| | 8.0 + 8.3 AB system | | | (2H) |
| | | | $J_{A-B} = 8.4$ Hz, the 8.0 peak being further split with the 1.9 Hz coupling | |

Deduce the structure of the photoproduct.

16. 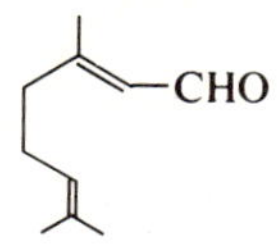

Citral

$\xrightarrow{h\nu}$ A + B (2 : 1)

photocitral A photocitral B

(1) ir 1720, 2715, 890, 1640, 3070 $cm^{-1}$

| | | | |
|---|---|---|---|
| nmr | $\tau$ 8.98 | (doublet) | 3H |
| | 8.34 | (singlet) | 3H |
| | 5.31 | (singlet) | 2H |
| | 0.36 | (doublet) | 1H |

(2) nmr

| | | |
|---|---|---|
| $\tau$ 9.27 | (s) | 3H |
| 8.9 | (s) | 6H |
| 8.44 | (m) | 4H |
| 7.80 | (m) | 1H |
| 7.45 | (m) | 1H |
| 0.42 | (d) | 1H |

Deduce the structures of (1) and (2).

17. The mass spectra of $\alpha$-ionone and of $\beta$-ionone,

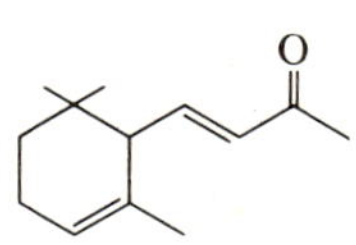

$\alpha$-Ionone

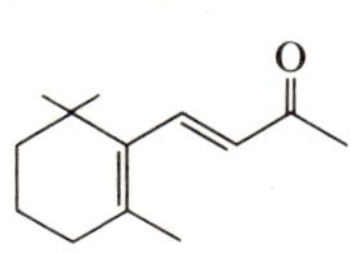

$\beta$-Ionone

differ markedly. That of $\beta$-ionone shows only two important peaks, at $m/e = 177$ and 43, whereas the mass spectrum of $\alpha$-ionone shows important peaks at $m/e = 43$, 77, 93, 121, 136, and 177. Explain this behavior.

## 15.6
### *APPENDIX*
*Summary of important infrared absorption frequencies for oxygen-containing molecules*

| Group | Band ($cm^{-1}$) | Comments |
|---|---|---|
| A. Alcohols, phenols | | |
| Primary—OH | 3640, 1050 | Free O—H stretching (only in dilute inert solution, $CCl_4$, $CHCl_3$) and C—O stretching; both bands strong |
| Secondary—OH | 3630, 1100 | |
| Tertiary—OH | 3620, 1150 | |
| Phenol | 3610, 1200 | |
| ROH, ArOH | 3600–3200 | Inter- or intramolecular hydrogen-bonded O—H |
| B. Ethers | | |
| ROR | 1150–1070 | C—O—C stretch |
| ArOR, C=C—OR | 1275–1200 | asym. C—O—C stretch |
| | 1075–1020 | sym. C—O—C stretch |
| $ROCH_3$ | 2830–2815 | sym. C—H stretch |
| $ArOCH_3$ | ~2850 | sym. C—H stretch |
|  | 3050–3000 | C—H stretch |
| | 1250, 950–810, 840–750 | ring stretching vibrations |
| C. Ketones | | |
| $-\overset{\text{O}}{\overset{\Vert}{\text{C}}}-$ | 1715 | C=O stretch, strong characteristic band |
| 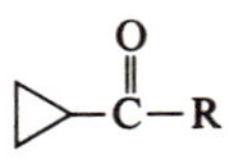 | 1695 | lower than normal C=O due to cyclopropyl overlap |
| $\text{Ar}\overset{\text{O}}{\overset{\Vert}{\text{C}}}\text{R}$ | 1690 | lower due to conjugation |
| α,β unsat. | 1675 | |
| α,β and γ,δ unsat.; α,β and α′,β′ unsat.; $\text{Ar}\overset{\text{O}}{\overset{\Vert}{\text{C}}}\text{Ar}$ | ~1665 | lower due to conjugation |
| Cyclic $C_7$, $C_8$ | ~1705 | see Section 15.2.1 |
| Cyclic $C_6$ | ~1715 | |
| Cyclic $C_5$ | ~1745 | |
| Cyclic $C_4$ | ~1780 | |
| $-\overset{\text{O}}{\overset{\Vert}{\text{C}}}-\overset{\text{O}}{\overset{\Vert}{\text{C}}}-$ | ~1720 | asym. stretch of two C=Os |
| D. Aldehyde | | |
| $\text{R}-\overset{\text{O}}{\overset{\Vert}{\text{C}}}-\text{H}$ | 2820, 2720 | C—H stretch |
| | ~1725 | C=O stretch |
| $\text{Ar}\overset{\text{O}}{\overset{\Vert}{\text{C}}}-\text{H}$ | ~1700 | C=O stretch |
| α,β unsat. | ~1685 | lower due to conjugation |
| α,β and γ,δ unsat. | ~1675 | |
| E. Carboxylic acid | | |
| RCOOH | 3000–2500 | H-bonded O—H stretch and combination bands |
| | 1760–1710 | strong C=O stretch |
| C=CCOOH, ArCOOH | 1720–1690 | conjugated C=O stretch |
| $-COO^-$ | 1610–1550 and ~1400 | asym. and sym. C=O stretch |
| F. Esters, lactones | | |
| $-\overset{\text{O}}{\overset{\Vert}{\text{C}}}-\text{O}-$ | 1735 | C=O stretch |
| | 1300–1050 | C—O—C stretch |
| $\text{C}=\text{C}-\overset{\text{O}}{\overset{\Vert}{\text{C}}}-\text{O}-$, $\text{Ar}\overset{\text{O}}{\overset{\Vert}{\text{C}}}-\text{O}-$ | ~1720 | conjugated C=O |

*(Continued)*

| Group | Band ($cm^{-1}$) | Comments |
|---|---|---|
| $-\overset{\overset{O}{\Vert}}{C}-O-C=C$ | ~1760 | |
| Six-member lactone | ~1735 | |
| Five-member lactone | ~1770 | |
| G. Anhydride | | |
| $-\overset{\overset{O}{\Vert}}{C}-O-\overset{\overset{O}{\Vert}}{C}-$ | 1820, 1760 | sym. and asym. C=O stretches |
| | 1300–1050 | C–O–C stretch |
| H. Acid halide | | |
| $R-\overset{\overset{O}{\Vert}}{C}-X$ | ~1800 | C=O stretch |
| $C=C-\overset{\overset{O}{\Vert}}{C}-X$ | 1780–1750 | conj. C=O stretch |

## 15.7 *SUGGESTED READINGS*

M. Tichy, "The Determination of Intramolecular Hydrogen Bonding by Infrared Spectroscopy and Its Application in Stereochemistry," in *Advances in Organic Chemistry*, R. A. Raphael, E. C. Taylor, and H. Wynberg, eds., Wiley-Interscience Publishers, Inc., New York, 1965, pp. 115–298.

R. R. Fraser, M. Kaufman, P. Morand, and G. Govil, "Stereochemical Dependence of Vicinal H—C—O—H Coupling Constants," *Can. J. Chem.*, *47*, 403 (1969).

H. Feltkamp and N. C. Franklin, "Conformational Analysis by Nuclear Magnetic Resonance. VI. The Conformations of Sixteen Stereoisomeric 2,5-Dialkyl Cyclohexanols of the Menthol Type," *Tetrahedron*, *21*, 1541 (1965).

F. G. Riddell, "Heterocyclic Conformational Analysis," *Quart. Rev.* (London), *21*, 364 (1967).

R. J. Ouellette, K. Liptak, and G. E. Booth, "Conformational Analysis. X. Determination of Intramolecular Hydrogen Bonding by Hydroxyl Proton Magnetic Resonance," *J. Org. Chem.*, *32*, 2394 (1967).

B. Galabov and D. Simov, "The Stretching Vibration of Carbonyl Groups in Cyclic Ketones," *Chem. Phys. Lett.*, *5*, 549 (1970).

# CHAPTER 16 NITROGEN-CONTAINING MOLECULES

Two aspects of the presence of nitrogen in a molecule are of concern: the exact identity of the nitrogen functionality, that is, $-NH_2$, $-CN$, $-NO_2$, and so on, and special properties such as nitrogen inversion and tautomeric equilibria. Quadrupolar relaxation of nitrogen has already been discussed. Let us describe first the qualitative identification of the nitrogen functionality.

## 16.1 *Identification of Nitrogen Functionality*

The presence of nitrogen can usually be discerned from the previous history of the sample or ascertained from elemental analysis or accurate molecular-weight determinations by high-resolution mass spectrometry (Chapter 10). The type of nitrogen functional group present is best determined by ir; Table 16.1 lists the most important nitrogen-containing functional groups and their characteristic ir absorption frequencies. Further information concerning nitrogen-containing molecules can be obtained from nmr and mass spectrometry. In the nmr the N—H proton resonates in the following regions:

| | |
|---|---|
| aliphatic amines | 0–2.0 ppm |
| aromatic amines | 2–5 ppm |
| amides | 5–9 ppm |

These resonances are often broad bands—a useful identifying feature—due to moderately slow chemical exchange and incomplete washing out of $^{14}N-H$ coupling. Groups such as $-CN$, and $-NO_2$ can be treated as electronegative substituents which

*TABLE 16.1*

*Nitrogen Functional-Group Frequencies in the Infrared*

| Group | Band ($cm^{-1}$) | Comments |
|---|---|---|
| Primary amine, $-NH_2$ | 3500–3400 | N—H stretch, usually two; sym. and asym. |
| | 1650–1550 | N—H bending |
| | 1350–1250 | C—N vibrations |
| Secondary amine N—H | 3400–3300 | N—H stretch, usually weak, only one band |
| | 1600–1490 | N—H bending, usually weak |
| | 1350–1250 | C—N vibrations |
| $-NH_3^+$ | 3100–3000 | sym. and asym. N—H, usually broad |
| | 1600–1500 | bending modes |
| $NH_2^+$, $NH^+$ | 2700–2200 | broad sym. and asym. N—H stretch |
| Imine C=N—H | 3400–3300 | N—H stretch |
| | 1700–1650 | C=N stretch |
| Nitrile —CN | 2260–2200 | C≡N stretch, very characteristic but sometimes lacking |
| Nitrate $-O-NO_2$ | 1650–1620 | |
| | 1290–1250 | |
| Nitrite —O—NO | 1700–1650 | |
| Nitro $C-NO_2$ | 1600–1500 | |
| Nitroso C—NO | 1600–1500 | |
| Amides —C(=O)—N | 1700–1650 | C=O stretch, see also N—H |
| Isocyanate —N=C=O | 2280–2250 | very strong |
| Azo compound —N=N— | 1630–1580 | N=N stretch |
| Diimide —N=N— | 2160–2130 | stretching |
| Azide $-N_3$ | 2160–2120 | |
| | 1350–1180 | |

cause a downfield shift. In the mass spectrometer parent peaks for amines are usually small due to easy fragmentation. Cleavage usually occurs $\beta$ to the heteroatom—in analogy to ethers—with the larger fragment eliminating more readily than the smaller one:

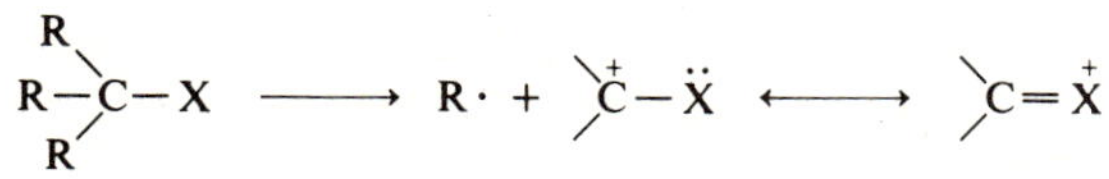

$$R{:}CH_2{:}\ddot{N}H_2 \longrightarrow R{:}CH_2{:}\overset{+\cdot}{N}H_2 \longrightarrow R\cdot + CH_2{=}\overset{+}{N}H_2 \quad m/e = 30$$

For primary aliphatic amines, for instance, such $\beta$-cleavage is dominant, resulting in a base peak at $m/e = 30$.

It is instructive to note that such cleavage $\beta$ to the heteroatom parallels closely the strength of the heteroatom as a Lewis base. The order of base strength of the various fragments is clearly reflected in the relative abundance of the various ions:

$$CH_2{=}\overset{+}{N}H_2 > CH_2{=}\overset{+}{S}H > CH_2{=}\overset{+}{O}H \gg CH_2{=}\overset{+}{C}l$$

| $H_2NCH_2$ | $CH_2OH$ | $HSCH_2$ | $CH_2OH$ | $ClCH_2$ | $CH_2OH$ |
|---|---|---|---|---|---|
| 57 percent | 2.5 percent | 13.7 percent | 8.5 percent | 6.9 percent | 53 percent |

The percentage indicates the relative abundance of fragments with respect to the base peak. In primary amides $\beta$ cleavage, $\alpha$ to the carbonyl, is prominent, resulting in a resonance-stabilized ion at $m/e = 44$:

$$[R{-}\overset{O}{\overset{\|}{C}}{-}NH_2]^{+\cdot} \xrightarrow{-R\cdot} O{=}C{=}\overset{+}{N}H_2 \quad m/e = 44$$

Long-chain amides containing $\gamma$ hydrogens are also capable of undergoing the McLafferty rearrangement discussed in Chapter 10.

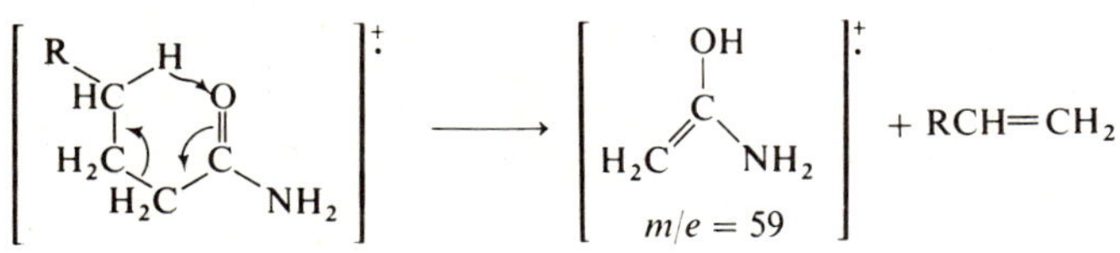

After this brief summary of the methods for the identification of nitrogen-containing compounds let us examine the special properties of some of these molecules.

## 16.2 *Nitrogen Inversion*

The possibility of inversion about pyramidal nitrogen is a well-recognized phenomenon. Such inversion results in the interconversion of enantiomers in the case of dissymetrically substituted tertiary amines and the interconversion of diastereotopic protons in the nmr spectrum:

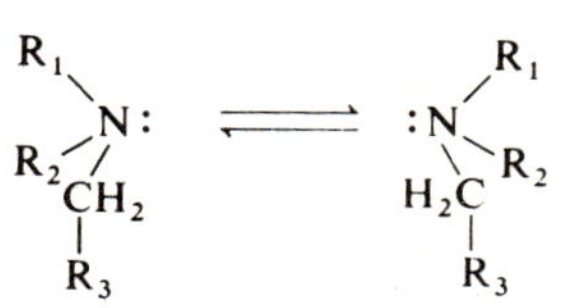

The configurational stability of such pyramidal nitrogen molecules depends upon the magnitude of the energy barrier to inversion at nitrogen. Two techniques in particular have been employed for the successful measurement of inversion barriers. The barrier in ammonia, where the two conformations are of course identical, has been determined by microwave spectroscopy. Because of quantum mechanical tunneling, a splitting of vibrational levels, known as "inversion doubling," occurs. From the observed splittings the barrier to inversion in ammonia has been determined as 5.9 kcal/mole.[1] This technique is complicated and difficult to apply to substituted amines because of the complexity of the vibrational and rotational interactions and therefore other, in particular nmr, techniques have evolved for the study of such systems.

The nmr method requires that the molecule contain diastereotopic protons which are distinguishable in the spectrum at sufficiently low temperature and become indistinguishable because of the rate of inversion at some higher temperature. An example is the study of the rate of inversion of dibenzylmethylamine.[2]

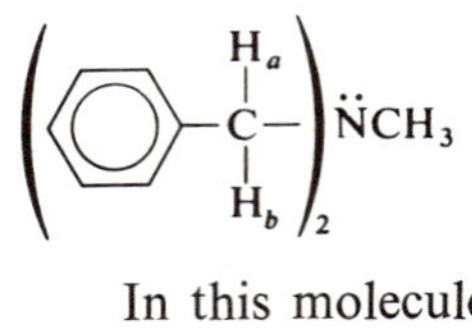

In this molecule $H_a$ and $H_b$ are diastereotopic, but inversion at nitrogen transforms the environment of $H_a$ into the mirror image of that of $H_b$. The chemical shifts of $H_a$ and $H_b$ are exchanged. In ordinary organic solvents inversion is too rapid and no value for the energy barrier was obtained. In concentrated aqueous hydrochloric acid, however, the nmr spectrum of the amine hydrochloride features a methyl doublet and displays nonequivalent methylene protons. Upon decreasing the acid concentration, the methyl doublet coalesces but the methylene protons remain nonequivalent until the pH is raised to 2.5. The rate of the exchange process (giving rise to coalescence of the methyl group) is calculated to be about 5 $sec^{-1}$ using the method of Meiboom.[3] A rough value for the second-order rate constant for proton transfer is calculated to be $6 \pm 3 \times 10^8$ liter/mole sec using the measured $pK_a$ of 7.5.

It was then argued that under conditions in which NH exchange was rapid enough to allow both the coalescence of the methyl group and the nonequivalence of the methylene protons, the amine salt must lose a proton and be reprotonated without the free amine suffering inversion. The measured $pK_a$ permitted calculation of the ratio [amine]/([amine] + [amine: HCl]) as a function of pH. Taking into account $J_{ab}$ and the natural linewidth, a value of 21 $sec^{-1}$ was calculated for the rate of interchange of the methylene protons at a pH of 2.5. Since $k_{\text{int.}} = k_{\text{inv.}}$ (fraction of substrate present as free amine),

$$k_{\text{int.}} = k_{\text{inv.}}\left(\frac{[\text{amine}]}{[\text{amine}] + [\text{amine: HCl}]}\right)$$

a value of $2 \pm 1 \times 10^5$ $sec^{-1}$ was calculated for the rate constant for inversion. A very rough estimation of the free energy of activation can be made with the usual relation $\Delta G^\ddagger = -RT \ln kh/KT$. In this case $\Delta G^\ddagger = (-1.98)(300)(2.303)\log(2 \times 10^5)(6\times 10^{-27})/(1.38 \times 10^{-10})(300) = 10.3$ kcal/mole.

An interesting system which has been extensively investigated is aziridine. At a sufficiently low temperature the parent system displays an AA′BB′ spectrum which

[1] G. Herzberg, *Molecular Spectra and Molecular Structure,* Van Nostrand Reinhold Co., New York, 1964, vol. II, pp. 220–227.

[2] M. Saunders and F. Yamada, *J. Amer. Chem. Soc., 85,* 1882 (1963); D. E. Leyden and W. R. Morgan, *Chem. Comm.,* 598 (1969); C. H. Bushweller and J. W. O'Neil, *J. Amer. Chem. Soc., 92,* 2159 (1970).

[3] A. Lowenstein and S. Meiboom, *J. Chem. Phys., 27,* 1067 (1967).

collapses to a single line as the temperature is raised and nitrogen inversion becomes rapid. A wide variety of aziridines have been investigated in order to elucidate the

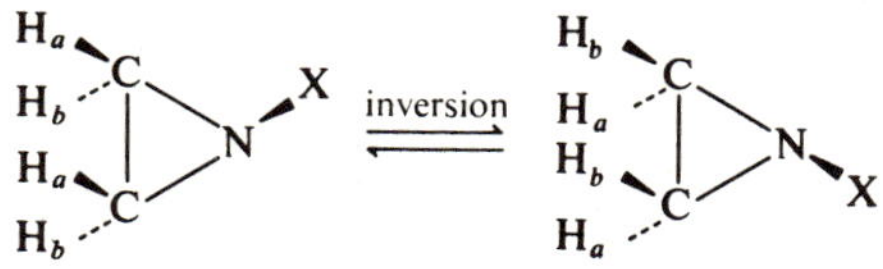


effect of substituents (X) on the rate of inversion and gain some understanding of the nature of the inversion barrier. From the organic chemist's point of view the following effects have been found to affect the rates of nitrogen inversion in aziridines[4]:

1. Steric effects. A large substituent on nitrogen is more compressed in the pyramidal ground state than in the planar transition state required for inversion and hence causes an increase in the frequency of inversion.
2. Inductive effects. Electronegative substituents on nitrogen decrease the rate of inversion since they increase the *s* character of the unshared electron pair on nitrogen which must be in a *p* orbital in a planar transition state.
3. Conjugative effects. Since conjugative overlap, particularly between two *p* orbitals, is greater in the planar state than in the pyramidal ground state, substituents capable of conjugation increase the rate of inversion.
4. Electrostatic or electron-repulsion effects. When a substituent attached to nitrogen possesses an unshared electron pair the electrostatic repulsion between this pair and the lone pair on nitrogen is less in the pyramidal ground state than in the planar transition state, resulting in a decrease in the rate of inversion. Such effects are particularly noticeable with halogens and in particular with fluorine.

The effects of substituents on the ring have been investigated as well. By taking advantage of the substituent effects configurationally stable aziridines (Figure 16.1) have been isolated recently.[5] Since the *trans* (1a) and *cis* (1b) isomers were configurationally stable even at 120°C a lower limit of 21 kcal/mole was calculated for the free-energy barrier to inversion in these molecules.

Another interesting system which has been investigated recently is piperidine. In piperidine and its derivatives there are two distinct processes that can take place: inversion at nitrogen (1) and inversion of the ring (2):

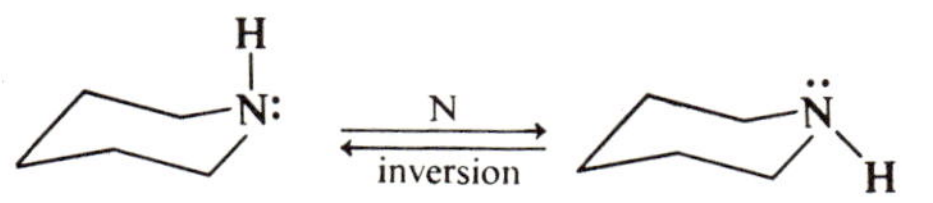


(1)

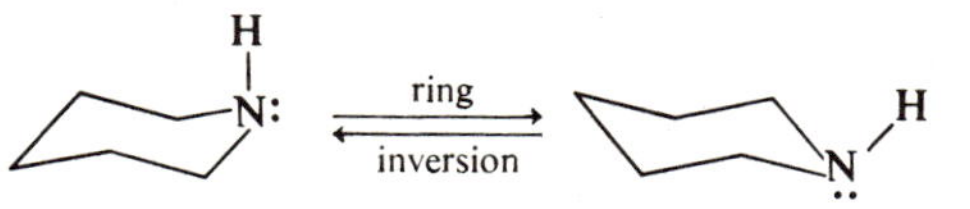


(2)

Yousif and Roberts[6] used *gem* fluorines in the 4 position to investigate the relative rates of these two processes by means of nmr. They found that below −40° they could slow the rate of ring inversion and observe different chemical shifts for the equatorial and axial fluorines (Figure 16.2) and below −60°C in acetone they could slow the rate of nitrogen inversion (3) and observe two separate isomers (2a and 2b). Their calculated activation parameter for ring inversion is $E_a \sim 14$ kcal/mole, whereas the nitrogen barrier is about 10.5 kcal/mole.

---

[4] F. A. L. Anet, R. D. Trepka, and D. J. Cram, *J. Amer. Chem. Soc.*, *89*, 352 (1968).
[5] S. J. Brois, *J. Amer. Chem. Soc.*, *90*, 506, 508 (1968); D. Felix and A. Eschenmoser, *Angew. Chem. Int. Ed.*, *7*, 224 (1968); J. M. Lehn and J. Wagner, *Chem. Comm.*, 148 (1968).
[6] G. A. Yousif and J. D. Roberts, *J. Amer. Chem. Soc.*, *90*, 6428 (1968).

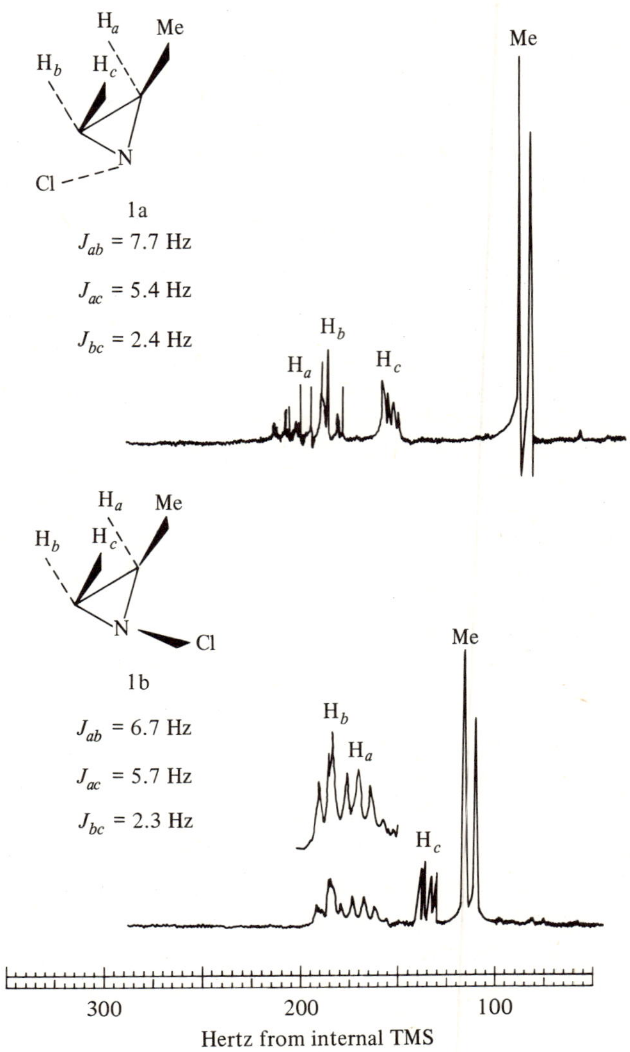


*FIGURE 16.1*

*Nuclear magnetic resonance spectra (100 MHz) of trans- and cis-1-chloro-2-methylaziridine (1a and 1b, respectively) as 25 percent (w/v) benzene solutions. The decoupled spectrum (insert) of (1b) was obtained by irradiation of the methyl protons.*

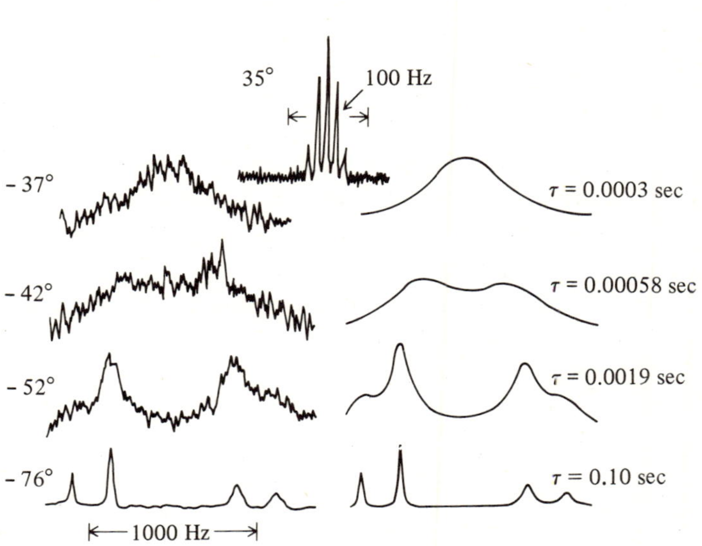


*FIGURE 16.2*

*Experimental and calculated fluorine nmr spectra of 4,4-difluoropiperidine as a function of temperature at 56.4 MHz. Note that at 35°C the fluorine resonance is a binomial quintet [arising from coupling with the adjacent hydrogens, a "deceptively simple" AA′(XX′)₂ system], showing no chemical-shift difference cetween the axial and the equatorial fluorines as a result of rapid equilibration.*

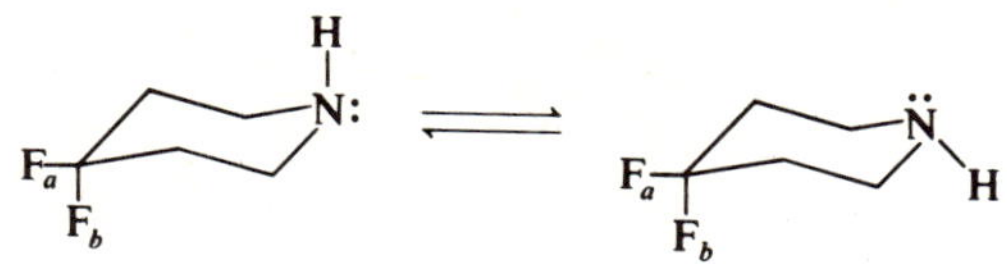


2a 2b (3)

Simultaneous inversion around two nitrogens has also been investigated. Anderson and Lehn[7] have studied the inversion process within the 2,3-diazabicyclo[2.2.1] heptane and 2,3-diazabicyclo[2.2.2]octane series with compounds 3, 4, and 5.

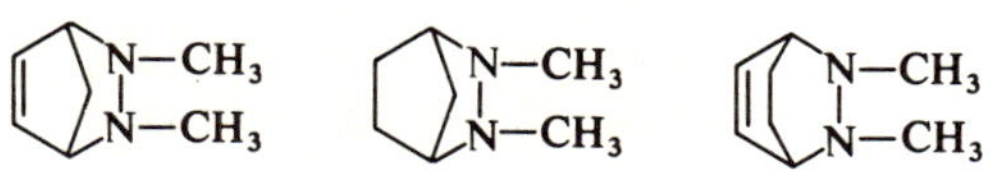


3 4 5

At higher temperatures each compound showed a single methyl resonance while at lower temperatures two distinct resonances were observed (Figure 16.3) corresponding to the process

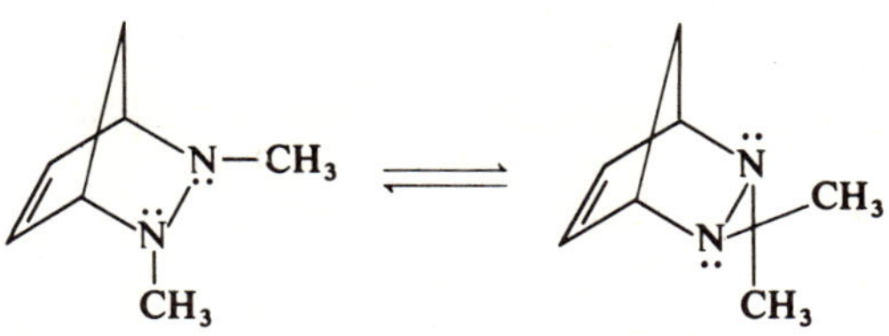


The observed activation parameters for the three compounds are shown in Table 16.2.

*TABLE 16.2*

*Activation Parameters for Compounds (3), (4), and (5)*

| Compound | Solvent | $\Delta G^\ddagger$ |
|---|---|---|
| 3 | $CDCl_3$ | 14.5 |
| 4 | $CDCl_3$ | 13.0 |
| 5 | $CDCl_3$ | 11.8 |

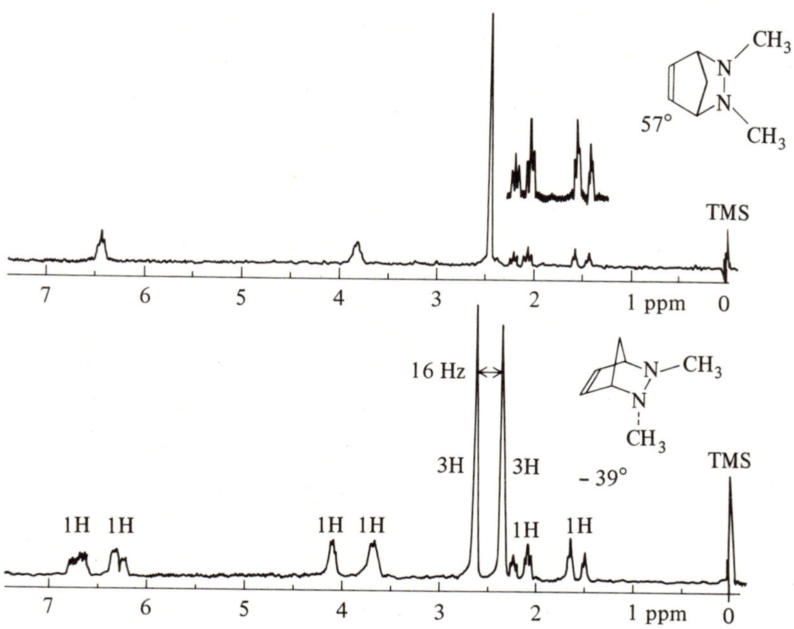


*FIGURE 16.3*
*The nmr spectrum of compound (3) at 57°C and at −39°C. Note in particular the two different methyl resonances at the lower temperature.*

[7] J. E. Anderson and J. M. Lehn, *J. Amer. Chem. Soc.*, *89*, 81 (1967).

## 16.3 *Hindered Rotation in Amides*

Because of the partial double-bond character about the C—N bond in amides due to resonance interaction of the type

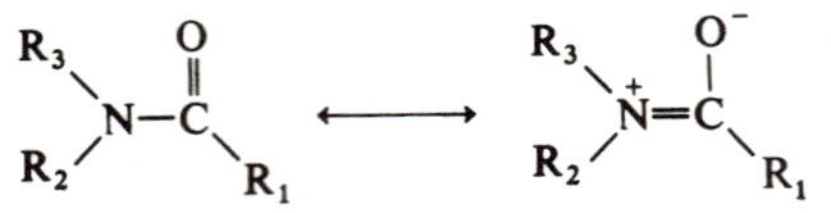

rotation about this bond is restricted. If $R_2$ and $R_3$ are different substituents and the barrier to rotation is sufficiently large this can lead to conformational isomers. The magnitude of the barrier has been determined for a number of amides, of which the work of Gutowsky and coworkers[8] is perhaps the most extensive. A particular example which has been investigated carefully is N-methyl-N-benzylformamide, by

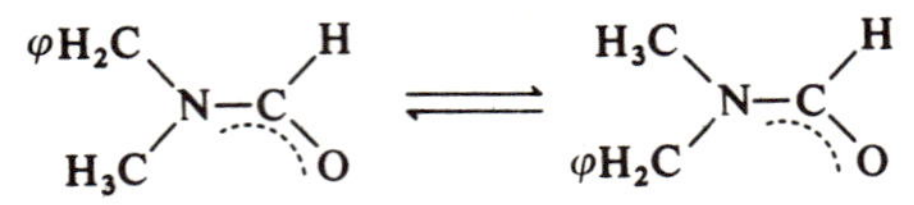

A B

looking at the temperature dependence in the nmr of the benzyl protons as well as the formyl protons (Figure 16.4) and also as the direct equilibration between the two isomers. The activation parameters observed for hindered rotation are as follows:

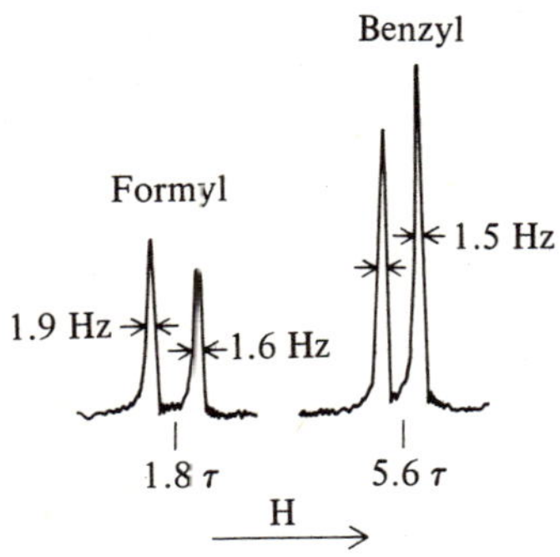


*FIGURE 16.4*
*Spectra of formyl and benzyl protons of N-methyl-N-benzylformamide at 40°C and 60 MHz. Note the two different resonances for both the formyl and benzyl protons, indicating two isomers A and B.*

| Activation parameter | Line shape analysis (nmr) Benzyl | Line shape analysis (nmr) Formyl | Equilibration |
|---|---|---|---|
| $\Delta H^\ddagger$ (kcal/mole) | 24.3 ± 1.5 | 22.0 ± 1.5 | 19.5 ± 3 |
| $\Delta G^\ddagger$ (kcal/mole) | 20.9 ± 0.4 | 21.0 ± 0.4 | 21.6 ± 0.7 |
| $\Delta S^\ddagger$ (e.u.) | 8.8 | 2.4 | −6.5 |

It is evident that the various techniques are in satisfactory agreement and that the barrier to rotation in amides is rather large. At this point one should remember the previous fallacious determinations of these barriers due to temperature-dependent chemical shifts as discussed in Chapter 11.

Hindered rotation in enamines has also been examined. In these compounds hindrance to free rotation about the C—N bond is dependent upon the importance of resonance hybrid (2):

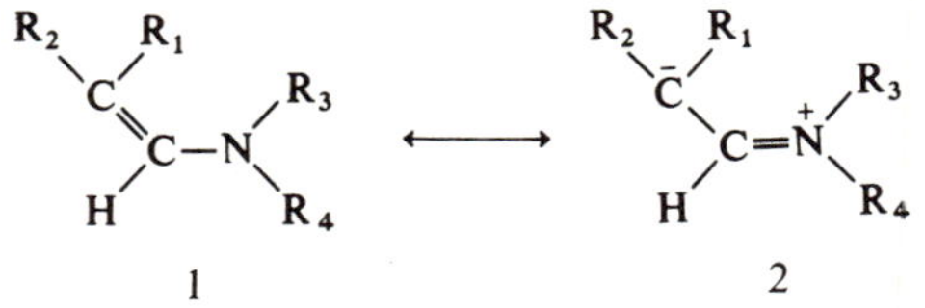

1 2

A number of compounds with differing $R_1$ and $R_2$ were investigated[9] and found to give activation energies $\Delta G^\ddagger$ for rotation of 12–20 kcal/mole, depending upon the substituents.

## 16.4 *Tautomeric Equilibria*

In some cases nitrogen-containing molecules can exist in a number of possible tautomeric forms. Spectroscopic methods are best suited for distinguishing between the various tautomeric forms. For instance, 2-hydroxypyridine can exist either in the hydroxy or keto form:

---

[8] H. S. Gutowsky, J. Jonas, and T. H. Siddall, *J. Amer. Chem. Soc.*, *89*, 4300 (1967).

[9] A. Mannschreck and U. Koelle, *Tetrahedron Lett.*, 863 (1967); Y. Shvo, E. C. Taylor, and J. Bartulin, *Tetrahedron Lett.*, 3259 (1967).

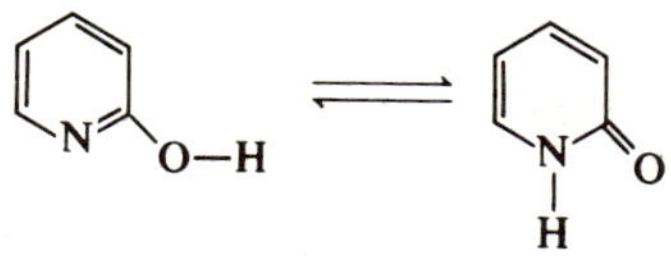

The uv spectrum of this compound shows absorption at $\lambda = 224$ nm ($\varepsilon = 7230$) and 293 nm ($\varepsilon = 5900$). Upon comparison with model compounds where tautomerism is prevented, such as in 2-methoxypyridine and N-methylpyrid-2-one,

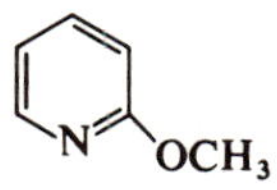

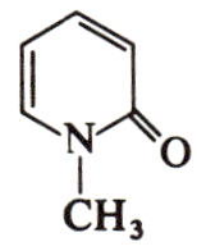

$\lambda_{max} = <205$ nm($\varepsilon > 5300$) 269 nm($\varepsilon = 3230$)    $\lambda_{max} = 226$ nm($\varepsilon = 6100$) 297 nm($\varepsilon = 5700$)

it is evident that the compound exists in the keto form rather than the hydroxy form.

Another example is maleic hydrazide, which can undergo the following tautomeric equilibrium:

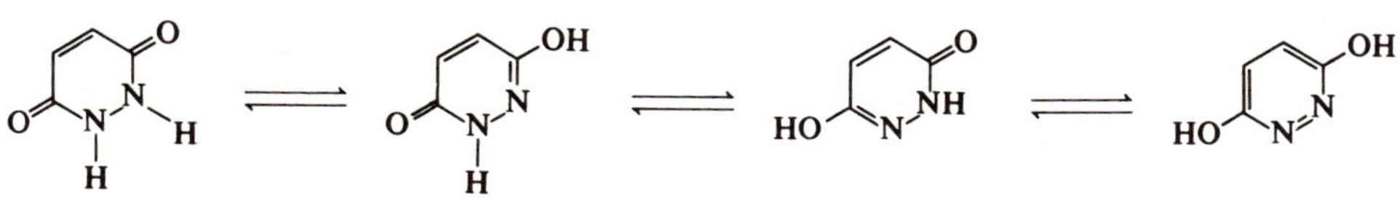

Experimentally a single peak for the ring protons is observed[10] at $\tau = 3.04$. The following model compounds were also examined:

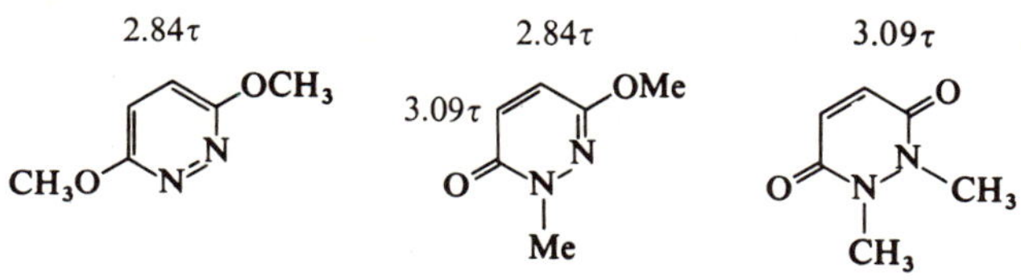


Since replacement of a methoxyl by hydroxyl shifts an *ortho* proton $0.11 \pm 0.04$ ppm and a *meta* proton $0.07 \pm 0.04$ ppm upfield the ring protons for the monomethyl derivative are predicted to occur at

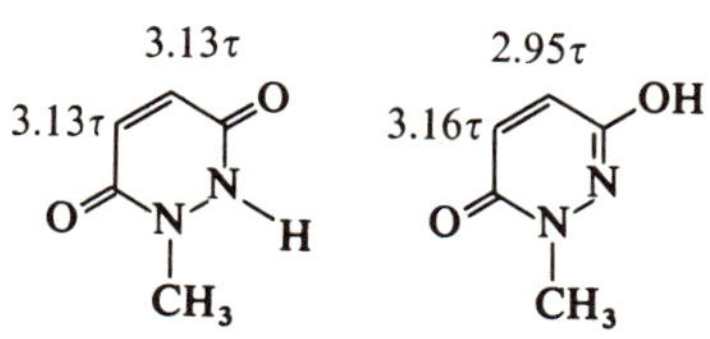


Since the experimental values are $\tau = 2.96$ and 3.13 the monomethyl derivative probably exists in the hydroxy form. For maleic hydrazide itself the estimated chemical shifts for the three possible forms are $\tau = 3.02$, 3.06, and 3.16 for the dihydroxy, monohydroxy, and diketo forms, respectively. Therefore the experimental value of $\tau = 3.04$ would indicate that maleic hydrazide does not exist in the diketo form.

## 16.5 *ILLUSTRATIVE EXAMPLES*

In Figure 16.5, note the absorption around 3400–3200 cm$^{-1}$ characteristic of N—H stretching; it is a doublet in primary amines such as *t*-butylamine corresponding to symmetric and asymmetric stretching, only a singlet in secondary amines, and of

[10] A. R. Katrizky and A. J. Waring, *J. Chem. Soc.*, 1523 (1964).

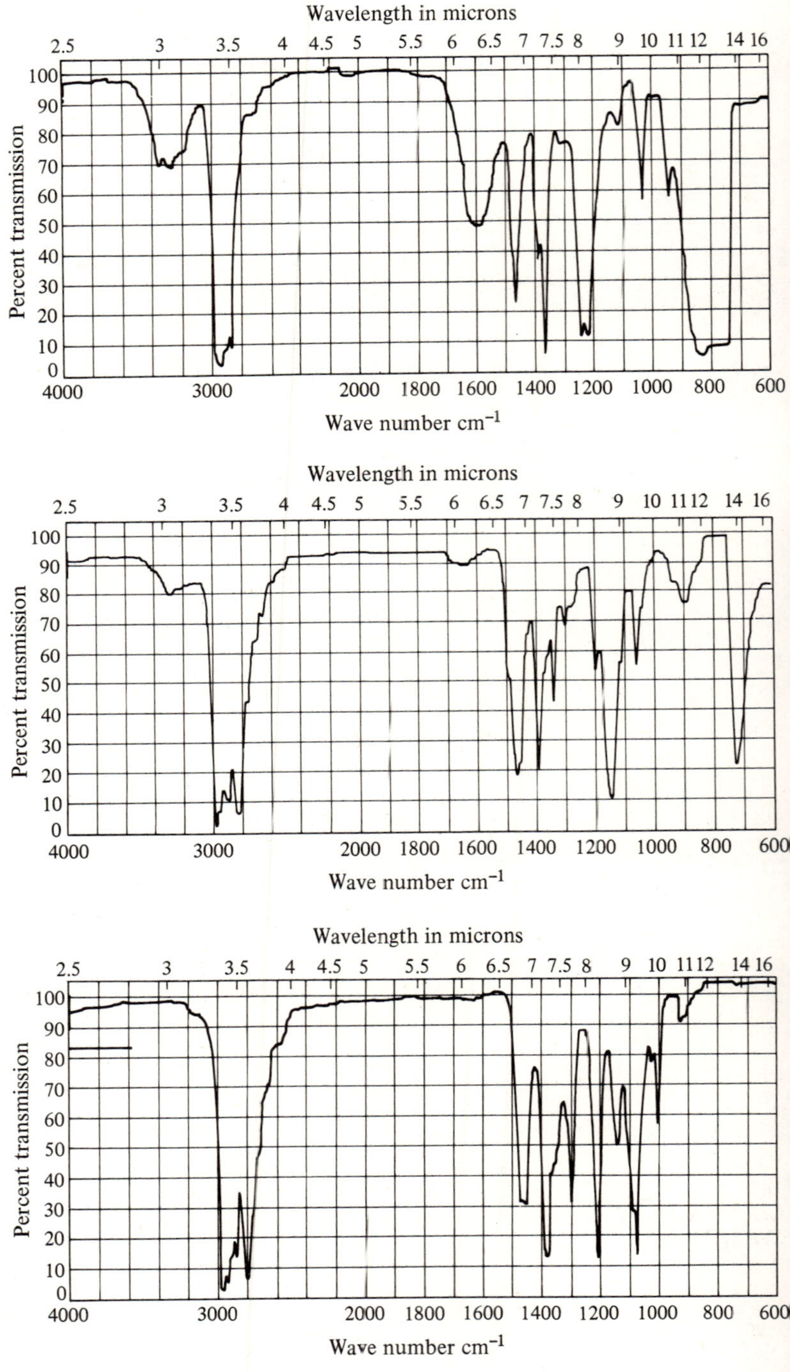


*FIGURE 16.5*
*Infrared spectrum of t-butylamine (top), diethylamine (middle), and triethylamine (bottom).*

course completely absent in tertiary amines. Note also the strong $NH_2$ bending modes in *t*-butylamine at 1600 $cm^{-1}$. Amines are further characterized by absorption in the 1230–1030 $cm^{-1}$ region due to C—N stretching.

The most characteristic absorption of a nitrile is in the region of 2260–2200 $cm^{-1}$ due to C≡N stretching (Figure 16.6). For benzonitrile it occurs at 2220 $cm^{-1}$ due to conjugation and is somewhat more intense than in unconjugated nitriles. Note also the absence of C—H absorption below 3000 $cm^{-1}$, indicating the absence of saturated hydrogens in this molecule. Furthermore, note the absorption pattern between 2000–1650 $cm^{-1}$ and below 1000 $cm^{-1}$ characteristic of substituted aromatic systems.

The bands characteristic of a $-NO_2$ group occur at 1515 and 1345 $cm^{-1}$ in the compound shown in Figure 16.7. The absorption at 860 $cm^{-1}$ is due to C—N stretching.

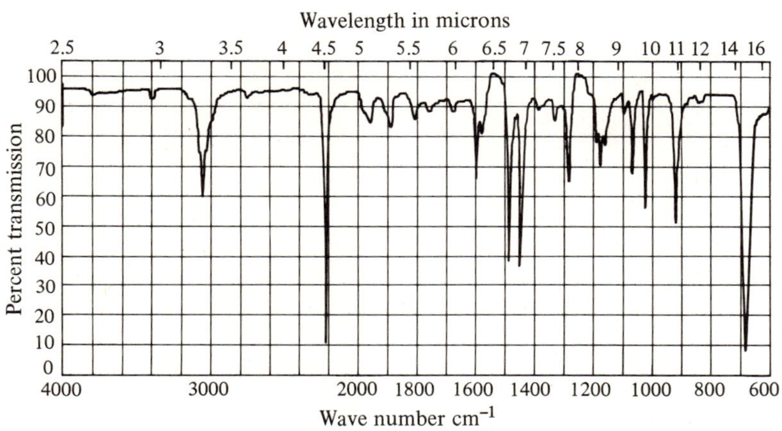


*FIGURE 16.6*
*Infrared spectrum of benzonitrile.*

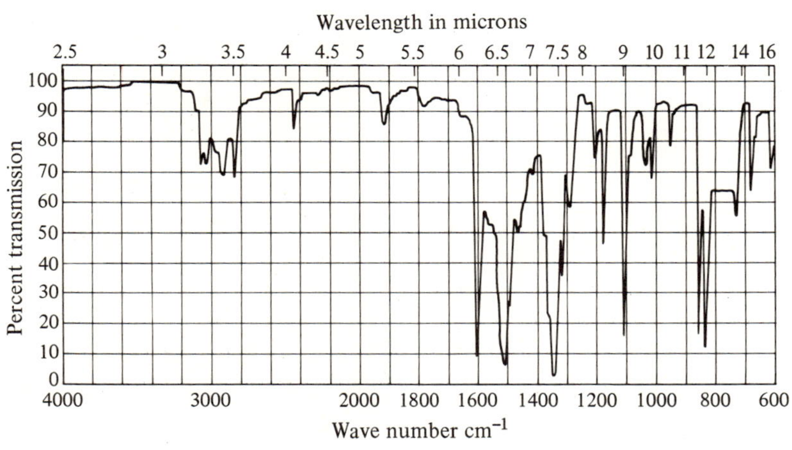


*FIGURE 16.7*
*Infrared spectrum of p-nitrotoluene.*

## 16.6 *PROBLEMS*

1. Nitriles do not show coupling of the $\alpha$ or the $\beta$ protons to nitrogen, but isonitriles do. How would you account for this difference in behavior?

$$\underset{\text{Nitrile}}{R{-}\underset{\beta}{CH_2}{-}\underset{\alpha}{CH_2}{-}C{\equiv}N} \qquad \underset{\text{Isonitrile}}{R{-}\underset{\beta}{CH_2}{-}\underset{\alpha}{CH_2}{-}\overset{+}{N}{\equiv}\overset{-}{C}}$$

2. The pmr spectrum of methylamine in water solution is pH dependent. At pH 10 one peak is seen for the rapidly exchanging water and amino protons, and the methyl resonance is a singlet. At pH 4.1 the methyl resonance is still a singlet, but that for the exchanging O—H and N—H from the solvent and from the amine is markedly broadened. At pH 0.87, and still at room temperature, the water line is again a sharp singlet, sandwiched between a low-field triplet (three broad lines of equal intensity separated from each other by $J = 52$ Hz), and a high-field triplet (1:2:1 relative intensities) characterized by $J = 7$ Hz. Explain these observations.
3. In the pmr spectrum of neat thiazole at 32°C, three distinct resonances are seen at $\delta = 7.44$ ($H_5$), 8.02 ($H_4$), and 9.01 ($H_2$); $H_2$ and $H_4$ show up as broadened doublets with $J_{25} = 1.8$ Hz and $J_{45} = 3.2$ Hz, and the $J_{24}$ coupling cannot be resolved. However, by cooling the sample to $-37$°C, both the $H_2$ and the $H_4$ resonances are transformed into doublets of doublets, displaying the long-range coupling $J_{24} = 0.6$ Hz. How do you explain these observations?

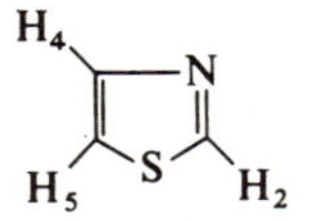


Thiazole

4. Give an explanation for the prominent fragmentations in these mass spectra:

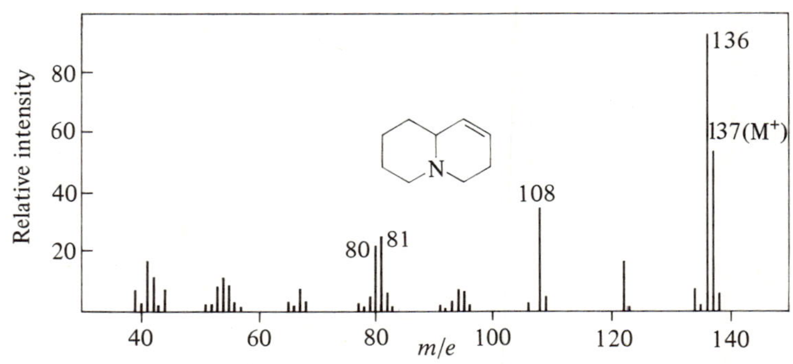


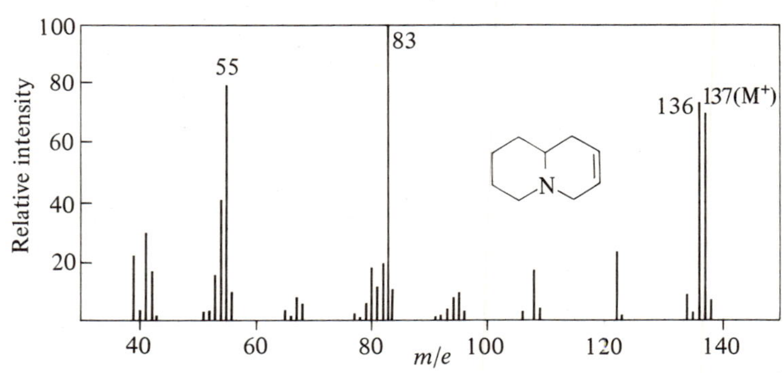


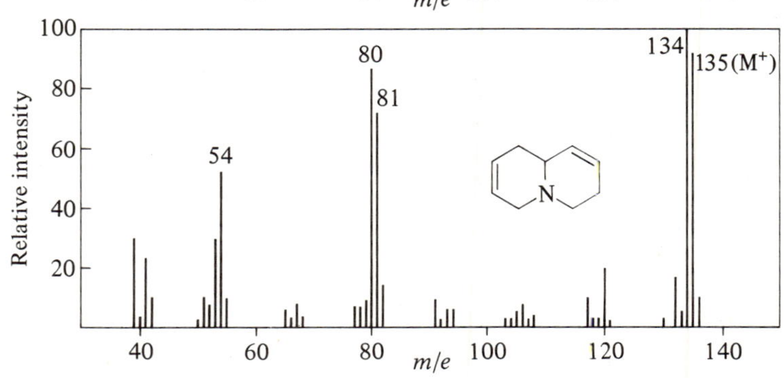

5. The mass spectrum below corresponds to a $C_6N_3H_5$ compound. Deduce its structure and explain the appearance of the spectrum.

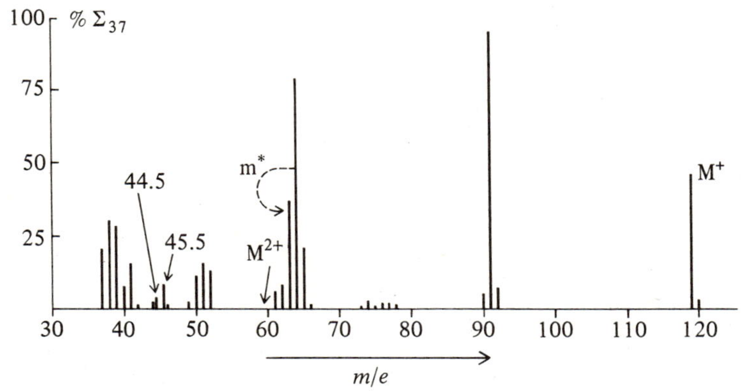


6. The labile hydrogen exchangeable with deuterium in the iminophenol below appears as a doublet, with a coupling constant of 21 Hz at 0°C and of 32.5 Hz at −50°C. Explain.

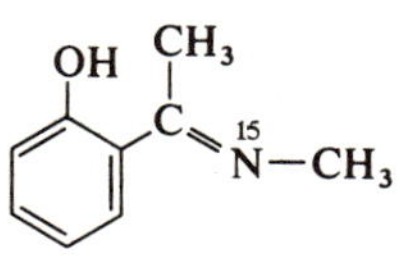


7. A natural product $C_{13}O_4NH_{15}$ has the following spectral properties:

| | |
|---|---|
| ir | strong bands at 1660 and 1620 $cm^{-1}$ |
| ms | 249 (M+) |
| nmr ($CDCl_3$) | $\delta$3.54 s, 3H; 3.93 s, 6H; 3.90 s, 3H; 6.91 d, 1H, $J = 12$ Hz; 6.58 d, 1H, $J = 12$ Hz; 7.56 s, 1H |
| uv | 247 nm ($\varepsilon = 42{,}000$), 270 nm ($\varepsilon = 4000$), 281 nm ($\varepsilon = 4800$), 293 nm ($\varepsilon = 5500$), 332 nm ($\varepsilon = 3000$), and 322 nm ($\varepsilon = 3200$). |

What is the structure of this compound?

8. Treatment of amine (1) with N-chlorosuccinimide in methylene chloride yielded a compound (2) with an elemental formula of $C_{24}H_{35}O_5N$. Whereas (1) shows in its nmr spectrum a sharp singlet at $\delta = 2.84$ ($CH_2$—$NH_2$ protons) and a vinylic proton resonance at $\delta = 5.21$, these have disappeared in the spectrum of (2). However, an AB-type quartet is seen in the spectrum of (2) with $J = 11.4$ Hz, at $\delta = 3.62$ and $\delta = 2.60$. Furthermore, (2) has a lower $pK_a$ value (7.17) than the 8.81 of (1). Write a possible structure for (2).

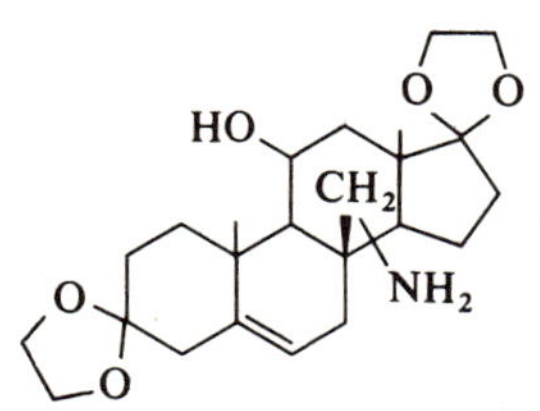


1

9. A compound $C_{13}H_{15}NOS$ results as product when bis (o-aminophenyl) disulfide (1) and cycloheptanone are reacted in the presence of air and an acid catalyst. Its spectral properties include

ir $3300 + 1700\ cm^{-1}$

nmr ($CDCl_3$) $\delta = 6.9$ (4H) multiplet

5.35 (1H) singlet

$\left.\begin{array}{l}2.65\ (2H)\\2.2\ (2H)\\1.6\ (6H)\end{array}\right\}$ multiplets

Also, Raney-nickel desulfurization of this product gives 2-anilinocycloheptanone.

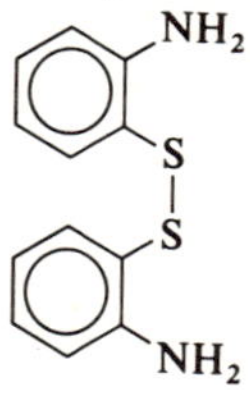

1

Deduce the structure of the product from the information given.

## 16.7 *SUGGESTED READINGS*

D. Y. Curtin, E. J. Grubbs, and C. G. McCarthy, "Uncatalyzed Syn-Anti Isomerization of Imines, Oxime Ethers, and Haloimines," *J. Amer. Chem. Soc.*, *88*, 2775 (1966).

M. Rabinovitz and A. Pines, "Hindered Internal Rotation and Dimerization of N,N-Dimethylformamide in Carbon Tetrachloride," *J. Amer. Chem. Soc.*, *91*, 1585 (1969).

H. S. Gutowsky, J. Jonas, and T. H. Siddall, "Comparison of Chemical Exchange Rates Determined by NMR Line-Shape and Equilibration methods. III. Internal Rotation of N-Methyl-N-Benzylformamide," *J. Amer. Chem. Soc.*, *89*, 4300 (1967).

M. Raban and F. B. Jones, Jr., "Stereochemistry at Trivalent Nitrogen. V. Origin of Sulfur-Nitrogen Torsional Barriers in N-Sulfenylsulfonamides," *J. Amer. Chem. Soc.*, *91*, 2180 (1969).

A. R. Katritzky, "Nuclear Magnetic Resonance: Some Applications to Organic Chemistry," in *The Techniques of NMR and EPR*, R. A. Y. Jones *et al.*, eds., United Trade Press, Ltd., London, 1965, pp. 8–11.

G. Maier, "Valence Tautomerism in Heterocyclic Systems," *Angew. Chem. Int. Ed.*, *6*, 812 (1967).

V. Bekarek, K. Rothschein, P. Vetesnik, and M. Vecera, "Estimation of Azo-Hydrazo Tautomeric Equilibrium in *Ortho*-hydroxyazo Compounds by NMR," *Tetrahedron Lett.*, 3711 (1968).

A. Rauk, L. C. Allen, and K. Mislow, "Pyramidal Inversion," *Angew. Chem. Int. Ed. 9*, 400 (1970).

J. M. Lehn, "Nitrogen Inversion," *Topics in Current Chemistry*, Springer-Verlag, Berlin, *15*, 311 (1970).

K. L. Henold, "Spin-Spin Coupling in Secondary Amines. The First Observation of Coupling by Amine Protons in Strongly Basic Amines," *Chem. Comm.*, 1340 (1970).

# CHAPTER 17 APPLICATIONS OF SPECTROSCOPIC TECHNIQUES TO BIOMOLECULES

What has been learned piecemeal in this book should now be applicable to the study of real molecules with a variety of interacting structural elements. The problems should already have helped the reader to familiarize himself with the kind of mental agility required to unravel these structures. A few of the applications to molecules of biological interest will now be presented. It has been often stated that the molecules of nature present the biggest challenge to the chemist. One of the reasons behind this statement is that the total is no longer necessarily equal to the sum of the parts. The particular folding of the native conformation of proteins is an example. Magnetic resonance techniques, in particular, have become very important to biophysicists and therefore we shall concentrate on this rapidly growing area (see the Appendix for the formulas of the amino acids mentioned in the text).

## 17.1 *Use of Chemical Shifts*

The conformations of some peptides and proteins have been investigated mainly through recourse to pmr. Even with the most powerful instruments (300 MHz at the time of writing), the resonances of the component amino acids do overlap extensively, except in a few " blank " regions of the spectrum. One of the possibilities for simplifying the spectrum is by specific deuteration of a number of amino acids. Carbon-13 spectroscopy is also extremely promising in this respect.

Another approach can be illustrated by the work of McDonald and Phillips[1] on proteins such as ribonuclease or lysosyme. The spectrum of ribonuclease is recorded at 220 MHz, in $D_2O$ in order to approach the natural conditions. The composition of ribonuclease in terms of its amino-acids components is well known. The spectra of the amino acids can also be recorded individually in $D_2O$ solution. When all these spectra are superimposed (this can best be achieved with a digital computer) a "synthetic" spectrum of ribonuclease which is the "sum of the parts" is obtained. It is very gratifying to find then that this synthetic spectrum is virtually identical to that of the protein which has been denatured by heating to 72.5°C. The spectrum of the native protein at 22°C, however, is different. By comparing it to that of the denatured protein, some of the changes become conspicuous (Figure 17.1). For instance, the denatured protein (denatured by heat, the action of trifluoroacetic acid, etc.) contains in its spectrum a sharp resonance at $\delta = 0.85$, due to the methyl protons of valine, leucine, and isoleucine; at lower field, between 6.8 and 7.8 ppm one observes the aromatic protons of the 4 histidine, 3 phenylalanine, and 6 tyrosine amino acids. In the folded native conformation, the 0.85-ppm resonance is replaced by at least four or five distinct resonances; in the 6.8–7.8 ppm spectral range new peaks appear which shows that the equivalencies of the 4 histidines, and so on, have been removed. Yet more remarkable is the example of lysozyme: The native conformation is characterized by the existence of peaks at extremely high fields (up to $\delta = -1$, i.e., above TMS) due to the alkyl residues of leucine, isoleucine, and valine which are being forced into

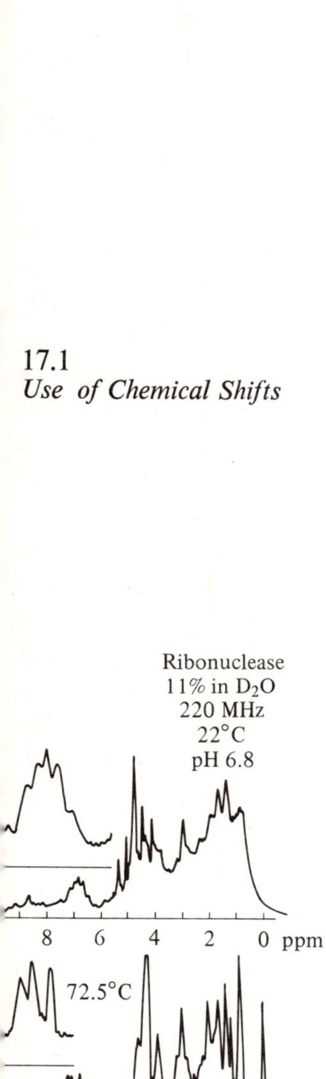


*FIGURE 17.1*

*The 220-MHz pmr spectra of ribonuclease in native (top) and denatured (bottom) conformations.*

[1] C. C. McDonald and W. D. Phillips, *J. Amer. Chem. Soc.*, *89*, 6332 (1967).

close contact with the aromatic rings of histidine, tyrosine, phenylalanine, and tryptophane, which are present in lysozyme. The magnetic anisotropy of these aromatic rings is responsible for these high-field shifts. Upon inspection of molecular models (based on the x-ray structural determinations of some of these macromolecules) it appears that such a very large shielding cannot be obtained with ribonuclease, which does not have the proper geometric arrangement of the $CH_3$ and $CH_2$ groups with respect to the aromatic rings. Also, one can count the number of protons involved from the integral curve and find that there are 128 protons in the 0.85-ppm resonance of denatured lysozyme, versus about 48 protons between $-0.7$ and $+0.7$ ppm in the native conformation of the protein. To account for these numbers, inspection of the model suggests that arginine-61 interacts with the adjacent tryptophane-62, methionine-105 with tyrosine-23, and so on. It can indeed be checked that the high-field resonances between $-0.3$ and 0 ppm are markedly affected by the interaction in solution of lysozyme with an inhibitor, such as N-acetyl glucosamine trimer. X-Ray studies show that this inhibitor predominantly affects the tryptophane residues No. 62, 63, and 108 in good accord with the nmr results.

Another example of the use of the chemical shifts due to the ring anisotropies of aromatic systems is provided by the elegant experiments of Chan and coworkers.[2] They have, for instance, studied the interaction of purine with the dinucleotides TpT, TpdU, and dUpT (where T = thymidine, and dU = 2′-deoxyuridine; p refers to the phosphate linkage). Upon addition of purine to solutions of these dinucleotides in water, the resonance positions of the protons in the dinucleotide are affected. For example, peaks for the thymine protons and for the $H'_1$ proton in TpT (see the Appendix for the formula of thymidine) are shifted towards higher fields. Of course, these shifts are diagnostic of the molecular interaction of the dinucleotide with the purine which has been added to the solution. Furthermore, they show that this interaction is very likely to be of the base-stacking type, the purine ring inserting itself between the two heterocyclic bases in the dinucleotide. As a check, the purine protons are also shifted upfield because of this interaction. Chan and coworkers could thus distinguish between base stacking (where the effects predicted are in general upfield shifts) and base pairing through hydrogen bonds (which correspond to downfield shifts due to the coplanarity of the magnetically anisotropic aromatic rings). Depending on the pH, the nature of the bases and of the dinucleotide, either base stacking or base pairing predominates.

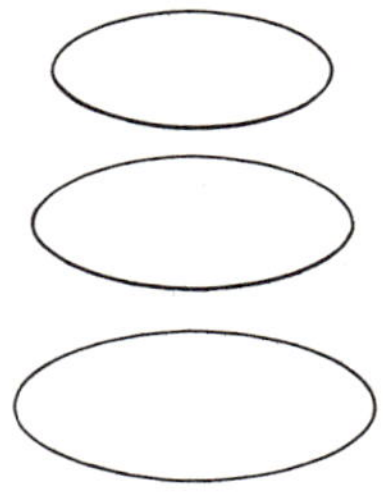

Base stacking

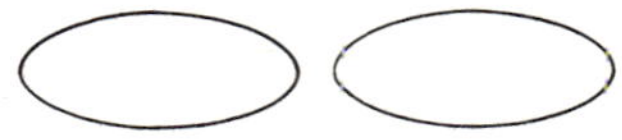

Base pairing

## 17.2 *Use of Dipolar Relaxation*

A large number of enzymatic reactions occur through binding of the enzyme (E) to the substrate (S) to form a complex (ES), which undergoes a conformational

[2] M. P. Schweizer, S. I. Chan, and P. O. P. Ts'O, *J. Amer. Chem. Soc.*, *87*, 5241 (1965); S. I. Chan, B. N. Bangerter, and M. M. Peter, *Proc. Nat. Acad. Sci. U.S.*, *55*, 720 (1966).

change (ES) → (ES)′ before going into products. The equilibrium constant $K_S$ is defined[3] by

$$K_S = \frac{[E][S]}{[ES]}$$

and it can be shown that ES = $S_{total}E_{total}/K_S + S_{total}$, as follows.

Given the equilibrium

$$E + S \rightleftharpoons ES$$

one can define the equilibrium constant $K_S$ = (E)(S)/(ES) where (E) = $E_{total}$ − ES and (S) = $S_{total}$ − ES
so that

$$K_S = \frac{(E_{total} - ES)(S_{total} - ES)}{(ES)}$$

If it is assumed that the concentration of the complex (ES) is small compared to that of the substrate

$$(ES) \ll S_{total}$$

then

$$K_S = \frac{(E_{total} - ES)S_{total}}{(ES)}$$

and

$$K_S(ES) = S_{total}E_{total} - (ES)S_{total}$$

$$K_S(ES) + S_{total}(ES) = S_{total}E_{total}$$

therefore

$$(ES) = \frac{S_{total}E_{total}}{K_S + S_{total}}$$

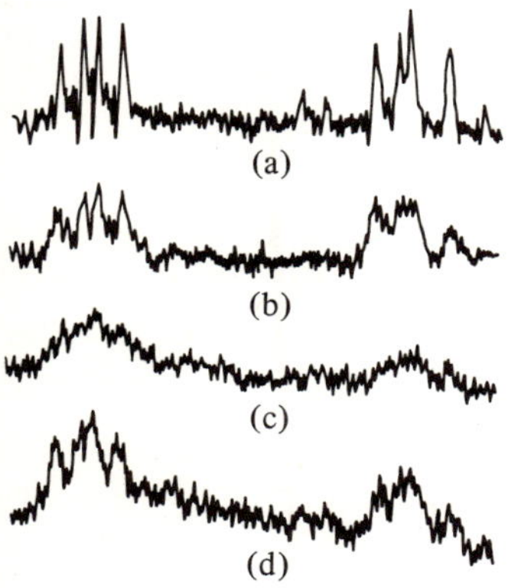

*FIGURE 17.2*
*Observed pmr spectra of the ABX region of (a) 39.8-mM* D-*tryptophan, (b) 49.5-mM* D-*tryptophan and 1.1-mM α-chymotrypsin, (c) 39.8-mM* D-*tryptophan and 2.6 mM α-chymotrypsin, and (d) 39.8-mM* D-*tryptophan and 3.3-mM DFP-chymotrypsin in 95 percent* $D_2O$*. The left side of each spectrum is at 0.50 ppm from internal water; the spectra are 100 Hz wide. The upfield wing of the AB resonance pattern is not shown.*

One example studied[4] is the reaction of the enzyme α-chymotrypsin with the substrate tryptophan. Appreciable line broadenings are observed with both the L and the D form of this amino acid in the presence of the enzyme (Figure 17.2). For instance, the following results are obtained with L-tryptophan.

| Conditions | Linewidth of the aromatic peaks |
|---|---|
| 46.0-m*M* Tryptophan | 1.4 Hz |
| 46.0-m*M* Tryptophan + 1.2-m*M* α-chymotrypsin | 2.5 Hz |
| 46.0-m*M* Tryptophan + 2.3-m*M* α-chymotrypsin | 4.0 Hz |

Since tryptophan can exist both in the free (S) and in the bound (ES) state the observed line broadening $\Delta\nu_{1/2}$ is related to the linewidth in the complex $\nu_{1/2}$ (ES) by

$$\Delta\nu_{1/2} = \frac{E_{total}}{S_{total} - K_S}\,\nu_{1/2}(ES)$$

[3] *Inter alia* T. C. Bruice and S. J. Benkovic, *Bioorganic Mechanisms*, W. A. Benjamin, Inc., New York, 1960, vol. 1, p. 212.
[4] J. T. Gerig, *J. Amer. Chem. Soc.*, *90*, 2681 (1968).

This relationship is obtained by expressing the observed linewidth $\nu_{1/2}$ as the weighted average of $\nu_{1/2}$ (free) and $\nu_{1/2}$ (bound), which we shall denote as $\nu_{1/2}^{S}$ and $\nu_{1/2}^{ES}$ respectively:

$$\nu_{1/2} = \frac{(ES)}{S_{total}} \nu_{1/2}^{ES} + \frac{S_{total} - ES}{S_{total}} \nu_{1/2}^{S}$$

Making the same assumption that (ES) $\ll S_{total}$, and making use of the expression derived above for the equilibrium constant $K_S$, it follows that

$$\Delta\nu_{1/2} = \nu_{1/2} - \nu_{1/2}^{S} = \frac{E_{total}}{K_S + S_{total}} \nu_{1/2}^{ES}$$

A plot of the observed linewidth $\nu_{1/2}$ against $(E_{total})(S_{total} - K_S)$ for several assumed values of $K_S$ in the range 3–12 m*M* yields the following linewidths for the tryptophan α-chymotrypsin complex:

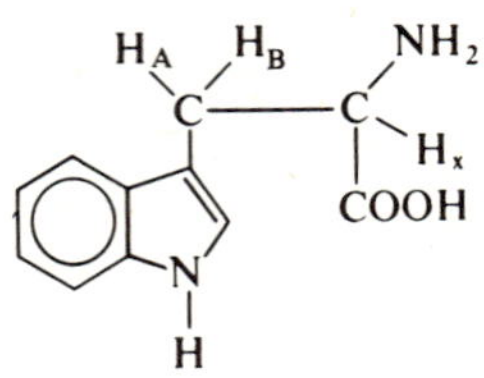


$\nu_{1/2}$ in the (ES) Complex (Hz)

| Proton type (see structure) | L-Tryptophan | D-Tryptophan |
|---|---|---|
| A, B | 60 | 85 |
| X | 20 | 33 |
| ArH | 28 | 43 |

Inhibition of the enzyme by phosphorylation of the serine-195 residue with diisopropylfluorophosphate diminishes considerably the observed binding and the attendant nmr resonance line broadening. Similarly, line broadening disappears upon alkylation of the methionine – 192 residue, in the region of the active site. One can use these line broadenings to estimate the correlation time $\tau_c$ for the tryptophan molecule. The relaxation time due to dipolar couplings between protons is given by

$$\frac{1}{T_1} \sim \frac{3}{2} \gamma^4 h^2 \sum_j \frac{1}{\langle r_{ij} \rangle^6} \tau_c$$

(see Chapter 5), where $1/r_{ij}$ is the average distance between nuclei *i* and *j*, which can be estimated from consideration of molecular models. For instance, the term $1/\langle r_{ij} \rangle^6 \times 10^2$ is equal to ~ 0.0080 for proton B. One thus can derive the following values for the correlation time:

$\tau_c = 3.6 \times 10^{-8}$ sec D-tryptophan

$\tau_c = 2.3 \times 10^{-8}$ sec L-tryptophan

This is in very nice agreement with the idea of a strong complex to α-chymotrypsin. This macromolecule has molecular dimensions of the order of 45 × 35 × 38 Å, corresponding to a correlation time $\tau_c = 4 \times 10^{-8}$ sec if Stokes behavior were strictly followed by it. Otherwise much shorter correlation times would be characteristic of the substrate because of its much smaller molecular volume.

## 17.3 *Use of Quadrupolar Relaxation*

Rather sharp lines ($\nu_{1/2} \sim 10$ Hz) are seen for the $^{35}Cl$ resonance of the spherically solvated chloride ion $Cl^-$ when sodium chloride is dissolved into water. Since the field gradient $q$ at the quadrupolar chlorine nucleus is zero, the $e^2qQ$ term is also zero, and quadrupolar relaxation is powerless to broaden the magnetic resonance of the $^{35}Cl$. However, when this $^{35}Cl$ nucleus is engaged in covalent bonding, the quadrupolar coupling constants $e^2qQ$ are then observed to be in the range of 40–90 MHz. The bandwidths are comparably large: for carbon tetrachloride $\nu_{1/2} = 14.5$ KHz, corresponding to a correlation time $\tau_c \sim 1.7 \times 10^{-12}$ sec. The idea behind the experiments to be presented now[5] is thus very simple: The range of values between $\sim 10$ and 14,500 Hz is enormous, and can be exploited if fast exchange between two such sites produces a $^{35}Cl$ resonance with a bandwidth which will be the weighted average of these two limiting numbers. Experiments have been devised, for instance, using mercurichloride —Hg—Cl) labeling of proteins. When sodium chloride NaCl is reacted with mercuric chloride $HgCl_2$, chlorine is present either as the chloride $Cl^-$ or engaged in a covalent molecule

$$3NaCl + HgCl_2 \rightarrow Cl^- + HgCl_4^= + 3Na^+$$

Many macromolecules of biological interest have thiol (S—H) groupings which react with $HgCl_2$ to give —S—Hg—Cl. For instance, the $^{35}Cl$ resonance of 0.5-*M* NaCl in $H_2O$ has a width $\nu_{1/2} = 16$ Hz, which is only increased to 22·Hz—same order of magnitude—upon addition of $1 \times 10^{-5}$-*M* hemoglobin. When $14 \times 10^{-5}$-*M* $HgCl_2$ is added to the solution the linewidth increases to 210 Hz. Therefore, as a rough estimate, the correlation time $\tau_c$ for the mercuric site on the hemoglobin is ca. 100 times that for $HgCl_4^=$:

$$\nu_{1/2} = K\tau_c \sim 210 \quad \text{hemoglobin}$$

$$\nu_{1/2} = K\tau_c' \sim 14{,}500 \quad HgCl_4^=$$

assuming equal electrical field gradients $q$ in the free $HgCl_4^=$ and in the S—Hg—Cl group. In another application of this chloride-probe technique, the helix-coil transition of polybenzyl-L-glutamate labeled at the terminal amide position was followed

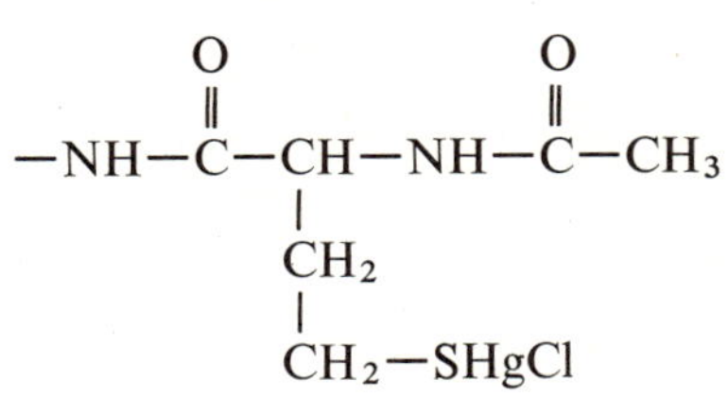

as a function of the pH. The chlorine linewidth drops from $\nu_{1/2} = 90$ Hz at low

[5] T. R. Stengle and J. D. Baldeschwieler, *Proc. Nat. Acad. Sci. U. S.*, *55*, 1020 (1966); R. P. Haugland, L. Stryker, T. R. Stengle, and J. D. Baldeschwielder, *Biochemistry*, *6*, 498 (1967); R. G. Bryant, *J. Amer. Chem. Soc.*, *89*, 2496 (1967); T. R. Stengle and J. D. Baldeschwieler, *J. Amer. Chem. Soc.*, *89*, 3045 (1967).

pH to $\nu_{1/2} = 75$ Hz at pH above 5.2, corresponding to the change from the more compact helix to the more loosely arranged random coil, which has a larger molecular volume and therefore a decreased correlation time (by roughly 15–20 percent).

## 17.4 *The McConnell Spin-Labeling Method*[6]

This method consists of reacting a nitroxide free radical

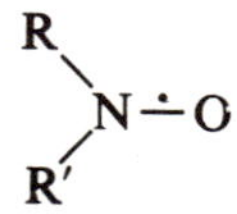


with a protein, to which it becomes attached. After the unreacted nitroxide radicals have been eliminated by dialysis the resultant esr spectrum is analyzed and yields information about the conformation of the protein. For instance, the following spin label has been prepared:

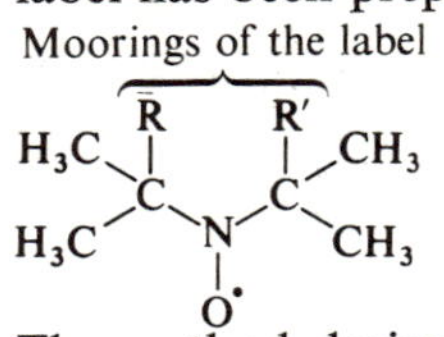


The method derives its elegance from the high sensitivity of esr, and from the fact that one looks at the spectrum of the unpaired spins and not at the intricate spectra for the nuclei of the protein itself. The esr spectrum of a nitroxide radical is a triplet and the magnitude of the hyperfine coupling indicates that the unpaired electron resides predominantly on the nitrogen. The appearance of the spectrum depends upon the bandwidths, which vary depending upon the experimental conditions. For the nitroxide amide the linewidth increases considerably in going from an aqueous

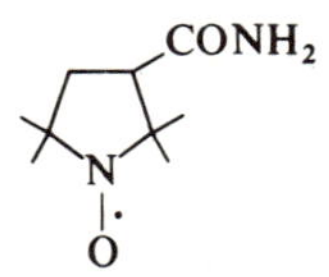


solution to a solution of 60 percent saccharose in water, and to a glycerol solution at −196°C. This increase with increasing viscosity is, of course, explained by the concomitant increase in the correlation time for molecular motion, which enters the expression for the relaxation time as indicated earlier (Chapter 5). When this particular spin label is attached to a protein, two cases can be experimentally observed. In one case the label has conformational mobility with respect to the protein and its spectrum is a regular three-line spectrum slightly broadened with respect to the esr of nitroxides in aqueous solution. In the other case the label is strongly immobilized in the particular conformation adopted by the protein and a broadened three-line spectrum similar to that obtained in glycerol at low temperature is observed.

It has been verified in a certain number of cases that the labeled protein has crystals isomorphous with those of the natural protein; that is, upon labeling the protein maintains its natural folding. As an example the isocyanate spin label,

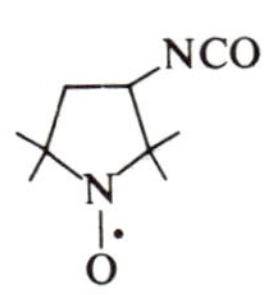


[6] C. L. Hamilton and H. M. McConnell, "Spin Labels," in *Structural Chemistry and Molecular Biology*, A. Rich and N. Davidson, eds., W. H. Freeman and Co., San Francisco, Calif., 1968, p. 115.

is attached to poly-L-lysine, 1.2 molecules of the free radical combining with 1 molecule of the polymer (M.W. = 50,000), presumably as

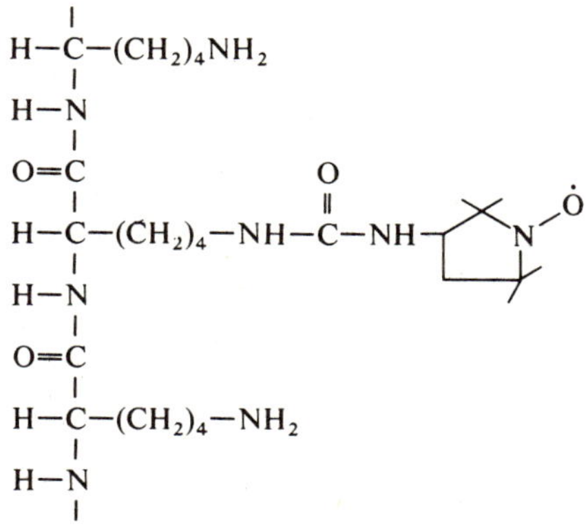

At pH = 8.0, sharp lines are observed while at pH = 11.0, broad lines are obtained, pointing to the existence of a helical structure at the higher pH value. Similarly, when the same isocyanate spin label is reacted with bovine serum albumin (one or two labels being attached on each of the macromolecules) two sites are found, one with a mobile spin label and the other with a strongly immobilized spin label. From the relative amounts of these distinct spectra, the percentage of mobile label can be measured as a function of the pH. It goes through a maximum at about pH 2, and the changes between pH 2 and 4 are reversible and parallel to those displayed by the viscosity or by the optical rotation. This observation suggests that at these pHs the protein undergoes a conformational change which liberates the rotation of the hitherto immobilized spin label. Many other spin labels have been devised and used to reveal information about other active sites of complex proteins.

## 17.5 *Structure of Some Amino Acids and Other Biomolecules*

| | |
|---|---|
| Leucine | $(CH_3)_2CHCH_2CH(NH_2)COOH$ |
| Isoleucine | $CH_3CH_2CH(CH_3)CH(NH_2)COOH$ |
| Valine | $(CH_3)_2CHCH(NH_2)COOH$ |
| Histidine | 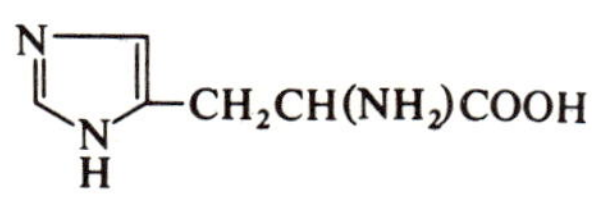 |
| Tyrosine | $p$-HO-$C_6H_4$-$CH_2CH(NH_2)COOH$ |
| Phenylalanine | $C_6H_5CH_2CH(NH_2)COOH$ |
| Tryptophan | 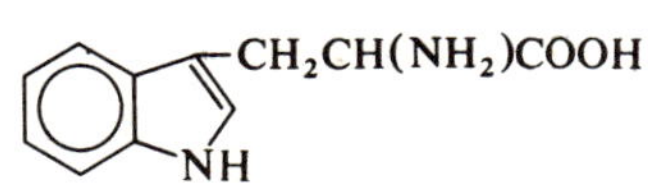 |
| Purine | 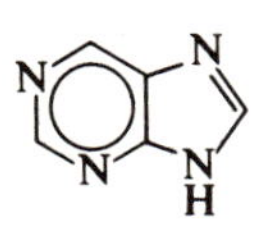 |
| Thymidine | 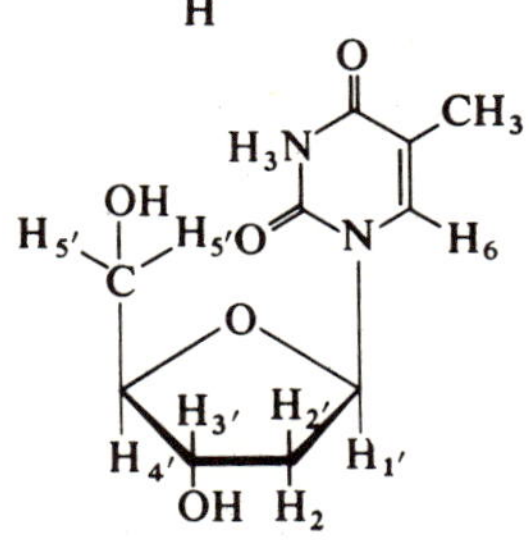 |

## 17.6 *SUGGESTED READINGS*


C. C. McDonald and W. D. Phillips, "Manifestations of the Tertiary Structure of Proteins in High-Frequency Nuclear Magnetic Resonance," *J. Amer. Chem. Soc.*, *89*, 6332 (1967).

M. P. Schweizer, S. I. Chan, and P. O. P. Ts'o, "Interaction and Association of Bases and Nucleosides in Aqueous Solutions. IV. Proton Magnetic Resonance Studies of the Association of Pyrimidine Nucleosides and Their Interactions with Purine," *J. Amer. Chem. Soc.*, *87*, 5241 (1965).

S. I. Chan, B. W. Bangerter, and H. H. Peter, "Purine Binding to Dinucleotides: Evidence for Base Stacking and Insertion," *Proc. Nat. Acad. Sci. U. S.*, *55*, 720 (1966).

J. T. Gerig, "Nuclear Magnetic Resonance Studies of the Interaction of Tryptophan with α-Chymotrypsin," *J. Amer. Chem. Soc.*, *90*, 2681 (1968).

T. R. Stengle and J. D. Baldeschwieler, "Halide Ions as Chemical Probes for NMR Studies of Proteins," *Proc. Nat. Acad. Sci. U. S.*, *55*, 1020 (1966).

T. R. Stengle and J. D. Baldeschwieler, "Halide Ions as Probes for Nuclear Magnetic Resonance Studies of Proteins. The Sulfhydryl Groups of Hemoglobin," *J. Amer. Chem. Soc.*, *89*, 3045 (1967).

R. P. Haughland, L. Stryer, T. R. Stengle, and J. D. Baldeschwieler, "Nuclear Magnetic Resonance Studies of Antibody-Hapten Interactions Using a Chloride Ion Probe," *Biochemistry*, *6*, 498 (1967).

H. L. Crespi, R. M. Rosenberg, and J. J. Katz, "Proton Magnetic Resonance of Proteins Fully Deuterated except for $^1$H-Leucine Side Chains," *Science*, *161*, 795 (1968).

C. L. Hamilton and H. M. McConnell, "Spin Labels," in *Structural Chemistry and Molecular Biology*, A. Rich and N. Davidson, eds., W. H. Freeman and Co., San Francisco, Calif., 1968, p. 115.

A. S. V. Burgen and J. C. Metcalfe, "The Application of Nuclear Magnetic Resonance to Pharmacological Problems," *J. Pharm. Pharmac.*, *22*, 153 (1970).

# IV

# REFERENCES
# ANSWERS TO PROBLEMS
# INDEXES

# REFERENCES


*RESONANCE SPECTROSCOPIES*

B. I. IONIN AND B. A. ERSHOV, *NMR Spectroscopy in Organic Chemistry*, Plenum Press, New York, 1970, 2nd ed.

E. F. MOONEY, ed., *Annual Reports on NMR Spectroscopy*, Academic Press, Inc., New York, 1968, vol. 1, 1969, vol. 2, and 1970, vol. 3.

R. M. LYNDEN-BELL AND R. K. HARRIS, *Nuclear Magnetic Resonance Spectroscopy*, Thomas Nelson and Sons Ltd., London, 1969.

D. G. E. INGRAM, *Biological and Biochemical Applications of Electron Spin Resonance*, Plenum Publishing Corp., New York, 1969.

I. P. BIRYUKOV, M. G. VORONKOV, AND I. A. SAFIN, *Tables of Nuclear Quadrupole Resonance Frequencies*, Israel Program for Scientific Translations, Jerusalem, 1969.

P. DIEHL, E. FLUCK, AND R. KOSFELD, eds., *NMR Basic Principles and Progress*, Springer-Verlag, Berlin, 1969, vol. 1, and 1970, vol. 2.

A. ABRAGAM, *The Principles of Nuclear Magnetism*, Oxford University Press, 1961.

J. S. WAUGH, ed., *Advances In Magnetic Resonance*, 1967, vol. 1, 1968, vol. 2, 1969, vol. 3, and 1970, vol. 4.

J. W. EMSLEY, J. FEENEY, AND L. H. SUTCLIFFE, eds., *Progress in Nuclear Magnetic Resonance Spectroscopy*, 1965, vol. 1, 1966, vol. 2, 1967, vol. 3, 1968, vol. 4, and 1969, vol. 5.

F. GERSON, *High Resolution Electron Spin Resonance Spectroscopy*, Wiley-Interscience, New York, 1970.

E. F. MOONEY, *Introduction to Fluorine 19 Nuclear Magnetic Resonance Spectroscopy*, Heyden and Sons Ltd., London, 1970.

L. M. JACKMAN AND S. STERNHELL, *Applications of Nuclear Magnetic Resonance Spectroscopy in Organic Chemistry*, Pergamon Press, Oxford, 1969, 2nd ed.

J. A. POPLE, W. G. SCHNEIDER, AND H. J. BERNSTEIN, *High-Resolution Nuclear Magnetic Resonance*, McGraw-Hill Book Co., New York, 1959.

J. W. EMSLEY, J. FEENEY, AND L. H. SUTCLIFFE, *High-Resolution Nuclear Magnetic Resonance Spectroscopy*, Pergamon Press, Oxford, 1965, vol. 1, and 1966, vol. 2.

L. M. JACKMAN, *Applications of Nuclear Magnetic Resonance Spectroscopy in Organic Chemistry*, Pergamon Press, London, 1959.

J. D. ROBERTS, *Nuclear Magnetic Resonance: Applications to Organic Chemistry*, McGraw-Hill Book Co., New York, 1959.

J. D. ROBERTS, *An Introduction to the Analysis of Spin-Spin Splitting in High-Resolution Nuclear Magnetic Resonance Spectra*, W. A. Benjamin, Inc., New York, 1962.

E. D. BECKER, *High Resolution NMR*, Academic Press, New York, 1969.

F. A. BOVEY, *Nuclear Magnetic Resonance Spectroscopy*, Academic Press, New York, 1969.

A. CARRINGTON AND A. D. MCLACHLAN, *Introduction to Magnetic Resonance*, Harper & Row, New York, 1967.

H. G. HECHT, *Magnetic Resonance Spectroscopy*, John Wiley and Sons, Inc., New York, 1967.

M. A. BERSOHN AND J. C. BAIRD, *An Introduction to Electron Paramagnetic Resonance*, W. A. Benjamin, Inc., New York, 1966.

P. B. AYSCOUGH, *Electron Spin Resonance in Chemistry*, Methuen, London, 1967.

H. M. ASSENHEIM, *Introduction to Electron Spin Resonance*, Plenum Press, New York, 1967.

R. S. ALGER, *Electron Paramagnetic Resonance: Techniques and Applications*, Wiley-Interscience Publishers, Inc., New York, 1968.

E. A. C. LUCKEN, *Nuclear Quadrupolar Coupling Constants*, Academic Press, New York, 1969.

D. CHAPMAN AND P. D. MAGNUS, *Introduction to Practical High Resolution Nuclear Magnetic Resonance Spectroscopy*, Academic Press, New York, 1966.

*MOLECULAR ORBITAL THEORY*

O. SINANOGLU AND K. B. WIBERG, *Sigma Molecular Orbital Theory*, Yale University Press, New Haven, 1970.

J. A. POPLE AND D. L. BEVERIDGE, *Approximate Molecular Orbital Theory*, McGraw-Hill Book Company, New York, 1970.

R. ZAHRADNÍK AND J. PANCÍŘ, *HMO Energy Characteristics*, Plenum Publishing Corporation, New York, 1970.

J. D. ROBERTS, *Notes on Molecular Orbital Calculations*, W. A. Benjamin, Inc., New York, 1961.

A. STREITWIESER, JR., *Molecular Orbital Theory for Organic Chemists*, John Wiley and Sons, Inc., New York, 1961.

K. HIGASI, H. BABA, AND A. REMBAUM, *Quantum Organic Chemistry*, Wiley-Interscience Publishers, Inc., New York, 1965.

L. SALEM, *The Molecular Orbital Theory of Conjugated Systems*, W. A. Benjamin, Inc., New York, 1966.

R. L. FLURRY, JR., *Molecular Orbital Theories of Bonding in Organic Molecules*, Marcel Dekker, Inc., New York, 1968.

J. D. MEMORY, *Quantum Theory of Magnetic Resonance Parameters*, McGraw-Hill Book Co., New York, 1968.

M. J. S. DEWAR, *The Molecular Orbital Theory of Organic Chemistry*, McGraw-Hill Book Co., New York, 1969.

*MASS SPECTROMETRY*

J. ROBOZ, *Introduction to Mass Spectrometry: Instrumentation and Techniques*, John Wiley and Sons, Inc., New York, 1968.

E. STENHAGEN, S. A. ABRAHAMSON, AND F. W. MCLAFFERTY, Eds., *Atlas of Mass Spectral Data*, John Wiley and Sons, Inc., New York, 1969, 3 vols.

R. I. REED, ed., *Aspects of Mass Spectrometry*, Plenum Publishing Corporation, New York, 1968.

P. F. KNEWSTUBB, *Mass Spectrometry and Ion-Molecule Reactions*, Cambridge University Press, Cambridge, 1969.

D. PRICE AND J. E. WILLIAMS, *Dynamic Mass Spectrometry*, Heyden-Sadtler, London, New York, 1970.

A. L. BURLINGAME, ed., *Topics in Organic Mass Spectrometry*, Wiley-Interscience, New York, 1970.

J. H. BEYNON, *Mass Spectrometry and Its Application to Organic Chemistry*, Elsevier Publishing Co., Amsterdam, 1960.

K. BIEMANN, *Mass Spectrometry, Organic Chemical Applications*, McGraw-Hill Book Co., New York, 1962.

C. A. MCDOWELL, *Mass Spectrometry*, McGraw-Hill Book Co., New York, 1963.

F. W. MCLAFFERTY, ed., *Mass Spectrometry of Organic Ions*, Academic Press, New York, 1963.

R. W. KISER, *Introduction to Mass Spectrometry and Its Applications*, Prentice-Hall, Inc., Englewood Cliffs, N.J., 1965.

F. W. MCLAFFERTY, *Interpretation of Mass Spectra*, W. A. Benjamin, Inc., New York, 1966.

J. H. BEYNON, R. A. SAUNDERS, AND A. E. WILLIAMS, *The Mass Spectra of Organic Molecules*, American Elsevier Publishing Co., Inc., New York, 1968.

*OPTICAL SPECTROSCOPIES*

G. SNATZKE, ed., *Optical Rotatory Dispersion and Circular Dichroism in Organic Chemistry*, Heyden-Sadtler, London, New York, 1967.

E. J. LOADER, *Basic Laser Raman Spectroscopy*, Heyden-Sadtler, London, New York, 1970.

J. H. VAN DER MAAS, *Basic Infrared Spectroscyy*, Heyden, London, 1969.

H. A. SZYMANSKI, *Correlation of Infrared and Raman Spectra of Organic Compounds*, Hertillon Press, 1969.

H. A. SZYMANSKI, ed., *Raman Spectroscopy: Theory and Practice*, Plenum Publishing Corporation, New York, 1961, vol. 1, and 1970, vol. 2.

A. FINCH, P. N. GATES, K. RADCLIFFE, F. N. DICKSON, AND F. F. BENTLEY, *Chemical Applications of Far Infrared Spectroscopy*, Academic Press, New York, 1970.

E. S. STERN AND C. J. TIMMONS, eds., *Gillam and Stern's Introduction to Electronic Absorption Spectroscopy in Organic Chemistry, Revised*, Edward Arnold, London, 1970, 3rd ed.

N. L. ALPERT, W. E. KEISER, AND H. A. SZYMANSKI, *Theory and Practice of Infrared Spectroscopy*, Plenum Publishing Corporation, New York, 1970, 2nd ed.

E. I. STEARNS, *The Practice of Absorption Spectrophotometry*, John Wiley and Sons, Inc., New York, 1969.

L. LANG, *Absorption Spectra in the Ultraviolet and Visible Region*, Academic Press, New York, 1961.

H. H. JAFFÉ AND M. ORCHIN, *Theory and Applications of Ultraviolet Spectroscopy*, John Wiley and Sons, Inc., New York, 1962.

R. P. BAUMAN, *Absorption Spectroscopy*, John Wiley and Sons, Inc., New York, 1962.

C. N. R. RAO, *Ultraviolet and Visible Spectroscopy*, Butterworths, London, 1961.

J. N. MURRELL, *The Theory of the Electronic Spectra of Organic Molecules*, Methuen, London, 1963.

A. I. SCOTT, *Interpretation of the Ultraviolet Spectra of Natural Products*, Pergamon Press, Oxford, 1964.

R. S. BECKER, *Theory and Interpretation of Fluorescence and Phosphorescence*, Wiley-Interscience Publishers, Inc., New York, 1969.

G. HERZBERG, *Infrared and Raman Spectra of Polyatomic Molecules*, Van Nostrand Reinhold Co., New York, 1945.

L. J. BELLAMY, *The Infrared Spectra of Complex Organic Molecules*, John Wiley and Sons, Inc., New York, 1958, 2nd ed.

A. D. CROSS, *Introduction to Practical Infrared Spectroscopy*, Butterworths, London, 1960.

W. BRÜGEL, *An Introduction to Infrared Spectroscopy*, Methuen, London, 1962.

K. NAKANISHI, *Infrared Absorption Spectroscopy: Practical*, Holden-Day, Inc., San Francisco, Calif., 1962.

C. N. R. RAO, *Chemical Applications of Infrared Spectroscopy*, Academic Press, New York, 1963.

R. T. CONELEY, *Infrared Spectroscopy*, Allyn and Bacon, Inc., Boston, 1966.

F. F. BENTLEY, L. D. SMITHSON, AND A. L. ROZEK, *Infrared Spectra and Characteristic Frequencies*, Wiley-Interscience Publishers, Inc., 1968.

M. ST. C. FLETT, *Characteristic Frequencies of Chemical Groups in the Infrared*, American Elsevier Publishing Co., New York, 1969.

C. DJERASSI, *Optical Rotatory Dispersion: Application to Organic Chemistry*, McGraw-Hill Book Co., New York, 1960.

L. VELLUZ, M. LEGRAND, AND M. GROSJEAN, *Optical Circular Dichroism: Principles, Measurements and Applications*, Verlag Chemie, Weinheim, 1965.

P. CRABBE, *Optical Rotatory Dispersion and Circular Dichroism in Organic Chemistry*, Holden-Day, Inc., San Francisco, Calif., 1965.

C. E. MEELOAN, *Elementary IR Spectroscopy*, Macmillan, New York, 1963.

*GENERAL*

D. R. BROWNING, ed., *Spectroscopy*, McGraw-Hill Publishing Co., Maidenhead, England, 1970.

E. G. BRAME, JR., ed., *Applied Spectroscopy Reviews*, Marcel Dekker, Inc., New York, 1969, vol. 2.

E. F. H. BRITTAIN, W. O. GEORGE, AND C. H. J. WELLS, *Introduction to Molecular Spectroscopy Theory and Experiments*, Academic Press, New York, 1970.

B. TROST, *Problems in Spectroscopy*, W. A. Benjamin, Inc., New York, 1967.

A. AULT, *Problems in Organic Structure Determination*, McGraw-Hill Book Co., New York, 1967.

D. H. WHIFFEN, *Spectroscopy*, John Wiley and Sons, Inc., New York, 1966.

K. MISLOW, *Introduction to Stereochemistry*, W. A. Benjamin, Inc., New York, 1965.

T. R. DYER, *Applications of Absorption Spectroscopy of Organic Compounds*, Prentice-Hall, Inc., Englewood Cliffs, N.J., 1965.

J. C. P. SCHWARTZ, ed., *Physical Methods in Organic Chemistry*, Oliver and Boyd, Edinburgh, 1965.

J. C. D. BRAND AND G. EGLINTON, *Applications of Spectroscopy to Organic Chemistry*, Oldbourne Press, London, 1965.

S. K. FREEMAN, ed., *Interpretative Spectroscopy*, Van Nostrand Reinhold, New York, 1965.

R. M. SILVERSTEIN AND G. C. BASSLER, *Spectrometric Identification of Organic Compounds*, John Wiley and Sons, Inc., New York, 1967, 2nd ed.

P. YATES, *Structure Determination: Commentaries on the Determination of the Structures of Some Natural Products*, W. A. Benjamin, Inc., New York, 1967.

D. PASTO AND C. JOHNSON, *Organic Structure Determination*, Prentice-Hall, Inc., Englewood Cliffs, N. J., 1969.

P. J. WHEATLEY, *Molecular Structure*, Oxford University Press, Oxford, 1968, 2nd ed.

T. CAIRNS, ed., *Spectroscopy in Education*, Heyden and Son, London. T. Cairns, *Spectroscopic Problems in Organic Chemistry*, 1964, vol. 1; A. J. Baker and T. Cairns, *Spectroscopic Techniques in Organic Chemistry*, 1965, vol. 2; H. C. Hill, *Introduction to Mass Spectrometry*, 1966, vol. 3; A. J. Baker, G. Eglinton, F. J. Preston, and T. Cairns, *More Spectroscopic Problems in Organic Chemistry*, 1967, vol. 4.

J. P. PHILIP, *Spectra-Structure Correlation*, Academic Press, New York, 1964.

F. SCHEINMANN, ed., *An Introduction to Spectroscopic Methods for the Identification of Organic Compounds*, 1970, vol. 1, *Nuclear Magnetic Resonance and Infrared Spectroscopy*, Pergamon Press, Oxford.

# ANSWERS TO PROBLEMS

*CHAPTER 1*

1. Carbon-13 has six protons and seven neutrons, one short of a closed shell. The protons contribute zero, the missing neutron counts for $\frac{1}{2}$ ($1p_{1/2}$) and the total spin is $\frac{1}{2}$. As for deuterium, the spin number is 1, as the simple composition of the spin numbers for one proton and one neutron. Nitrogen-14 can be thought of in a similar manner, by reference to the closed shell of eight protons and eight neutrons, since both the seventh proton and the seventh neutron go into the $1p_{1/2}$ shell.
2. Spin-lattice relaxation is synonymous with thermal agitation. It is all the more effective as intermolecular fluctuations increase in frequency.
3. $E(\nu_0) = C/(\Delta\nu)^2$, where $C$ is a proportionality constant. $E_{1/2} = C/2(\Delta\nu)^2$ is the ordinate at half-height. Setting $(\nu - \nu_0)^2 + (\Delta\nu)^2 = 2(\Delta\nu)^2$

   $$|\nu - \nu_0| = |\Delta\nu|$$

   the width at half-height $\nu_{1,2}$ is twice $\Delta\nu$:

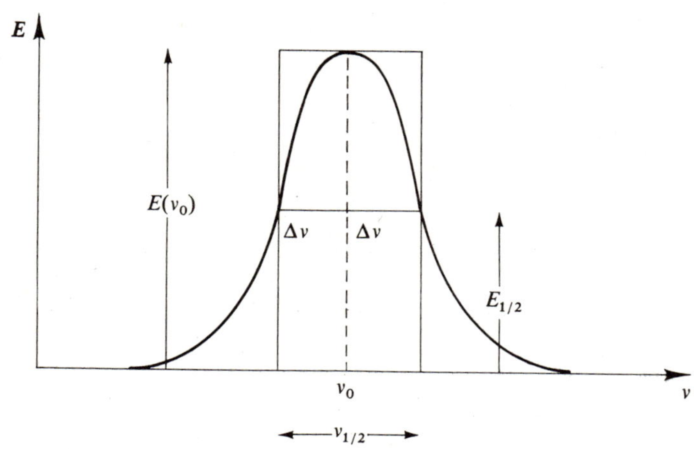


*CHAPTER 2*

1. Since $h\nu = eqQ/2$ and $Q$ is negative, the electrostatic field gradient $q$ at the chlorine nucleus is greater for isopropyl chloride than for cyclopropyl chloride. Hence the C—Cl bond is more polar in the former compound than in the latter, indicating more *s* character for the cyclopropyl carbon.
2. The results are due to an increase in *s* character in going from ethyl chloride to vinyl chloride and to chloroacetylene, the C—Cl bond becoming less polar (with more double bond character). [R. Bersohn, *J. Chem. Phys.*, *22*, 2078 (1954).]
3. The major effect is due to inductive effects decreasing with distance. The mesomeric effect has been found to be unimportant in $^{35}$Cl for benzene derivatives. [D. Biedenkapp and A. Weiss, *J. Chem. Phys.*, *49*, 3933 (1968).]
4. Since $CO_2$ is a linear molecule it has $3n - 5$ or (3)(3) − 5 = 4 normal modes. Two of these are stretching modes and two bending as follows:

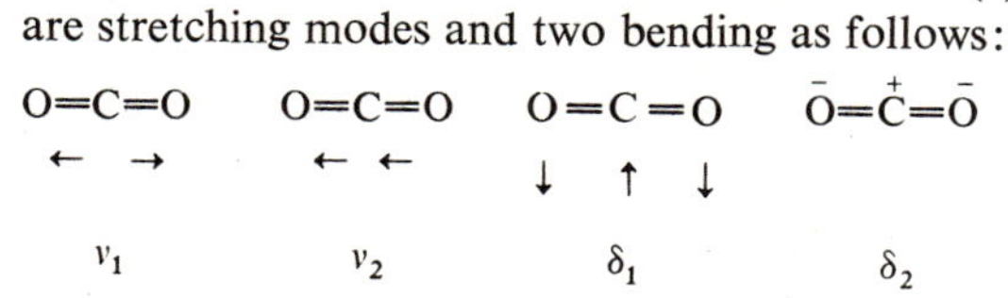


   Of these $\nu_1$ is ir inactive and Raman active; the remaining three are ir active and Raman inactive.
5. Both acetylene and dimethylacetylene are symmetric molecules, so there is no change in dipole moment upon bond stretching and hence no ir absorption.
6. The triplet state is lower in energy, since in the triplet state the two electrons are in different orbitals hence further apart, decreasing the unfavorable electron-electron repulsion energy.

CHAPTER 3

1. The correct structure is represented by (2). Structure (1) includes a cyclopropane ring which should have a $^1J_{C-H}$ value of the order of 160 Hz instead of the observed 152 Hz.

2.

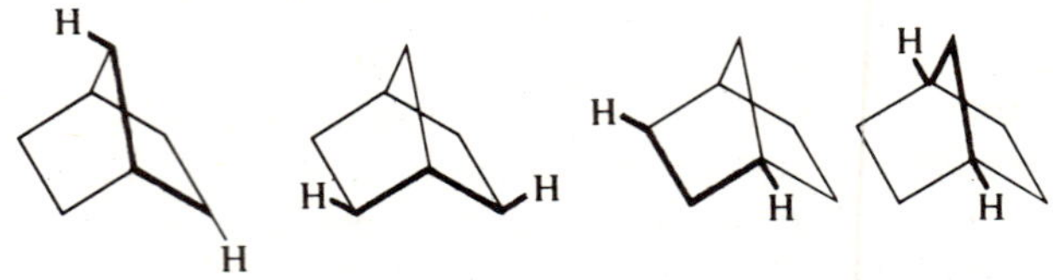

3. In the *trans* isomer there are three protons suitably disposed for long-range coupling interaction whereas in the *cis* isomer there is only one proton in the correct orientation according to the "M" or "W" rule. [K. L. Williamson, T. Howell, and T. A. Spencer, *J. Amer. Chem. Soc.*, *88*, 325 (1966).]

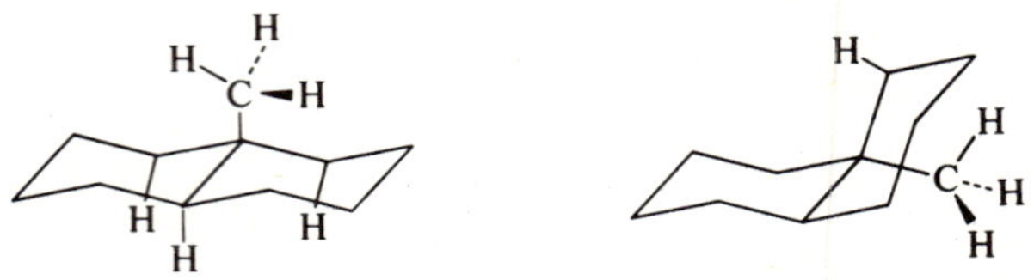

4. The six-membered ring dearomatizes to avoid making sulfur tetravalent, or having a dipolar resonance form. Hence it can be represented with the predominant resonance form:

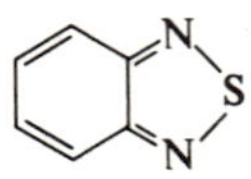

[N. M. D. Brown and P. Bladon, *Spectrochim. Acta*, *24A*, 1896 (1968); N. M. D. Brown, D. G. Lister, and J. K. Tyler, *Spectrochim. Acta*, *26A*, 2133 (1970).]

5. Three lines of equal intensity due to coupling with $^{14}N$ ($I = 1$).
6. The anthraceneonium ion shown has been generated, and it corresponds to $sp^3$ hybridization at $C_9$ and $sp^2$ hybridization at $C_{10}$, and $sp^2$ C—H bonds result in larger $^1J_{C-H}$ coupling.

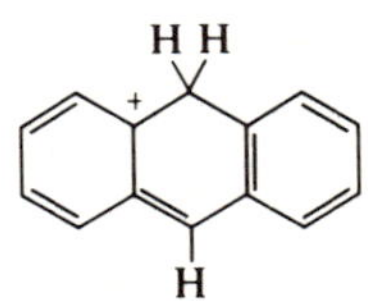

$Al_2Cl_7$ (and other resonance forms)

[V. A. Koptyug, I. S. Isaen, and A. I. Rezvukhin, *Tetrahedron Lett.*, 823 (1967).]

CHAPTER 4

1. Compounds (1) have $4n + 2 = 14$ $\pi$ electrons and are aromatic, whereas the molecules (2) are antiaromatic $4n$ systems with $n = 4$. Note the magnitude of the paramagnetic and of the diamagnetic shifts. Also note that they do *not* mirror each other, a feature in line with theoretical predictions. [R. H. Mitchell, C. E. Klopfenstein, and V. Boekelheide, *J. Amer. Chem. Soc.*, *91*, 4931 (1969).]

2. $\Delta\sigma_{\text{total}} = \Delta\sigma_{\text{anisotropy}} + \Delta\sigma_{\text{electric field}} + \Delta\sigma_{\text{steric compression}}$

(a) Anisotropy

$$\Delta\sigma = \frac{1}{3r^3}\Delta x(1 - 3\cos^2\theta) \qquad r \sim 2.5\text{Å} \qquad \theta \sim 90°$$

$$\Delta\sigma = \frac{-2 \times 5 \times 10^{-30}}{3 \times 32 \times 10^{-24}} \sim 0.1 \text{ ppm}$$

(b) Electric field

$$\Delta\sigma = 3 \times 10^{-12} E_z$$

$$E_z = \frac{2\mu\cos\phi}{4\pi r^3} \qquad \mu = 1.1\text{D} = 1.1\ 10^{-18} \text{ e.u.}, \phi \sim 45°, r \sim 2.5\text{Å}$$

$$= \frac{2 \times 1.1 \times 10^{-18}}{4 \times \pi \times 32 \times 10^{-24}} \sim \frac{1.5}{4} \times 10^4 \rightarrow \Delta\sigma = \frac{3 \times 10^{-12} \times 1.5 \times 10^4}{4}$$

$\sim 10^{-2}$ ppm (negligible)

(c) Steric compression

$\Delta\sigma = -600r^{-6} \sim -0.4$ ppm

$\Delta\sigma_{total} \sim -0.5$ ppm, which is in the correct direction and even though only $\frac{1}{2}$ the experimental value is nevertheless considered as rather good agreement.

3. 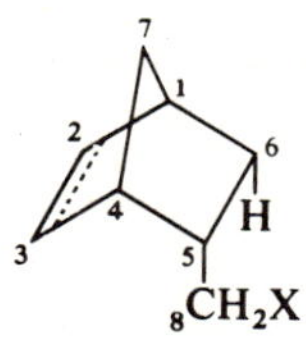


The high-field resonance is due to the $C_6$-endo H, which is subject to *two* positive shielding influences: the diamagnetic anisotropy of the double bond and also the ($C_5-C_8$) carbon-carbon single bond (dominant). [R. G. Foster and M. C. McIron, *Chem. Comm.*, 380 (1967).]

4. There are two distinct complexes:
   1. Hydrogen-bonded complexes, which are formed by fluoroform, chloroform or bromoform $Cl_3C{-}H{\cdots}X^-$, resulting in a downfield shift due to electron withdrawing by the halogen.
   2. Halogen-bonded or charge-transfer complexes, which can occur with iodoform or bromoform, resulting in an upfield shift due to electron donation by the halogen.

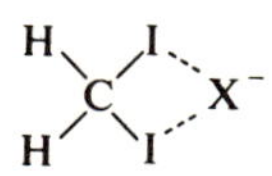


Approximately 0.7 ppm are due to van der Waals effects, the remainder to electrostatic effects. [R. D. Green and J. S. Martin, *J. Amer. Chem. Soc.*, *90*, 3659 (1968).]

5. No correlation. [B. P. Dailey, *J. Chem. Phys.*, *34*, 1099 (1961); *Chem. Abst.* *69*, 23523m (1968).]

6. In the trans isomer the olefinic protons are in the plane of both benzene rings. The cis isomer has twisted out-of-plane aromatic rings providing less shielding and also the rings are more distant from the $\beta$ proton than in the trans isomer.

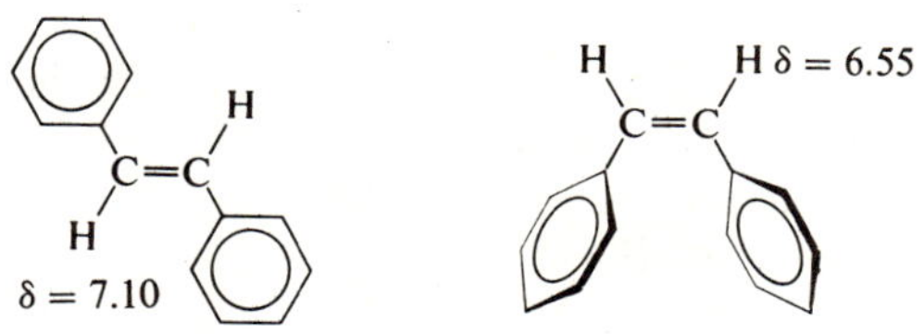


7.

| | | | |
|---|---|---|---|
| 1 | 7.35 | 2 | 6.85 |
| 3 | 5.60 | 4 | |

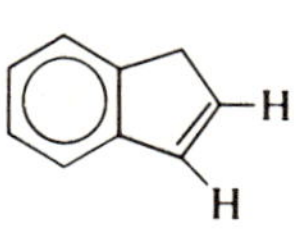


6.50

6.80

| | | | |
|---|---|---|---|
| 5 | 6.28 | 6 | 6.83 |
| 7 | 1.88 | 8 | 2.14 |

8. Upon protonation the homotropylium ion is formed with the indicated chemical shifts. [R. Pettit *et al.*, *J. Amer. Chem. Soc.*, *84*, 2842 (1962); *88*, 604 (1966); P. Warner, D. L. Harris, C. H. Bradley and S. Winstein, *Tetrahedron Lett.*, 4013 (1970).

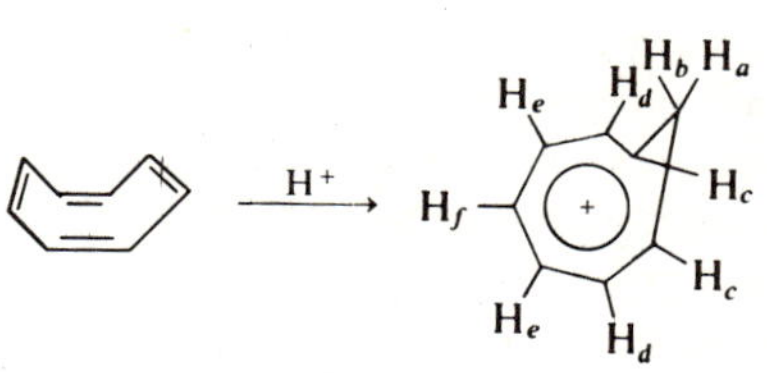


| Homotropylium ion | |
|---|---|
| $H_b$ | 10.6τ |
| $H_a$ | 4.8τ |
| $H_c$ | 3.4τ |
| $H_{d,e,f}$ | 1.4τ |

9. This is an example of London–van der Waals compression shift (deshielding of the protons). In the case of the methyl substituent (III), it causes the 0.38-ppm downfield shift. In the case of the phenyl substituent (II), it reduces to 0.18 ppm the shielding expected from $H_a$ being very close to the surface of the aromatic ring. The latter is forced by the geometry of the molecule into a plane perpendicular to that of the naphthyl moiety (examine a molecular model).

10. $\Delta\delta(^{13}C) = 123$ ppm; $123/170 \sim 0.75$ unit of positive charge on the central carbon. This result can be checked from $\Delta\delta(^1H) = 2.85$ ppm. Let us assume (as a zero-order approximation) a unit charge on the central carbon:

$$E = \frac{q}{r^2} = \frac{4.8 \times 10^{-10}}{(1.55 \times 10^{-8})^2} \sim 2 \times 10^6$$

$$E_z = E \cos (109.5°) = 0.18 \times 2 \times 10^6$$

$$E^2 = 4 \times 10^{12}$$

$$\Delta\delta \text{ (calc.)} = -3 \times 0.36 \times 10^6 \times 10^{-12} - 4 \times 10^{12} \times 10^{-18}$$

$$\sim -(1 + 4) \times 10^{-6} \sim -5 \text{ ppm (too big)}$$

Refining this calculation, by starting for instance with 0.75 unit of charge on the central carbon as indicated by the carbon-13 shift, one finally comes up with a distribution such as

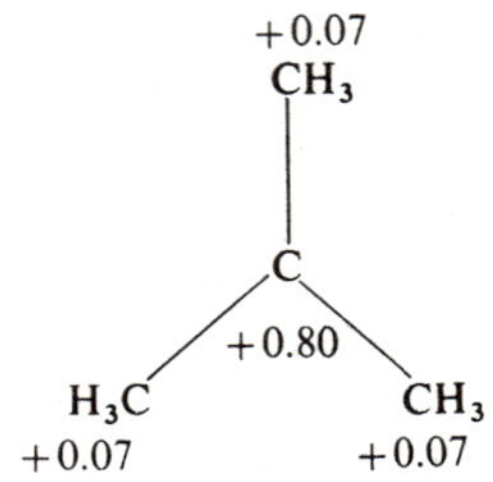


[G. Fraenkel and D. G. Farnum, "Nuclear Magnetic Resonance Spectra," in *Carbonium Ions*, G. A. Olah and P. von R. Schleyer, eds., Wiley-Interscience, New York, 1968, vol. 1, p. 237.]

## CHAPTER 5

1. Dissolved oxygen is paramagnetic, and would relax nuclei much more effectively than they would relax each other. Likewise, it is preferable to use solvents having only negligible magnetic moments, so that only intramolecular relaxation processes through direct dipolar coupling are available.
2. Since $\gamma_H/\gamma_D \sim 6.52$ and $\gamma_H/\gamma_{^{13}C} \sim 4$ the theoretical limits for the NOE enhancements are therefore 320 percent and 200 percent respectively. The experiments must be performed by saturating the proton resonance, since protons have the greater magnetic moment.
3. The two interactions are [R. H. Martin and J. C. Nouls, *Tetrahedron Lett.*, 2727 (1968)]:

33 percent (1,10-phenanthrene interaction)

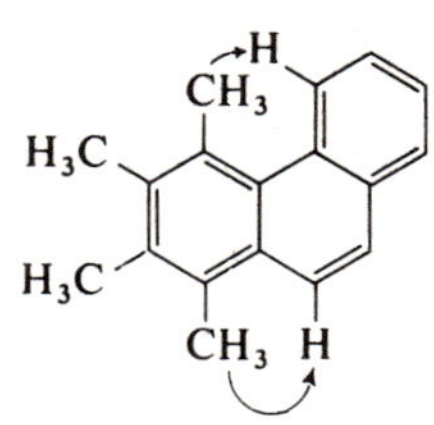


11 percent (*peri* interaction)

4. Molecule (2) has a bigger molecular volume than (1), and tumbles more slowly, hence it has a larger correlation time. [C. Brevard, J. P. Kintzinger, and J. M. Lehn, *Chem. Comm.*, 1193 (1969).]
5. Protons within the methyl group are much closer to each other—the contacts are ca. 1.8 Å—than to protons elsewhere in the molecule. Hence they relax each other by dynamic nuclear coupling much more effectively than they are relaxed by the olefinic proton. Since the olefinic proton does not intervene in the methyl relaxation to a significant extent, no NOE will be seen by irradiating it with a strong RF field. Of course, the converse statement is not true: The methyl protons provide a relaxation mechanism for the otherwise isolated olefinic proton, hence the possibility of an observable NOE.

6.

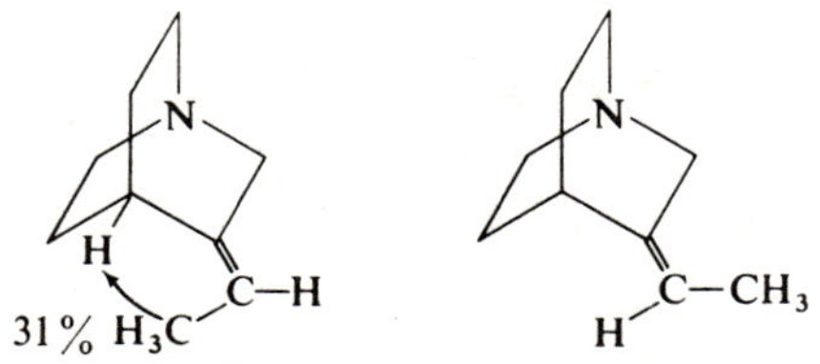


[J. C. Nouls, G. Van Binst, and R. H. Martin, *Tetrahedron Lett.* 4065 (1967).]

CHAPTER 6

1. The Hückel charge densities are

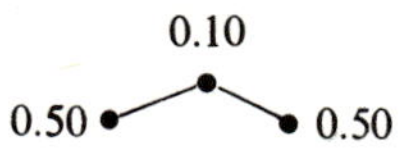


The unpaired electron is in $\psi_2 = (1/\sqrt{2})(\phi_1 - \phi_3)$. The experimental charge densities from McConnell relationship are

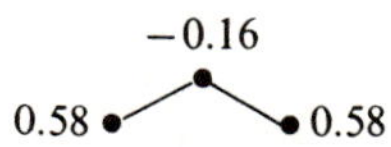


[Note: More sophisticated theoretical treatments, such as self-consistent field methods, indeed predict negative electronic density at position 2.]

2. Qualitative consideration of the following resonance forms:

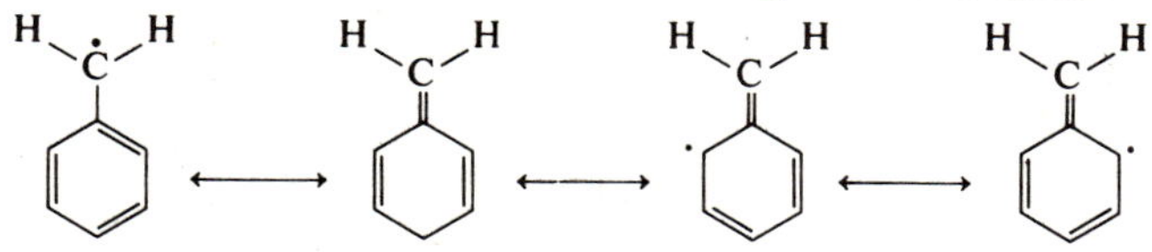

leads to the expectation of no unpaired spin delocalization to the meta position, hence no coupling. Quantitatively, the Hückel mo coefficients are calculated to be

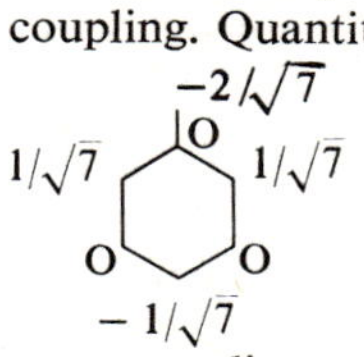


corresponding to the following distribution of electronic densities: 4/7 at $C_\alpha$, 1/7 at $C_2$, $C_4$, $C_6$, and 0 at $C_1$, $C_2$, $C_3$. The prediction based upon these and the McConnell relationship is thus $a_H = 14$ Oe($C_\alpha$); $a_H = 3.5$ Oe ($C_2$, $C_4$, $C_6$); and $a_H = 0$ ($C_3$, $C_5$). Compare with the observed values $a_H = 16.4$ Oe ($C_\alpha$); $a_H = 6.2$ Oe ($C_4$); $a_H = 5.2$ Oe ($C_2$, $C_6$); $a_H = 1.8$ Oe ($C_3$, $C_5$).

3. The molecule is forced out of coplanarity by the bulky *t*-butyl groups. Its electronic spectrum is intermediate between those of ethylene and butadiene.

4. The MOs are arranged according to the following energy diagram:

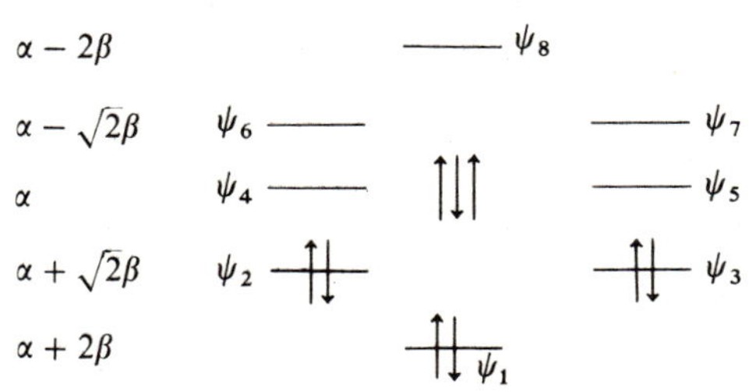


Three electrons must be partitioned between $\psi_4$ and $\psi_5$:

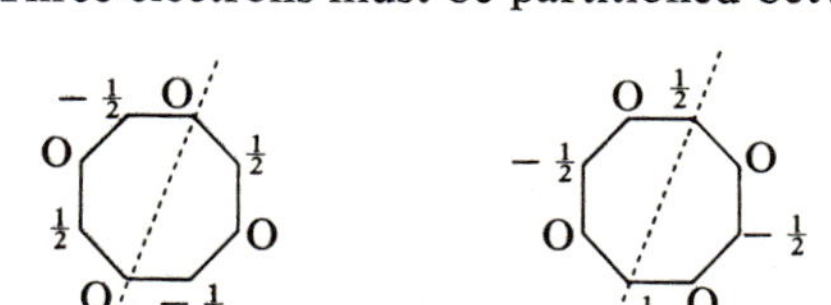


$\psi_4$(antisymmetric) $\psi_5$ (symmetrical)

The methyl substituent will seek the $C_1$ position of lowest electron density, hence will stabilize $\psi_4$ relative to $\psi_5$. Thus $\psi_4$ will hold the electron pair, leaving $\psi_5$ for the unpaired spin. Hence one predicts $a_H = 0$ at positions 3, 5, 7, which again agrees qualitatively with the experimental values. Quantitatively, the data suggest 75 percent $\psi_5$, 25 percent $\psi_4$.

Note the difference in the case of the toluene radical anion where the unpaired spin was inserted into the stabler of the two degenerate MOs. For the methyl cyclooctatetraene anion, a $9\pi$-electron system, the unpaired spin goes into the leftover MO predetermined by the other two electrons occupying the MO of least energy (*Aufbau* principle).

5. $\lambda = \frac{c}{\nu} = 162.5 \times 10^{-9}$ cm

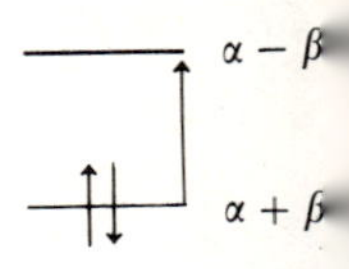


$$\nu = \frac{3 \times 10^{10}}{162.5 \times 10^{-9}} = 61{,}500 \text{ cm}^{-1}$$

$h\nu = 2\beta$

$\beta = 60.5$ kcal/mole ($H_2C = CH_2$)

Benzene R. E. = 36 kcal/mole = $2\beta$. Therefore $\beta = 18$ kcal/mole. The difference is the compression energy: In benzene, one must compress single bonds from 1.50 Å, $sp^3$-$sp^3$ to 1.39 Å; and the $sp^2$-$sp^2$ double bonds are stretched from 1.34 to 1.39 Å.

6. If one prepares a graph of the corresponding $h\nu$ versus the calculated Hückel MO transition energies (1.41, 1.00, 0.765, and $1.69\beta$, respectively for the above compounds), the correlation is linear as it should be and excellent. From the slope, one calculates $\beta = 48$ kcal/mole, in good agreement with the value of 54 kcal/mole derived from the photoelectronic spectrum of benzene.

7. Protonation will occur at the most electron-rich centers. One has therefore merely to look up the coefficients of the lowest vacant MO which will house the additional electron(s): one for the monoanion, two for the dianion. Upon examination of Figure 6.5, it appears that $\psi_5$ is the best choice, corresponding to electron densities of $\frac{1}{3}$ at positions 1 and 4: the product is kinetically controlled. This Hückel treatment has met with considerable success in accounting for the nature of the Birch products observed for a large number of hydrocarbons. (J. March, *Advanced Organic Chemistry*, McGraw-Hill Book Co., New York, pp. 597–598.)

8. The favored structure is (1). From the MO coefficients one can calculate a bond order of $p_{1,2} = 0.778$ and $p_{2,3} = 0.520$ which in a manner analogous to six-membered rings allows a prediction of the experimental coupling constants. (P. Yates, in *Advances in Alicyclic Chemistry*, H. Hart and G. J. Karabatsos, eds., Academic Press, New York, vol. 2, 1968.)

## CHAPTER 7

1. The higher-wavelength band is due to the $n \rightarrow \pi^*$ transition and shifts to lower wavelength due to increased solvent polarity and hydrogen bonding as the solvent is changed from hexane and ether to ethanol and water. The lower-wavelength transition is due to the $\pi \rightarrow \pi^*$ transition and undergoes a red shift as the solvent polarity increases. In the excited state the molecule is more polar due to the greater contribution of resonance forms such as $^+C—C=C—C^-$ in the case of butadiene or

$$(CH_3)_2\overset{+}{C} - CH = \underset{}{\overset{O^-}{\overset{|}{C}}}CH_3$$

in mesityl oxide. Such polar forms are more stable in polar solvents so there is a lowering of the energy of the excited state and a red shift as one goes from the nonpolar hexane to polar water.

2. The octant diagrams for these bicyclo[2.2.2]octanones bear out the data. For example, for (1):

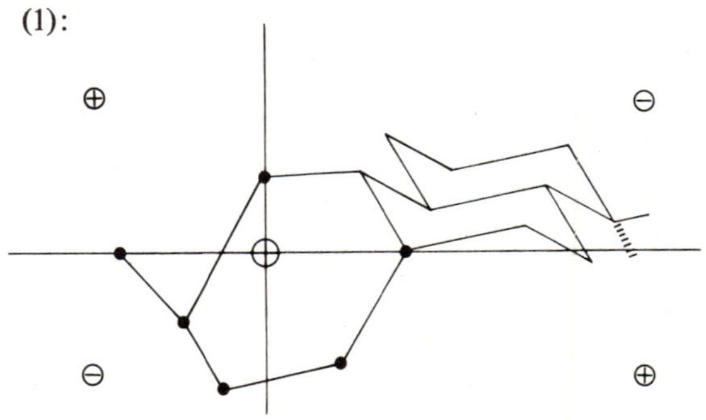

3. Consideration of the A and B rings suffices. In the octant diagram:

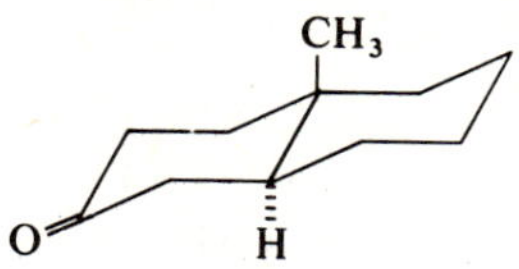


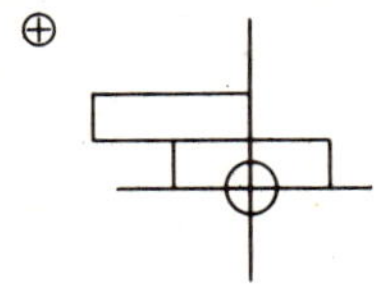

1

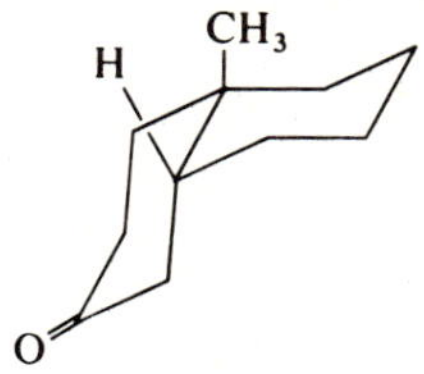


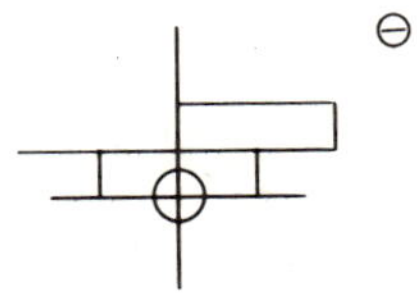

2

[C. Djerassi and D. Marshall, *J. Amer. Chem. Soc.*, *80*, 3986 (1958).]

4. The configuration of the methyl group is α, as shown by the octant diagram. [S. W. Pelletier *et al.*, *J. Org. Chem.*, *30*, 4234 (1965).]

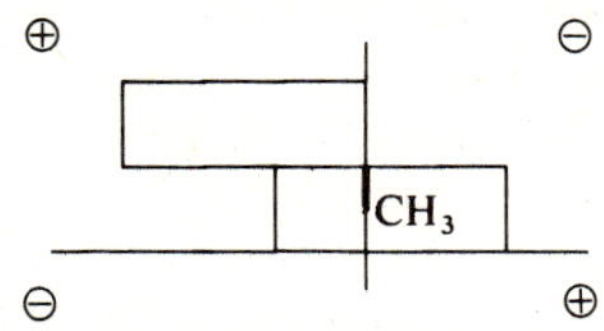


5. The $C_3$ configuration is *R*.

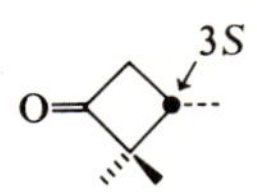


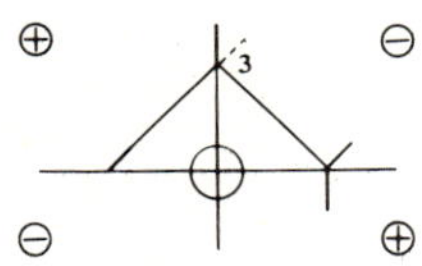


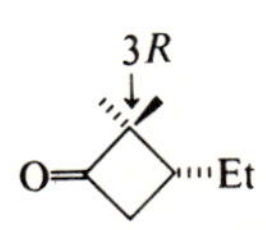


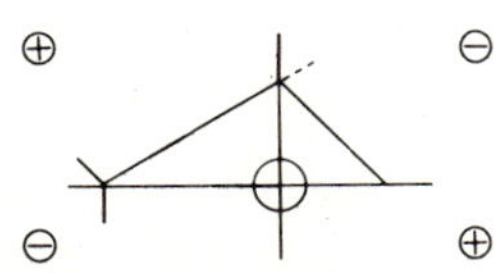

6. The sign of the Cotton curve is positive as seen from the octant diagram:

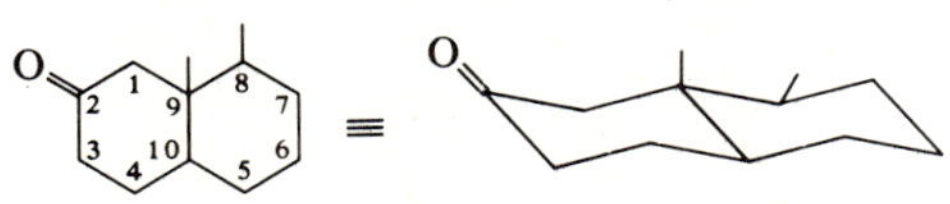


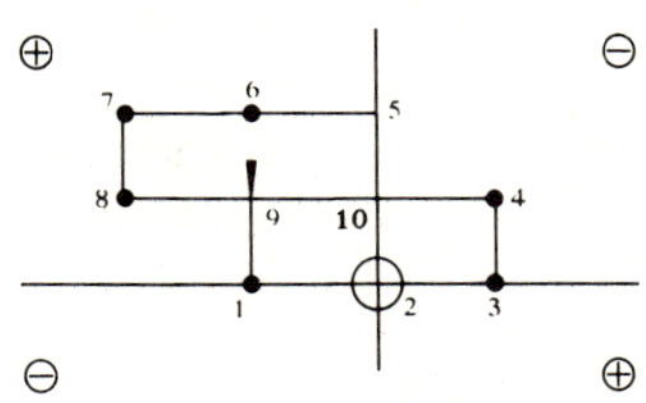


7. The normal chair conformation which would correspond to a weakly negative Cotton effect is destabilized by the severe 1,3-diaxial interaction between the methyl and isopropyl groups. The cyclohexanone ring distorts into a twist conformation in order to relieve this repulsion, hence the strongly negative effect observed.

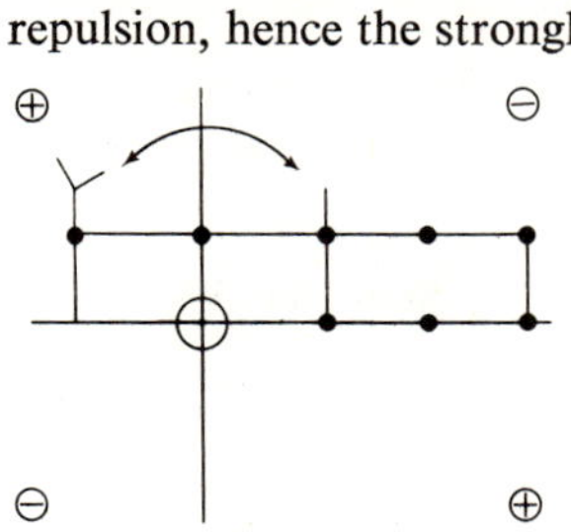

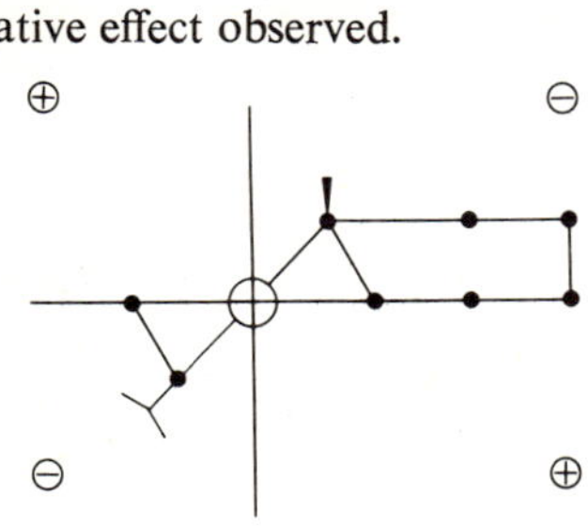

*CHAPTER 8*

1. $CH_2F_2$ is an $A_2$–$X_2$ case.
   $H_2C{=}CF_2$ is an AA′–XX′ case ($J_{AX}$ cis different from $J'_{AX}$ trans).
   $H_2C{=}C{=}CF_2$ is another $A_2$–$X_2$ case.
   HFC=CHF and HFC=C=CHF are AA′–XX′.
2. Inspection of the energy-level diagram shows that this is not feasible; identical frequencies and intensities are obtained for the four lines whether $J_{AX}$ is positive or negative.
3. The stereochemistry corresponds to minimization of the cis N-R/C-R and N-R/N-R interactions between substituents on adjacent centers:

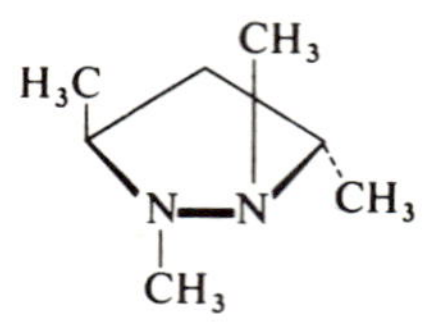


   The other conformations would lead to $ABX_2$ spectra even with fast nitrogen inversions. [J. Elguero, C. Marzin, and D. Tizane, *Tetrahedron Lett.*, 513 (1969).]
4. No: The isopropyl methyls are diastereotopic when adjacent to a chiral center or asymmetric carbon irrespective of the rate of rotation of the isopropyl group. [G. M. Whitesides, D. Holtz, and J. D. Roberts, *J. Amer. Chem. Soc.*, *86*, 2628 (1964); J. I. Kroschwitz, M. Winokur, H. J. Reich, and J. D. Roberts, *J. Amer. Chem. Soc.*, *91*, 5927 (1969).]
5. Diastereotopic for the meso isomer; enantiotopic for the dl isomer. [see T. Sato, Y. Saito, M. Kainosho, and K. Hata, *Bull. Chem. Soc. Jap.*, *40*, 391 (1967).]
6. Due to the pyramidal structure of the carbanion the ortho protons are in diastereotopic environments. At low temperatures, the rotations round the C-aryl bonds have become restricted on the nmr time scale. [H. Kloosterziel, *Chem. Comm.*, 1330 (1968).]
7. This observation is consistent only with the bisected conformation in which the plane of the

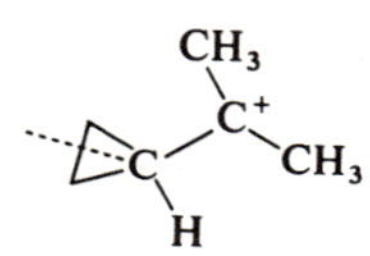


   cyclopropane ring is parallel to the vacant *p* orbital of the positively charged carbon. The anisotropy of the cyclopropane ring is the main component of the rather large chemical-shift difference. [G. A. Olah, "Stable, Long-Lived Carbonium Ions," *Chem. Eng. News*, March 27, 1967, p. 77–88.]

*CHAPTER 9*

2. 1. $J_{AB} = 4.5$ Hz. $\nu_A - \nu_B = 3.2$ Hz, $\delta_A = 1.94$ ppm; $\delta_B = 1.89$ ppm.
   2. $J_{AB} = 3.0$ Hz. $\nu_A - \nu_B = 4$ Hz, $\delta_A = 1.95$ ppm; $\delta_B = 1.88$ ppm.
   3. $J_{AX}$ and $J_{BX}$ of like signs:
      $J_{AB} = 3.0$ Hz $\quad \nu_A = 2.95$ Hz
      $J_{AX} = 2.1$ Hz $\quad \nu_B = 7.55$ Hz
      $J_{BX} = 0.9$ Hz
      $J_{AX}$ and $J_{BX}$ of opposite signs:
      $J_{AB} = 3.0$ Hz $\quad \nu_A = 4.95$ Hz
      $J_{AX} = 6.1$ Hz $\quad \nu_B = 5.55$ Hz
      $J_{BX} = 3.1$ Hz $\quad$ (other solutions are possible)
4. The compound is $C_6H_5CHBr{-}CH_2Br$ and the coupling constants and chemical shifts are
   $J_{AB} = 10$ Hz
   $J_{AX} = 4.7$ Hz } both positive (vicinal couplings)
   $J_{BX} = 11.3$ Hz }
   $\nu_A = 243.6$ Hz
   $\nu_B = 240.3$ Hz

*CHAPTER 10*

1.

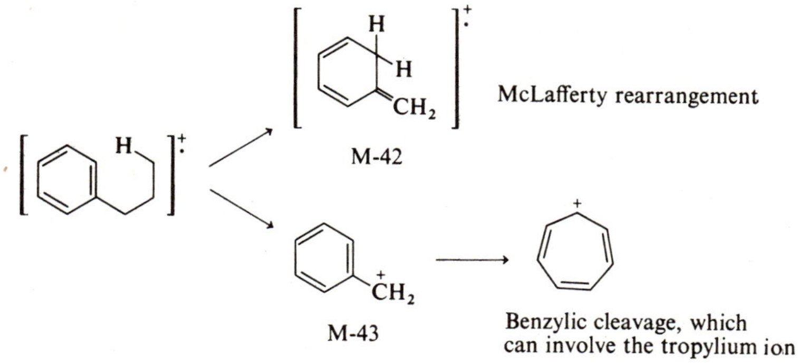


[R. Nicoletti and D. A. Lightner, *Tetrahedron Lett.*, 4553 (1968).]

2. Steric inhibition of resonance

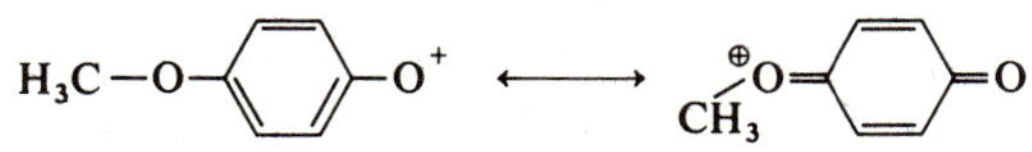


[M. M. Bursey, *J. Amer. Chem. Soc.*, *91*, 1861 (1969).]

3. The $m/e = 300$ peak is coronene which is less strained than hexahelicene since it has more planar (as opposed to twisted) rings. It arises via the elimination of ethylene. [R. C. Dougherty, *J. Amer. Chem. Soc.*, *90*, 5788 (1968).]

4.

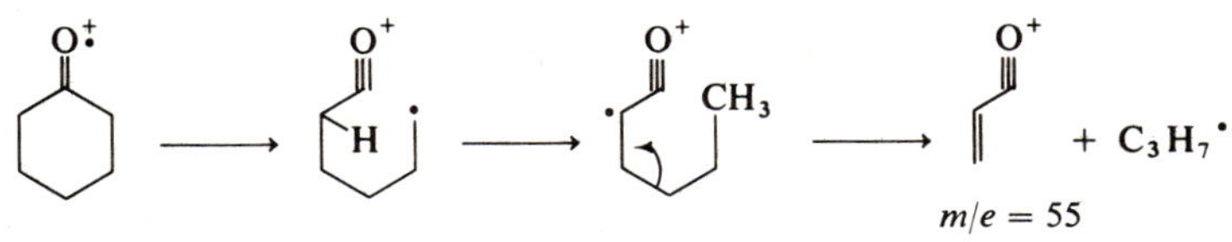


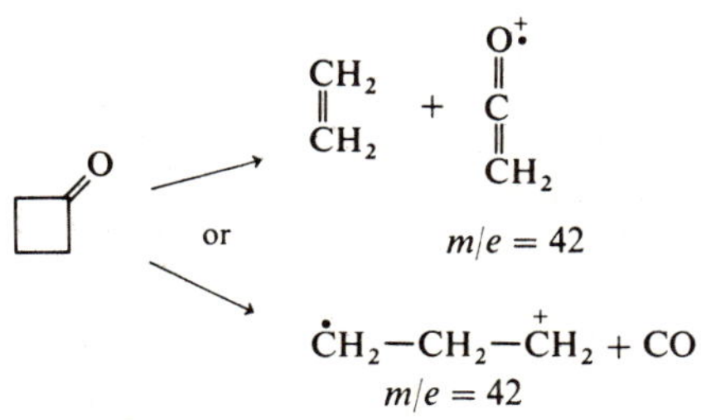


The transition state for H-transfer involves an unfavorable 4-membered ring in the case of cyclobutanone.

[D. H. Williams, H. Budzikiewicz, Z. Pelah, and C. Djerassi, *Monatsh. Chem.*, *95*, 166 (1964); H. J. Hofman, *Tetrahedron Lett.*, 2329 (1964).]

5. The amino group is electron releasing, the nitro group is electron withdrawing. [M. M. Bursey and F. W. McLafferty, *J. Amer. Chem. Soc.*, *89*, 1 (1967).]

6. These peaks are important because they suggest the formation of the cyclobutadiene singly charged and doubly charged ion after loss of one nitrogen molecule. A number of studies have postulated the intermediacy of cyclobutadiene in relevant mass spectral reactions. [M. H. Benn, T. S. Sorensen, and A. M. Hogg, *Chem. Comm.*, 574 (1967); M. M. Bursey *et al.*, *J. Amer. Chem. Soc.*, *90*, 1557 (1968).]

*CHAPTER 11*

1. $\ln\left(\frac{\mathrm{AK} - \mathrm{B}}{\mathrm{A_0K} - \mathrm{B_0}}\right) = -(k_1 + k_{-1})t, k_1 = 2.46 \times 10^{-5}\ \mathrm{sec}^{-1}$ and $k_{-1} = 1.56 \times 10^{-5}\ \mathrm{sec}^{-1}$.

[A. Speert (private communication).]

2.
$$k = k_0 \exp\left(\frac{-E_a}{RT}\right) \quad \text{Arrhenius}$$
$$= \frac{KT}{h} \exp\left(\frac{-\Delta G^*}{RT}\right) \quad \text{Eyring}$$

$E_a$ = 10.3 kcal/mole
$\Delta H^*$ = 10.8 kcal/mole
($E_a$ and $\Delta H^*$ are related by $\Delta H^* = E_a + RT$.)
$\Delta S^*$ = 2.8 e.u.

3. [F. A. L. Anet and A. J. R. Bourn, *J. Amer. Chem. Soc.*, *89*, 760 (1967).]

*CHAPTER 12*

1. Yes, mechanism A applies, not B. [B. G. Derendyaev *et al.*, *Tetrahedron Lett.*, 5 (1969).]
2. *n*-Pentane: $3^{4-2} = 9$; *n*-hexane: $3^{5-2} = 27$; *n*-heptane: $3^{6-2} = 81$; *n*-dodecane: $3^{11-2} = 19{,}683$.

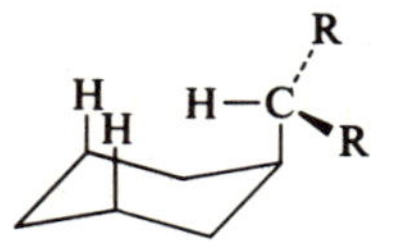


R = H,CH$_3$
Small 1,3 repulsion

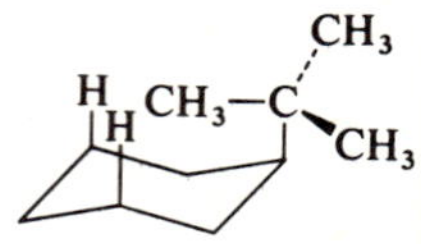


Large 1,3 repulsion

3.

1 → 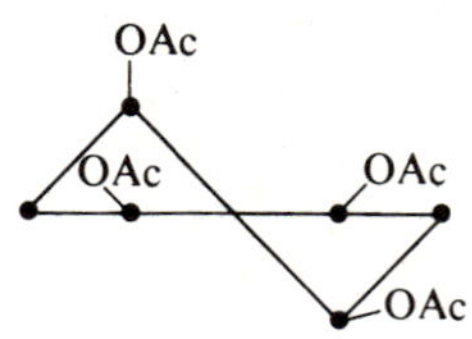


2 → 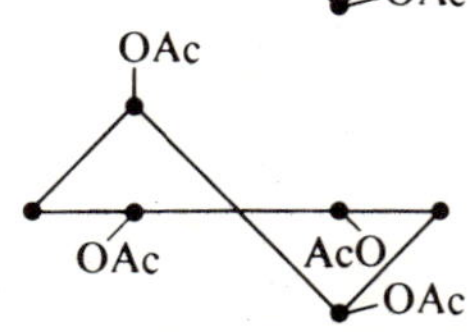


The transition states are virtually equienergetic; the difference between the free energies of activation reflects the relative stabilities of the initial states. Hence, molecule (1) suffers from the 1,3-diaxial interaction of the two acetoxyl functions, which is approximately 0.9 kcal/mole. [S. Wolfe and J. R. Campbell, *Chem. Comm.*, 877 (1967).]

4. Transannular participation is the explanation:

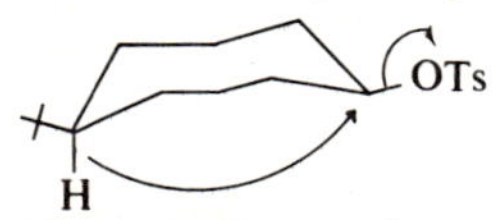


The trans isomer has much less participation, as witnessed by the smaller percentage of the 1-olefin; the parent compound is intermediate. [E. L. Eliel, N. L. Allinger, S. J. Angyal, and G. A. Morrison, *Conformational Analysis*, Wiley-Interscience Publishers, Inc., New York, 1965, pp. 224–225.]

5. The trans isomer has the large dipole moment and is chiral: this molecule is very rigid (hence the chirality) and strained (hence the rehybridization at the trigonal centers which increases the C—C group moment).

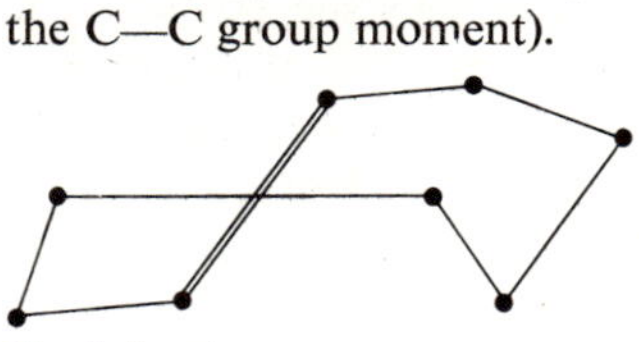

6. Cycloheptane:

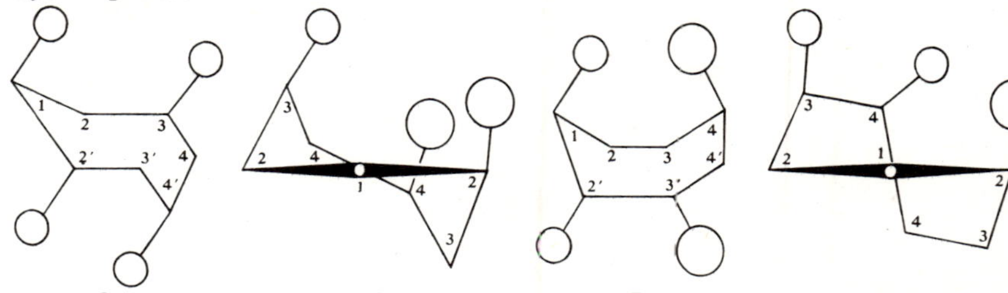

C TC B TB

Cyclooctane:

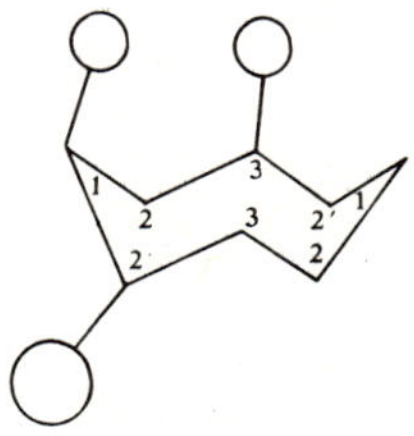

CC

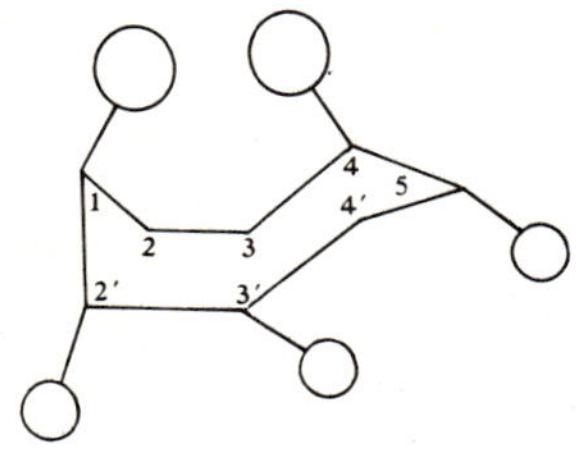

BC

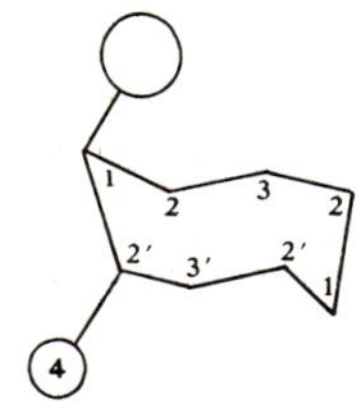

TC

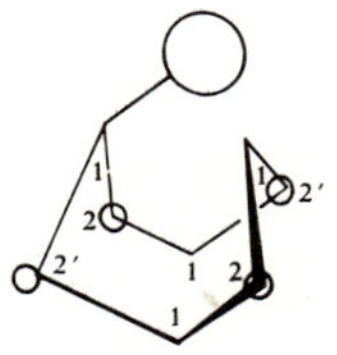

BB

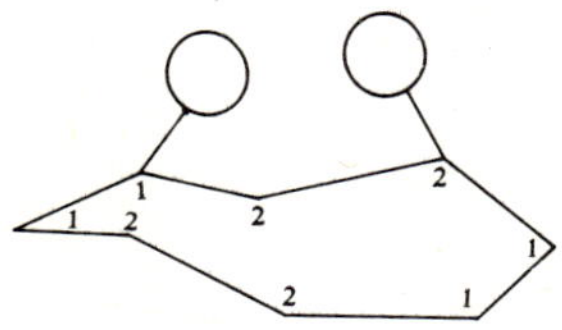

TCC

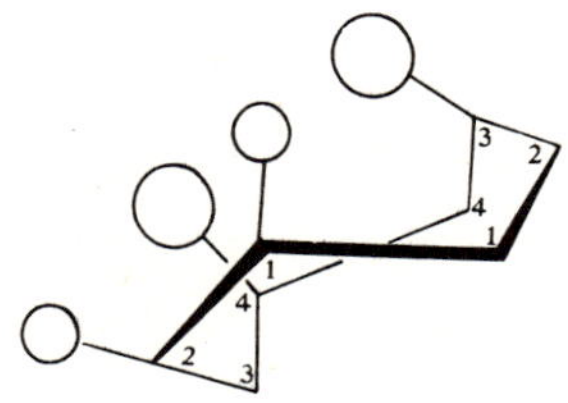

TBC

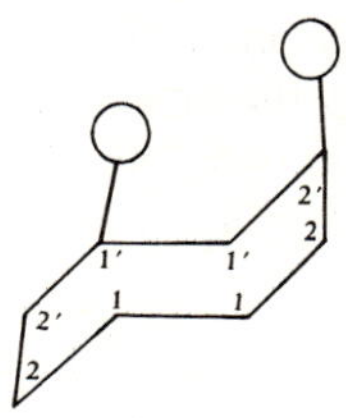

C

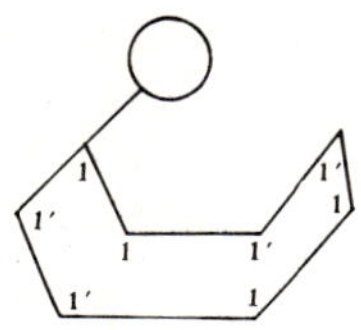

B

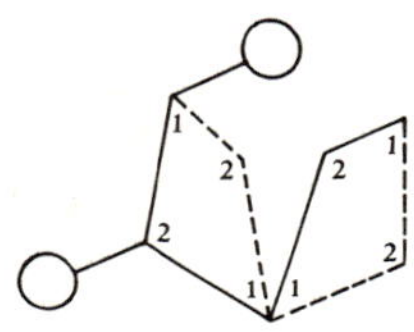

$S_4$

Cyclononane:

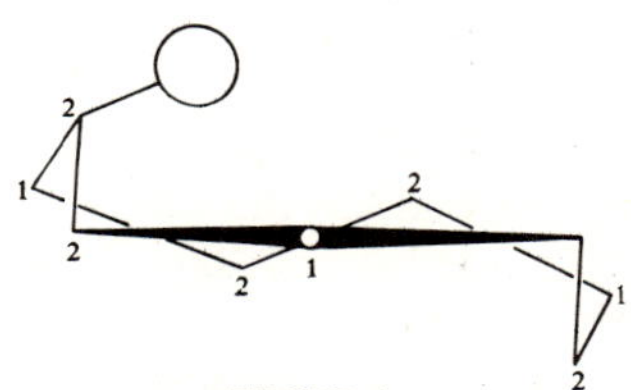

TBC($D_3$)

[J. B. Hendrickson, *J. Amer. Chem. Soc.*, *89*, 7043 (1967).]

7. Two possible approaches are indicated as follows:

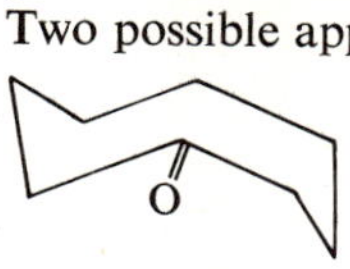


Introduction of the carbonyl group in this position relieves the greatest number of nonbonded interactions and offers the best torsional arrangements. [F. A. L. Anet, in *Conformational Analysis: Scope and Present Limitations*, G. Chiurdoglu, ed., Academic Press, New York, 1970.]

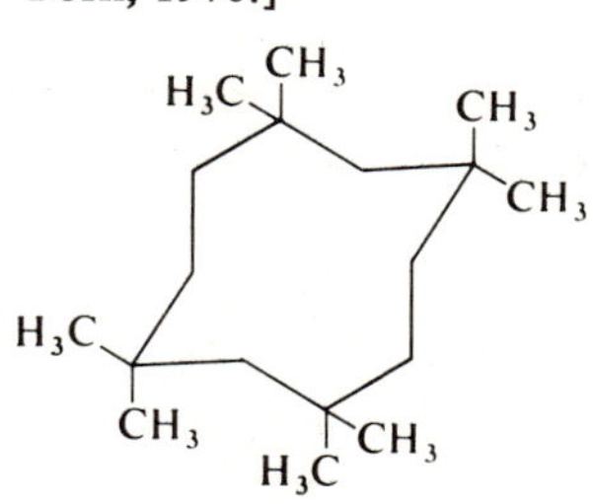


Methyl groups in all the unhindered "corner" positions. [J. Sicher, *Chimia 16*, 300 (1960); J. Sicher *et al.*, *Tetrahedron Lett.*, 4009 (1967).]

8. 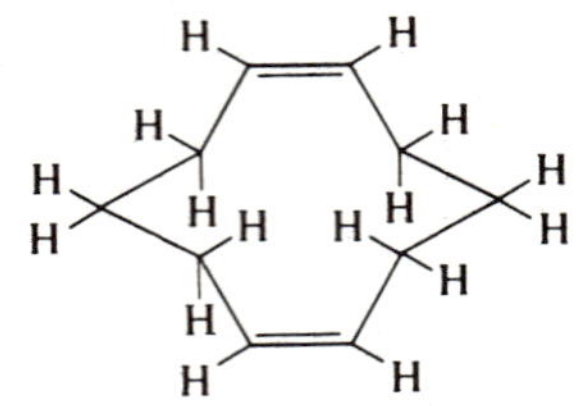

This conformation removes the greatest number of nonbonded interactions. It is the only strain-free conformation for this hydrocarbon. [J. Dale, T. Ekeland, and J. Schaug, *Chem. Comm.*, 1477 (1968).]

9. Triasterane:

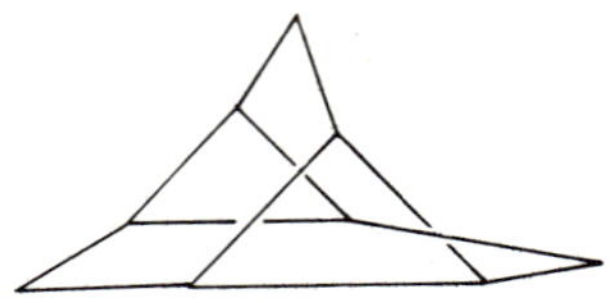

The key to the structure is the cyclopropyl protons indicated by their high-field position in the nmr and confirmed by the high-frequency ir absorption centered about 3025 $cm^{-1}$. The ir spectrum further suggests, by its very simplicity, a highly symmetrical molecule. The methylene groups are characterized both by the nmr and by the ir rocking vibration at 755 $cm^{-1}$, together with the stretching bands around 2900 $cm^{-1}$. The major peaks in the mass spectrum are due to loss of methyl (105), ethyl (91), and allyl (79). [H. Musso and U. Biethan, *Chem. Ber.*, *100*, 119 (1967).]

## *CHAPTER 13*

1.

| Calculated (nm) | Observed (nm) |
|---|---|
| (1) 303 | 306 |
| (2) 353 | 355 |
| (3) 288 | 285 |

2. In each of the two series, the double bonds are about equally strained; the added interaction with the exocyclic C—C bonds attached has the effect of raising the frequency. In fact, the C=C stretch is constant at ~1685 $cm^{-1}$ for 1,2-dimethylcyclobutene, -pentene, and -hexene, because the endocyclic C—C bonds are pulled together as the exocyclic C—C bonds are spread apart in these molecules, and there is almost exact mutual compensation of the two factors.

3. 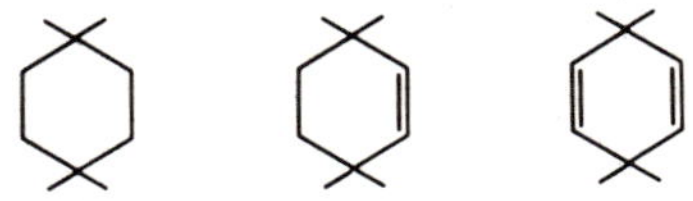

1 2 3

1,1,4,4-tetramethylcyclohexane displays slowed ring inversion at 78°C; two signals are seen together with a complex AA′BB′ multiplet for the ring methylenes. The mass spectra of these hydrocarbons are characterized by loss of methyl and by the retro Diels–Alder cleavage of the cyclohexene. [W. Reusch and D. F. Anderson, *Tetrahedron*, *22*, 583 (1966); *J. Org. Chem.*, *29*, 2446 (1964).]

4. Relatively slow ring inversion exchanges inner and outer protons of the crown conformation of cyclononatriene. [K. B. Untch and R. J. Kurland, *J. Amer. Chem. Soc.*, *85*, 346 (1963).]

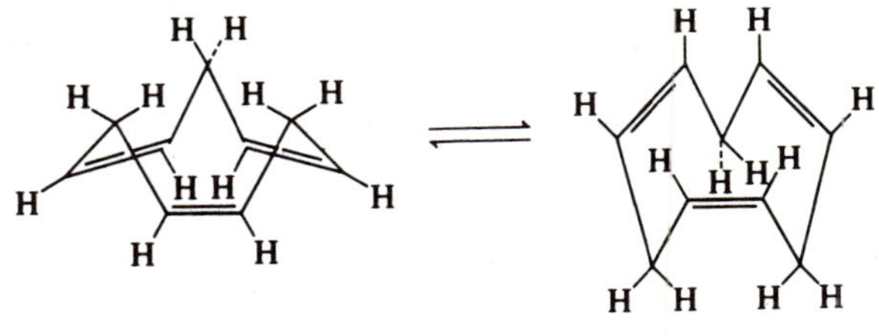

5. The conclusion is that the molecule is aromatic; it is also verified by other facts, such as x-ray determination of the C—C bond lengths (2 − 3 = 1.37 Å; 3 − 4 = 1.41 Å), or the high-field shift of the C=CH$_2$ resonance at δ = 3.19 instead of the usual olefinic regions. [H. Günther and H. H. Hinrichs, *Tetrahedron, 24*, 7033 (1968).]

6.

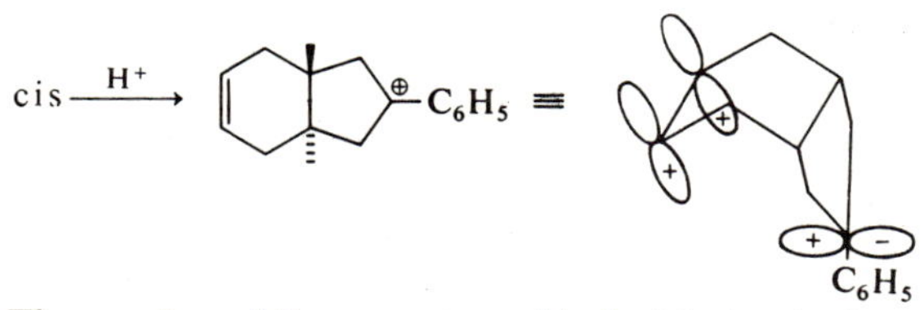


The overlap of the vacant *p* orbital with the double bond increases delocalization resulting in a shift to higher wavelength. Such an overlap is geometrically impossible in the trans isomer. [G. Leal and R. Pettit, *J. Amer. Chem. Soc., 81*, 3160 (1959).]

7. Twenty-one percent naphthalene + 50 percent tetralin + 29 percent decalin; the aromatic protons of tetralin are accidentally isochronous, and upfield from those of naphthalene since hydrogenation has removed one ring current.

8.

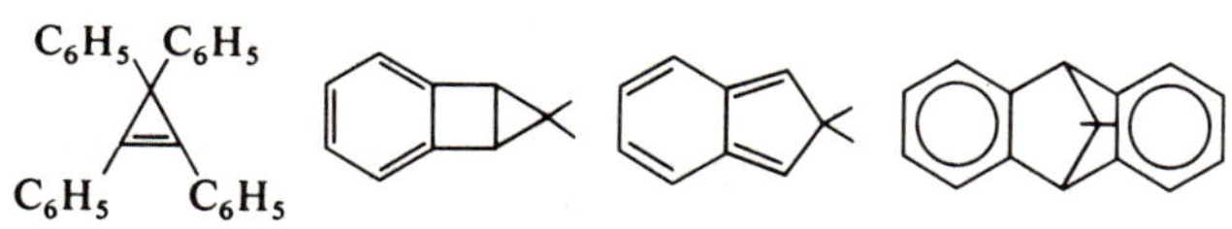


[J. E. Berson *et al., J. Amer. Chem. Soc., 86*, 3896 (1964).]

9.

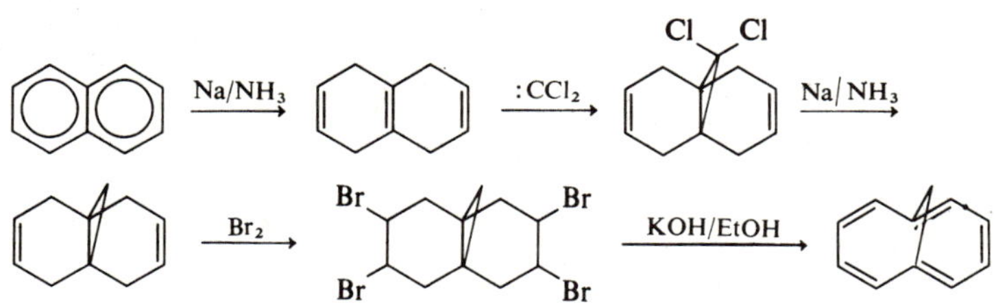


1, 6-Methanocyclodecapentaene
(aromatic)

Note: 1,6-ethanocyclodecapentaene is constrained to exist in the nonaromatic "cyclobutane" formulation below, as evidenced by its nmr spectrum, with peaks centered at δ5.5 (8H) and δ = 2.55 (4H). Build models of these two molecules to appreciate the structural difference. [E. Vogel "Aromatic 10π Electron Systems," *Chem. Soc. (London), Special Publication 21*, 113 (1967).]

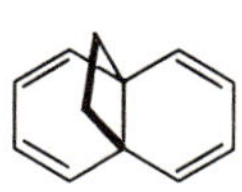

## CHAPTER 14

1.

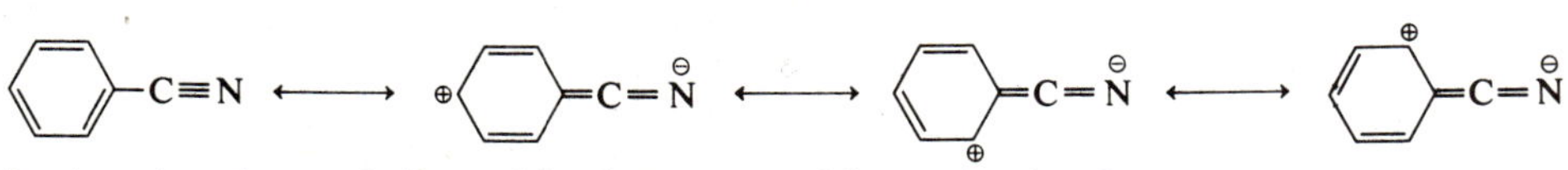


In 4-methoxybenzonitrile and in the cyanopyridines, the ring has an electronegative-withdrawing substituent, thus making the nitrile group less ionic. 3-Cyanopyridine is like benzonitrile because no resonance is possible. [E. Schempp and P. J. Bray, *J. Chem. Phys., 69*, 3450 (1968).]

2.

| | | |
|---|---|---|
| 1.187 | 1.299 | 2-cyanopyridine |
| 1.219 | 1.322 | 2-methylpyridine |
| 1.227 | 1.308 | 2-chloropyridine |
| 1.334 | 1.343 | 2-aminopyridine |

3. The cyano group is electron attracting both inductively and by resonance:

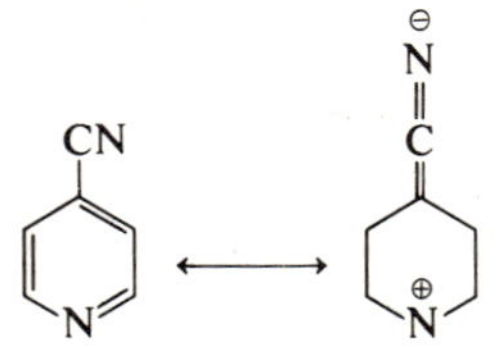


*CHAPTER 15*

1. Evidently a $\pi \rightarrow \pi^*$ transition. The Woodward rules give

$\lambda_{max}$ (calc.) = 227 nm for (1)

$\lambda_{max}$ (calc.) = 254 nm for (2)

hence structure (2) appears to be very likely. This conclusion can be confirmed by independent chemical and physical tests.

2.

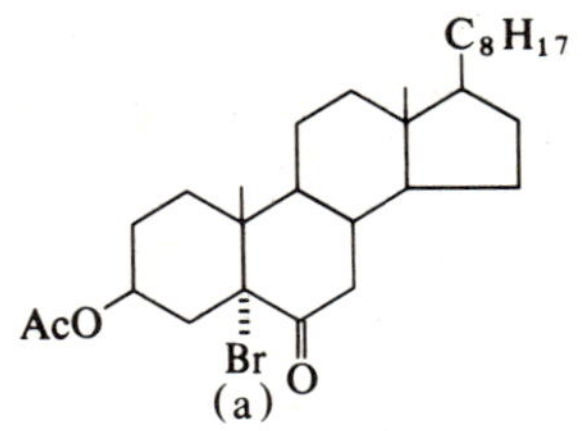


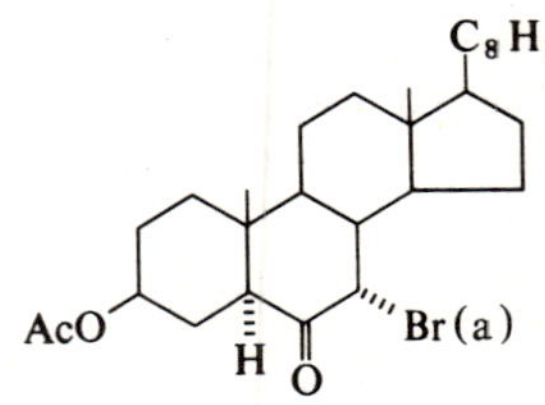


308 nm($\varepsilon$ = 125)    310 nm($\varepsilon$ = 160)

[R. C. Cookson, *J. Chem. Soc.*, 282 (1954). See E. J. Corey, *J. Amer. Chem. Soc.*, 76, 175 (1954) for the determination of the configuration of these $\alpha$-bromo ketones by ir spectroscopy.]

3.

| Calculated (nm) | Observed (nm) |
|---|---|
| (1) 227 | 222 |
| (2) 237 | 240 |
| (3) 249 | 253 |
| (4) 286 | 290 |
| (5) 280 | 279 |
| (6) 317 | 315 |

4. These chemical shifts are characteristic of hydrogen-bonded protons. Their differences clearly reflect the variations in the hydrogen bond strength. 2-Acetyl-3-hydroxynaphthalene has the weakest hydrogen bond because the corresponding $p_{23}$ bond order is low:

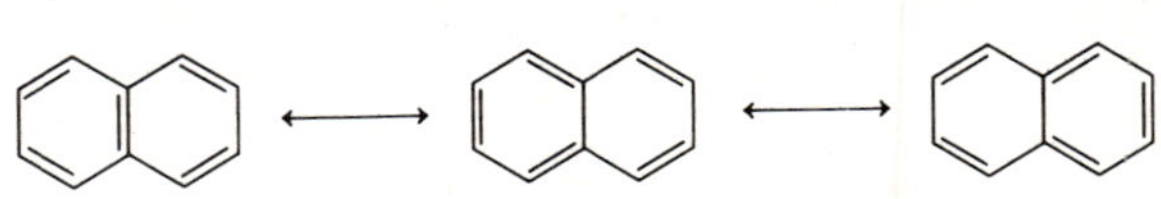

1-Hydroxy-2-acetylnaphthalene may have a stronger hydrogen bond than 1-acetyl-2-hydroxynaphthalene because the peri hydrogen either twists the acetyl group slightly out of coplanarity in the latter compound, or forces the phenolic hydrogen towards the carbonyl group in the former compound. Also, small differences due to the magnetic anisotropy of the naphthalene skeleton would have to be considered in a comprehensive treatment. [G. O. Dudek, *Spectrochim. Acta*, *19*, 691 (1963).]

5.

$\delta$ = 13.47    $\delta$ = 14.60

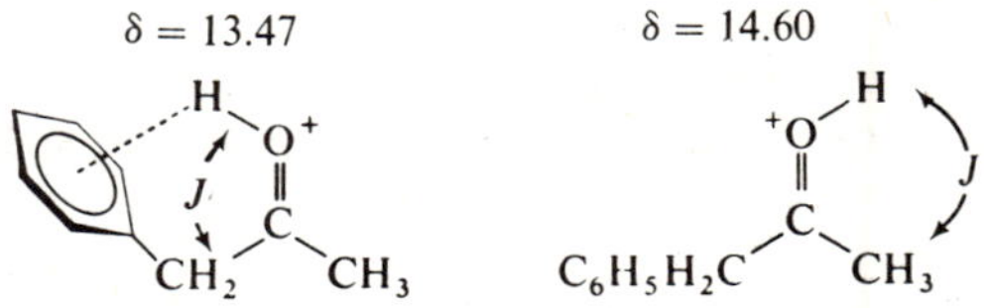


Intramolecular OH---$\pi$ hydrogen bonding places the syn C=O—H proton in the shielding zone of the aromatic ring; conversely, the anti C=O—H proton is assuredly in a les shielding region of space. Furthermore, cisoid allylic coupling constants are usually slightly greater in magnitude than transoid allylic coupling constants. [G. C. Levy and S. Winstein, *J. Amer. Chem. Soc.*, *90*, 3574 (1968).]

6. 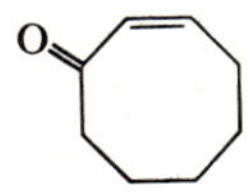

1

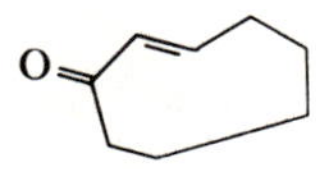

2

[P. E. Eaton and K. Lin, *J. Amer. Chem. Soc.*, *86*, 2087 (1964).]

7. From the elemental analysis, it is apparent that this product contains two dimethylcyclopropanone, and two dimethoxyethylene elements. Equivalence by symmetry, however, reduces the number of distinct resonances. For instance, the cyclopropylmethyl groups occur in two pairs, hence in each pair the two methyls are either homotopic or enantiotopic. Likewise for the methoxyl groups. The simplest working hypothesis for the constitution is then:

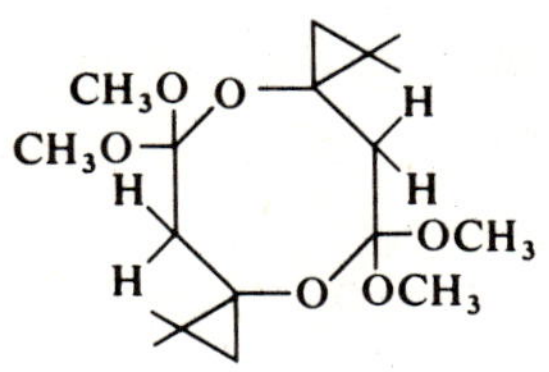


The observation of the AB quartets then dictates the stereochemistry as

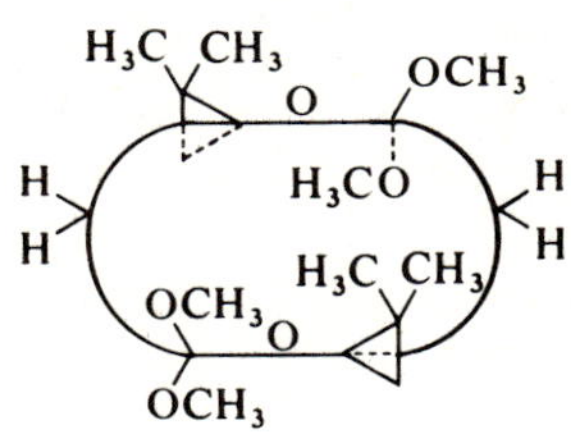


rather than the anti isomer which would have a center of inversion. [N. J. Turro and J. R. Williams, *Tetrahedron Lett.*, 321 (1969).]

8. 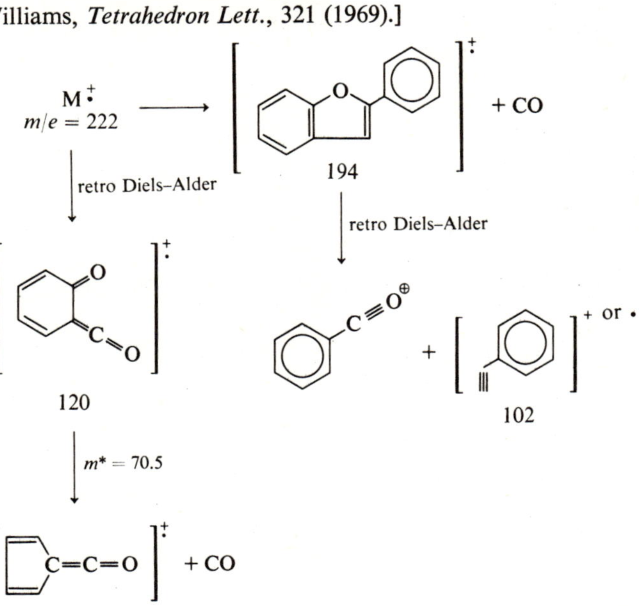


The peak at 97 mass units corresponds to a doubly charged fragment from the initial cleavage ($m/e = 194 \rightarrow m/2e = 97$), a rather common occurrence with aromatic molecules.

9. The McLafferty rearrangement

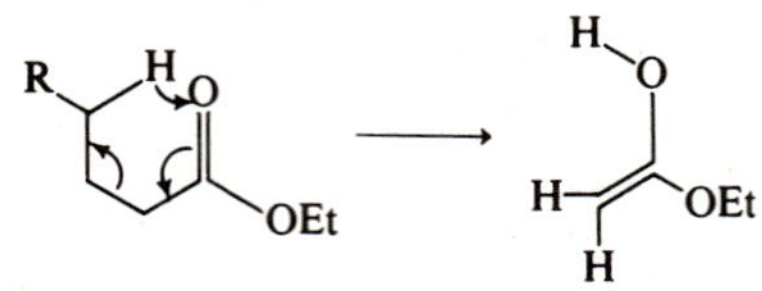

$m/e = 88$

[R. Ryhage and E. Stenhagen, *Ark. Kemi*, *14*, 483 (1959).]

10.

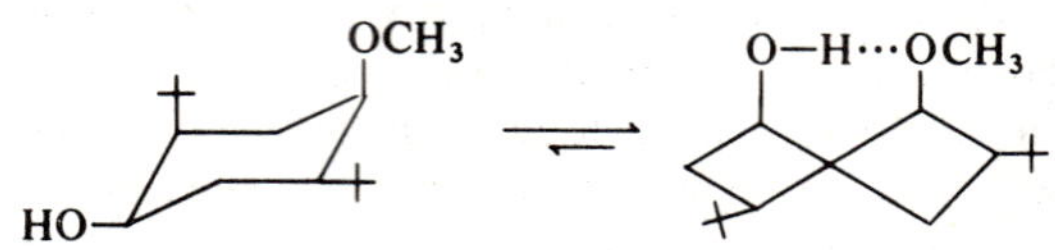

3620 $cm^{-1}$
(axial *t*-butyl)

3463 $cm^{-1}$

[R. D. Stolow and M. M. Bonaventura, *J. Amer. Chem. Soc.*, *85*, 3636 (1963).]

11.

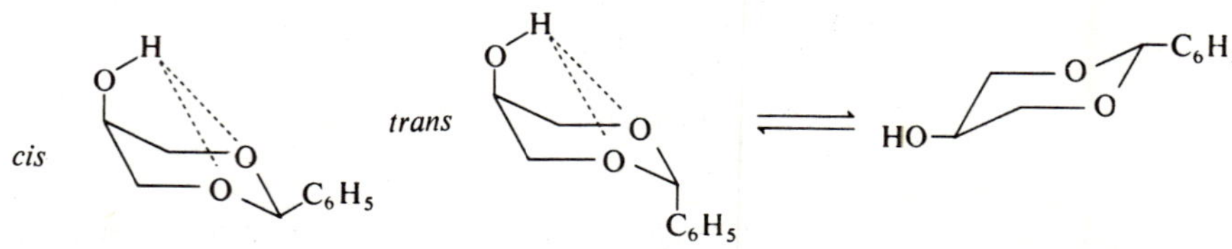

3602 $cm^{-1}$

3633 $cm^{-1}$

[N. Baggett *et al.*, *J. Amer. Chem. Soc.*, *82*, 2574 (1960).]

12.

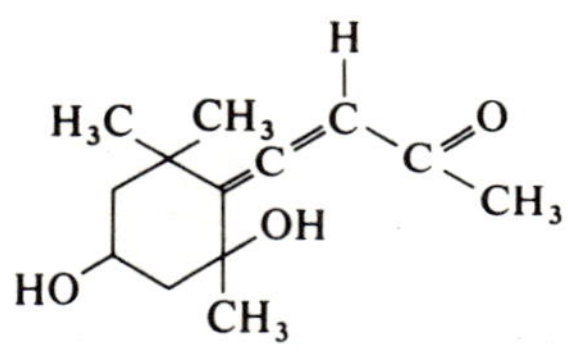

[J. Meinwald, K. Erickson, M. Hartshorn, Y. C. Meinwald, and T. Eisner, *Tetrahedron Lett.*, 2959 (1968).]

13.

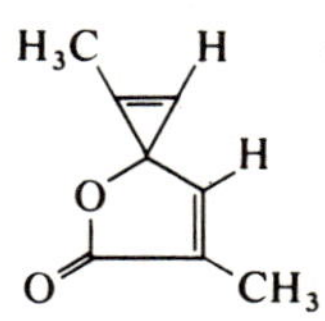

[R. Breslow and L. J. Altman, *J. Amer. Chem. Soc.*, *88*, 504 (1966).]

14.

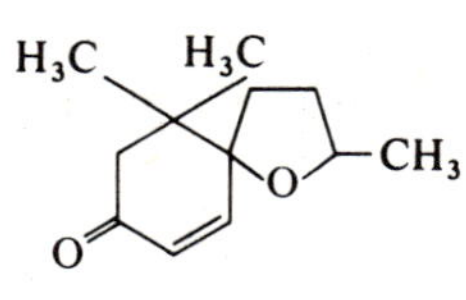

[K. Ina, Y. Sakato, and H. Fukami, *Tetrahedron Lett.*, 2777 (1968).]

15.

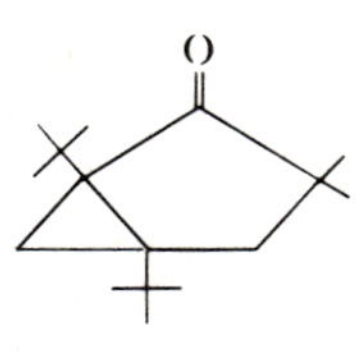

[E. E. Van Tamelen and T. H. Whitesides, *J. Amer. Chem. Soc.*, *90*, 3894 (1968).]

16.

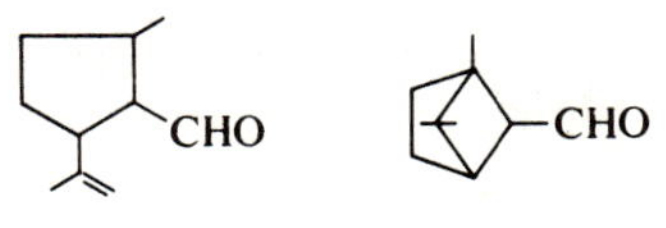

1 2

[R. C. Cookson *et al.*, *Tetrahedron Lett.*, 1995 (1963).]

17.

*β*-Ionone ⟶ 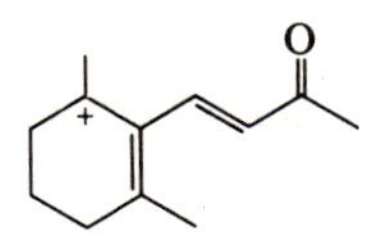 ⟶ $CH_3C{\equiv}O^+$

$m/e = 177$ very stable ion (allylic) $m/e = 43$

*α*-ionone M = 192 ⟶ M-15 loss of methyl

↓

$m/e = 136$ retro Diels–Alder ⟶ 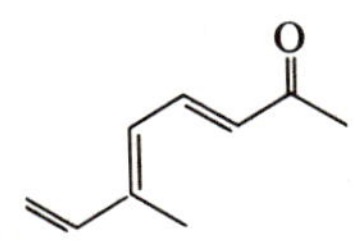

↓ *108.2

$m/e = 121$ cleavage *α* to the carbonyl 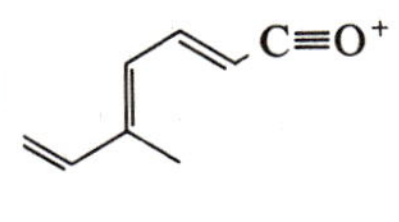


↓ *71.8

$m/e = 93$ loss of carbon monoxide

and $CH_3{-}C{\equiv}O^+$ at $m/e = 43$

## *CHAPTER 16*

1. The electrical field gradient $q$ at nitrogen is large in nitriles due to the presence of a lone pair on nitrogen, but extremely small in the isonitriles, which have tetravalent nitrogen. [I. D. Kuntz, Jr., P. von R. Schleyer, and A. Allerhand, *J. Chem. Phys.*, *35*, 1533 (1961).]
2. $H_2N{-}CH_3$; ${}^1J_{N{-}H} = 52$ Hz; ${}^3J_{HCNH} = 7$ Hz. [J. R. Dyer, *Application of Absorption Spectroscopy of Organic Compounds*, Prentice-Hall, Inc., Englewood Cliffs, N. J., 1965, p. 95.]
3. $H_2$ and $H_4$ are both adjacent to the quadrupolar heterocyclic nitrogen. Their resonances are broadened due to the incomplete washing out, by quadrupolar relaxation, of their coupling to the $^{14}N$. Cooling renders the quadrupolar relaxation yet more efficient, principally because of the increase in viscosity, and the $^{14}N$ is effectively decoupled from the adjacent $H_2$ and $H_4$ protons: the $J_{24}$ coupling constant becomes visible. [J. P. Kintzinger and J. M. Lehn, *Mol. Phys.*, *14*, 1333 (1968).]
4. 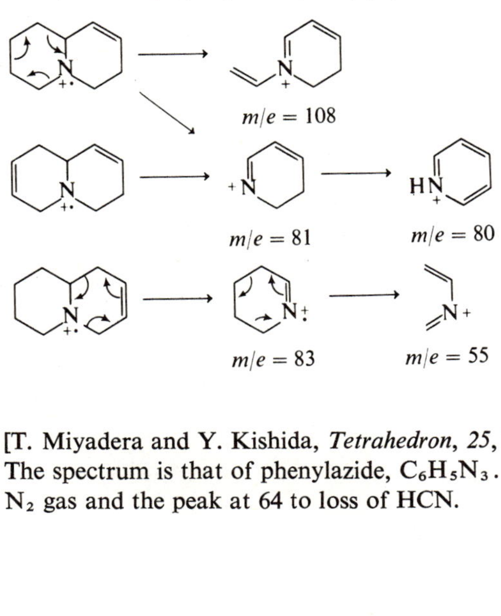

5. [T. Miyadera and Y. Kishida, *Tetrahedron*, *25*, 209 (1969).]
The spectrum is that of phenylazide, $C_6H_5N_3$. The base peak at 91 corresponds to loss of $N_2$ gas and the peak at 64 to loss of HCN.

6. Coupling is to $^{15}N(I = \frac{1}{2})$. This observation unambiguously demonstrates the existence of an iminophenol-enaminoketone tautomeric equilibrium:

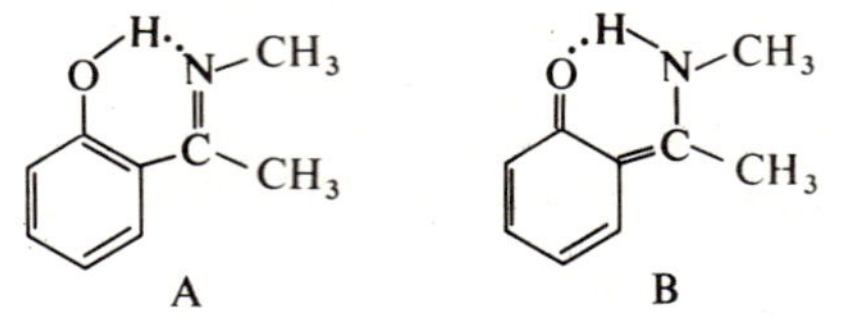


$^1J_{N-H}$ in the *pure* enaminoketone form can be independently estimated at ca. 8.9 Hz. Hence molecule (1) consists of approximately 76 percent phenol at 0°C and 64 percent phenol at −50°C and $^1J_{N-H}$ is zero or close to zero in the phenol. The enthalpy difference $\Delta H°$ for the equilibrium can be calculated as 1.5 kcal/mole. [G. O. Dudek and E. P. Dudek, *Chem. Comm.*, 464 (1965).]

7.

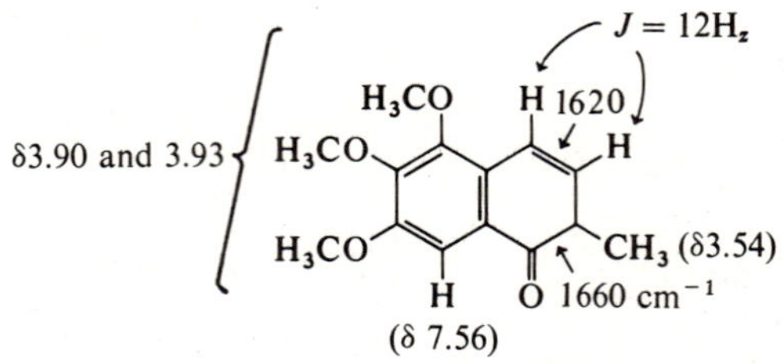


[N. M. Mollov and H. B. Dutschewska, *Tetrahedron Lett.*, 1951 (1969).]

8.

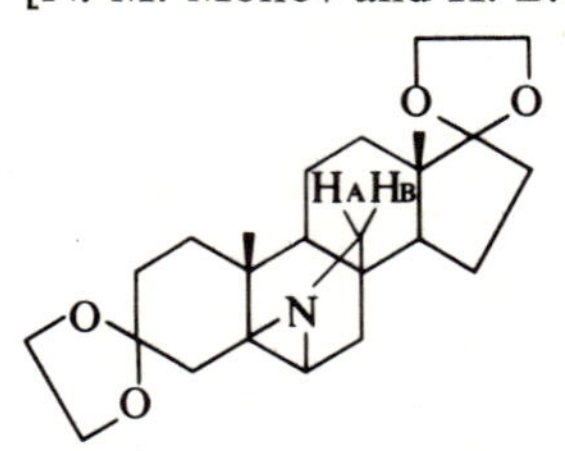


[W. Nagata, S. Hirai, K. Kawata, and T. Aoki, *J. Amer. Chem. Soc.*, *89*, 5045 (1967).]

9. The spectral evidence points to one N—H proton, a cycloheptanone, and an ortho-disubstituted benzene ring. The compound is

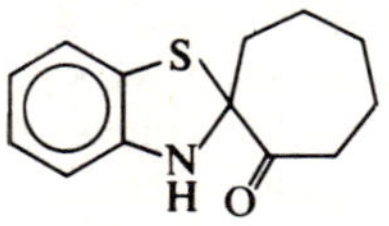


[V. Carelli, P. Marchini, F. M. Moracci, and G. Liso, *Tetrahedron Lett.*, 3421 (1967).]

# NAME INDEX

# SUBJECT INDEX

71 72 73 7 6 5 4 3